S0-BLM-044

Modern Polymer Spectroscopy

Edited by Giuseppe Zerbi

Modern Polymer Spectroscopy

Edited by
Giuseppe Zerbi

Weinheim · New York · Chichester
Brisbane · Singapore · Toronto

Prof. G. Zerbi
Dipartimento di Ingegneria Chimica e Chimica Industriale
Politecnico di Milano
Piazza Leonardo da Vinci 32
20132 Milano
Italy

Cover illustration: Dr. Erica Mannucci

Library of Congress Card No. applied for.

A catalogue record for this book is available from the British Library

Deutsche Bibliothek Cataloguing-in-Publication Data

Modern polymer spectroscopy / ed. by Giuseppe Zerbi. – 1. Aufl. – Weinheim; New York; Chichester; Brisbane; Singapore; Toronto: Wiley-VCH, 1999
ISBN 3-527-29655-7

Printed on acid-free and chlorine-free paper.

Composition: Asco Typesetters, Hong Kong.
Printing: Strauss Offsetdruck Gm bH., D-69509 Mörlenbach.
Bookbinding: Wilhelm Osswald 8 Co, D-67433 Neustadt.
Printed in the Federal Republic of Germany.

Preface

For unfortunate reasons the success of vibrational (infrared and Raman spectroscopy) in industrial and university laboratories seems to fade quickly in favor of other physical techniques which aim at chemical or structural diagnosis of unknown samples. On the other hand, engineers and instrument manufacturers in the field of vibrational spectroscopy keep producing magnificent and very sophisticated new instruments and accessories which enable the recording of vibrational spectra of samples under the most awkward experimental conditions which can never be attained by the other very popular new physical techniques.

At present, infrared spectra can be obtained with fast and very fast FTIR interferometers with microscopes, in reflection and microreflection, in diffusion, at very low or very high temperature, in dilute solutions, etc. Raman scattering can span a very large energy range in the excitation lines, thus reaching off-resonance or resonance conditions; spectrometers and interferometers can be used, microsampling is common, a few scattered photons can be detected with very sensitive CCD detectors and optical fibers provide a variety of new sampling procedures.

Parallel to the technological development we watch the invasion (even on a commercial level) of theoretical and computational techniques (both *ab initio* or semiempirical) which, by simply pushing a button, provide vibrational frequencies and intensities and even show the animated wiggling of molecules on the screen of any personal computer.

In spite of the wealth of experimental and theoretical data which are easily available for the study of molecules and of their behavior, other fields of physics and chemistry seem to have the priority in the teaching at Universities. Vibrational spectroscopy is, in general, no longer taught in detail and, at most, students are quickly exposed to molecular vibrations in some courses of analytical chemistry or structural organic chemistry. It is curious to notice that molecular dynamics is considered a very complicated mathematical machinery which must be avoided by simply mentioning briefly that molecules 'wiggle' in a curious way. Then, the traditional old-fashioned structure-group frequency spectral correlations (mostly in infrared) are considered the only useful tool for the interpretation of spectra. This very limited spectroscopic culture is obviously transferred to industrial laboratories who rely on the use of vibrational spectra only for very simple chemical diagnosis and routine analytical determinations.

The conclusion of this analysis is that the ratio: (number of information)/(capability of the experimental and theoretical techniques) turns out to be very small, in spite of the great potential offered by vibrational spectroscopy.

The above analysis, shared by many spectroscopists in the field of small molecules, can be further expanded when vibrational spectroscopy is considered in the field of polymers and macromolecules in general. The wiggling of polymers adds new flavor to physics and chemistry. The translational periodicity of infinite polymers with perfect structure generates phonons and collective vibrations which give rise to absorption or Raman scattering bands that escape the interpretation based on the traditional spectroscopic correlations. The concept of collective motions forms the basis for the understanding of the vibrations of finite chain molecules which form a nonnegligible part of industrially relevant materials. On the other hand, real polymer samples never show perfect chemical, strereochemical, and conformational structure. Symmetry is broken and new bands appear which become characteristic of specific types of disorder.

If a few simple theoretical concepts of the dynamics of ordered and disordered chain molecules are taken into account it can be easily perceived that the vibrational infrared and Raman spectra contain a wealth of information essential to analytical polymer chemistry, structural chemistry, and physics.

The content of this book has been planned to rejuvenate the vibrational spectroscopy of polymers. At present, the classes of polymeric materials are very many in number and range from classical bulk polymers of great industrial and technological relevance to highly sophisticated functional polymers which reach even the interest of photonics and molecular electronics. Also, the whole world of biopolymers requires great attention from spectroscopy.

This book touches on a very few classes of polymeric materials which we consider representatives for introducing problems, spectroscopic techniques, and solutions prototypical of many other classes of polymers and plastics.

Chapters 1 and 2 introduce new experimental techniques that provide new sets of relevant data for the study of the local and overall mobility of polymer chains. In Chapter 1, the development of two-dimensional infrared spectroscopy is described with a discussion of the mathematical principles, the description of the instrumental technique, and a detailed analysis of a few cases. In Chapter 2 the success of Fourier transform infrared polarization spectroscopy is shown for the study of segmental mobility in a polymer or a liquid-crystalline polymer under the influence of an external directional perturbation such as electric, electromagnetic, or mechanical forces. Industrial research laboratories should pay much attention to the information which can be acquired with this technique for technologically relevant polymeric materials.

Chapter 3 is a guided tour of molecular dynamics and infrared and Raman frequency and intensity spectroscopy of polymethylene chains, from oligomers to polymers, in their perfect and disordered states. The reader is exposed in the easiest possible way to the basic theoretical concepts and to the numerical techniques which can be applied to such studies. Many references are provided for the spectroscopist who wants to develop his or her own independent skill and judgment.

However, theory and calculations yield concepts and data to be used also in polymer characterization, even for routine analytical work. After discussing several oligomers or polymers the reader is guided step-by-step in the practical exercise to derive the detailed scenario of the mechanism of phase transition and melting of n-alkanes.

Some of the theoretical tools laid in Chapter 3 are fully exploited in Chapter 4, which exposes the reader to the actual problems (and proposed solutions) of the chemistry and physics of modern and technologically relevant polyconjugated polymers in their intact (insulating) and in the doped (electrically conducting) states.

Chapter 5 is an up-to-date review of the spectroscopic and structural problems and solutions reached by a modern approach to the dynamics of polypeptides. A clear and comprehensive discussion is presented on the force fields necessary for a reliable structural analysis through the vibrational spectra. Tools are thus becoming available for a systematic study of the structure even of complex polypeptides.

The message we wish to send to the reader of this book is that modern machines provide beautiful infrared and Raman spectra of polymers full of specific, unique and detailed information which can be extracted, not just merely and lazily using group frequency correlations. We will be very pleased to find that this book has provided the reader with enough motivation to overcome the potential barrier of some theoretical technicalities in order to enjoy fully the wealth of information inherently contained in the vibrational spectra of ordered or disordered chain molecules.

Milan, June 1998

Giuseppe Zerbi

Contents

Contributors

M. Del Zoppo
Dipartimento di Chimica Industriale
Politecnico di Milano
Piazza L. Da Vinci 32
20133 Milano
Italy

A. E. Dowrey
The Procter and Gamble Company
Miami Valley Laboratories
P.O. Box 398707
Cincinnati, OH 45239-8707
USA

Y. Furukawa
Department of Chemistry
School of Science and Engineering
Waseda University
Shinjuku-ku
169 Tokyo
Japan

U. Hoffmann
Bruins Instruments
D 82178 Puchheim
Germany

S. Krimm
Biophysics Research Division and
Department of Physics
University of Michigan
Ann Arbor, MI 48109
USA

Ch. Kulinna
Bayer AG
D 40789, Monheim
Germany

C. Marcott
The Procter and Gamble Company
Miami Valley Laboratories
P.O. Box 398707
Cincinnati, OH 45239-8707
USA

I. Noda
The Procter and Gamble Company
Miami Valley Laboratories
P.O. Box 398707
Cincinnati, OH 45239-8707
USA

S. Okretic
Fachhochschule Niederrhein
Fachbereich 04
D 7805 Krefeld
Germany

S. Shilov
Institute of Macromolecular
Compounds
199004 St. Petersburg
Russia

H. W. Siesler
Department of Physical Chemistry
University of Essen
D 45117 Essen
Germany

M. Tasumi
Department of Chemistry
Faculty of Science
Saitama University
Urawa
Saitama 338
Japan

I. Zebger
Department of Physical Chemistry
University of Essen
D 45117 Essen
Germany

G. Zerbi
Dipartimento di Chimica Industriale
Politecnico di Milano
Piazza L. Da Vinci 32
20133 Milano
Italy

1 Two-Dimensional Infrared (2D IR) Spectroscopy

I. Noda, A. E. Dowrey, and C. Marcott

1.1 Introduction

Two-dimensional infrared (2D IR) correlation spectroscopy [1–8] is a relatively new addition to the spectroscopic methods used for the characterization of polymers. In 2D IR, infrared spectral intensity is obtained as a function of two independent wavenumbers as shown in Figure 1-1. The well-recognized strength of IR spectroscopy arises from the specificity of an IR probe toward individual molecular vibrations which are strongly influenced by the local molecular structure and environment [9–12]. Because of such specificity toward the submolecular state of a sample, surprisingly useful information about a complex polymeric system is provided by expanding IR analysis to the second spectral dimension.

The initial idea of generating two-dimensional correlation spectra was introduced several decades ago in the field of NMR spectroscopy [13–16]. Since then, numerous successful applications of multidimensional resonance spectroscopy techniques have been reported, including many different types of studies of polymeric materials by 2D NMR [17–19]. However, until now the propagation of this powerful concept of multidimensional spectroscopy in other areas of spectroscopy, especially vibrational spectroscopies such as IR and Raman, has been surprisingly slow.

One of the reasons why the two-dimensional correlation approach as applied in NMR spectroscopy was not readily incorporated into the field of IR spectroscopy was the relatively short *characteristic times* (on the order of *picoseconds*) associated with typical molecular vibrations probed by IR. Such a time scale is many orders of magnitude shorter than the relaxation times usually encountered in NMR. Consequently, the standard approach used so successfully in 2D NMR, i.e., multiple-pulse excitations of a system followed by the detection and subsequent double Fourier transformation of a series of free induction decay signals, is not readily applicable to conventional IR experiments. A very different experimental approach, therefore, needed to be developed in order to produce 2D IR spectra useful for the characterization of polymers using an ordinary IR spectrometer.

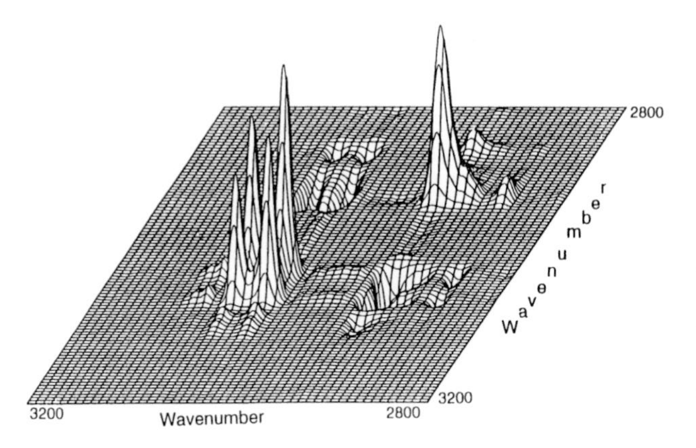

Figure 1-1. A fishnet plot of a synchronous two-dimensional infrared (2D IR) correlation spectrum of atactic polystyrene in the CH-stretching vibration region at room temperature.

1.2 Background

1.2.1 Perturbation-Induced Dynamic Spectra

The basic scheme adapted for generating 2D IR spectra [3, 8] is shown in Figure 1-2. In a typical optical spectroscopy experiment, an electromagnetic probe (e.g., IR, X-ray, UV or visible light) is applied to the system of interest, and physical or chemical information about the system is obtained in the form of a *spectrum* representing a characteristic transformation (e.g., absorption, retardation, and scattering) of the electromagnetic probe by the system constituents. In a 2D IR experiment, an external physical perturbation is applied to the system [2, 3] with the incident IR beam used as a probe for spectroscopic observation. Such a perturbation often induces time-dependent fluctuations of the spectral intensity, known as the *dynamic spectrum*, superposed onto the normal *static* IR spectrum.

There are, of course, many different types of physical stimuli which could induce such dynamic variations in the spectral intensities of polymeric samples. Possible sources of perturbations include electrical, thermal, magnetic, acoustic, chemical, optical, and mechanical stimuli. The waveform or specific time signature of the perturbation may also vary from a simple step function or short pulse, to more complex ones, including highly multiplexed signals and even random noises. In this chapter, dynamic spectra generated by a simple sinusoidal mechanical perturbation applied to polymers will be discussed.

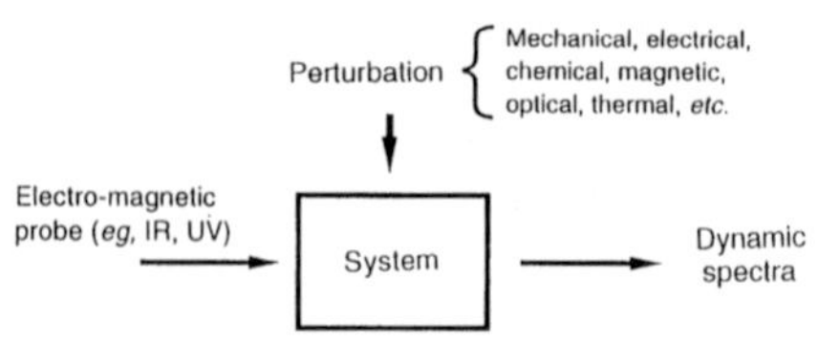

Figure 1-2. A generalized experimental scheme for 2D correlation spectroscopy based on perturbation-induced dynamic spectral signals [8].

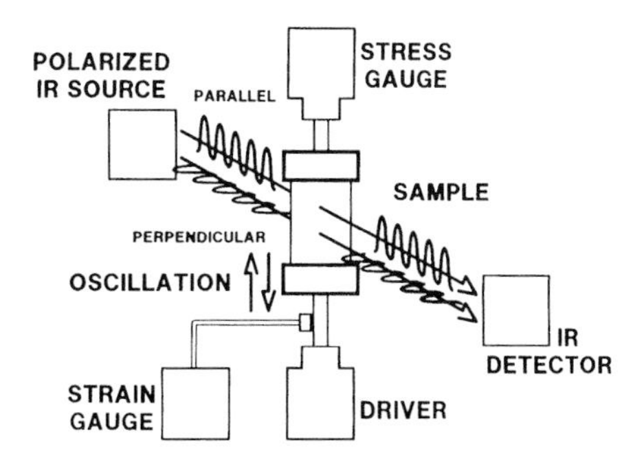

Figure 1-3. Schematic diagram of the dynamic infrared linear dichroism (DIRLD) experiment [23]. A small-amplitude sinusoidal strain is applied to a sample, and submolecular level reorientation responses of chemical moieties are monitored with a polarized IR beam.

1.2.2 Dynamic IR Linear Dichroism (DIRLD)

Figure 1-3 shows a schematic diagram of a dynamic IR linear dichroism (DIRLD) experiment [20–25] which provided the foundation for the 2D IR analysis of polymers. In DIRLD spectroscopy, a small-amplitude oscillatory strain (ca. 0.1% of the sample dimension) with an acoustic-range frequency is applied to a thin polymer film. The submolecular-level response of individual chemical constituents induced by the applied dynamic strain is then monitored by using a polarized IR probe as a function of deformation frequency and other variables such as temperature. The macroscopic stress response of the system may also be measured simultaneously. In short, a DIRLD experiment may be regarded as a combination of two well-established characterization techniques already used extensively for polymers: *dynamic mechanical analysis* (DMA) [26, 27] and *infrared dichroism* (IRD) spectroscopy [10, 11].

The optical anisotropy, as characterized by the difference between the absorption of IR light polarized in the directions parallel and perpendicular to the reference axis (i.e., the direction of applied strain), is known as the IR linear dichroism of the system. For a uniaxially oriented polymer system [10, 28–30], the *dichroic difference*, $\Delta A(\nu) \equiv A_{\parallel}(\nu) - A_{\perp}(\nu)$, is proportional to the average orientation, i.e., the second moment of the orientation distribution function, of transition dipoles (or electric-dipole transition moments) associated with the molecular vibration occurring at frequency ν. If the average orientation of the transition dipoles absorbing light at frequency ν is in the direction parallel to the applied strain, the dichroic difference ΔA takes a positive value; on the other hand, the IR dichroism becomes negative if the transition dipoles are perpendicularly oriented.

If a sinusoidally varying small-amplitude dynamic strain is applied to a polymeric system, a similar sinusoidal change in IR dichroic difference, as shown in Figure 1-4, is usually observed [3, 23]. The dynamic variation of IR dichroism arises from the time-dependent reorientation of transition dipoles induced by the applied strain. Interestingly, however, the dichroism signal often is not fully in phase with the strain. There is a finite phase difference between the two sinusoidal signals representing the externally applied macroscopic perturbation and the resulting dynamic molecular-level response of the system. This phase difference is due obviously to

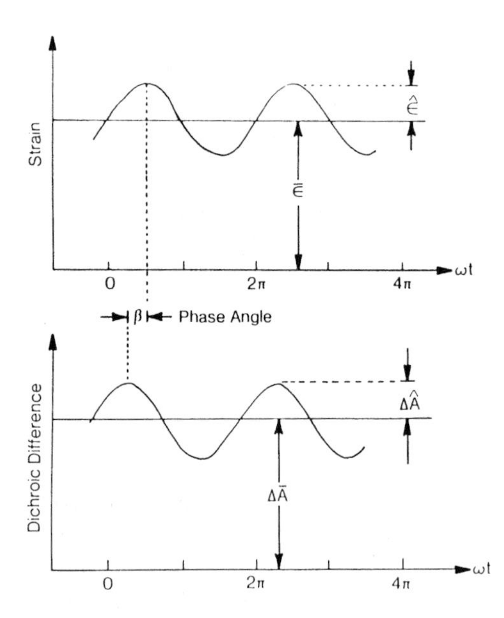

Figure 1-4. A small-amplitude sinusoidal strain applied to a sample and resulting dynamic IR dichroism (DIRLD) response induced by the strain. The two sinusoidal signals are not always in phase with each other.

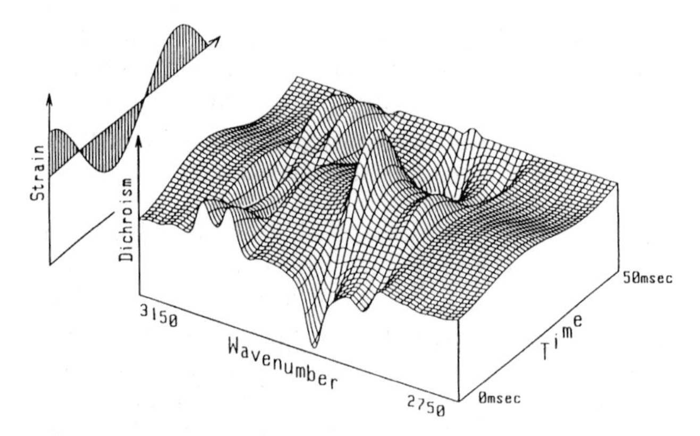

Figure 1-5. A time-resolved dynamic IR dichroism (DIRLD) spectrum of an atactic polystyrene film under a small-amplitude (ca. 0.1%) sinusoidal (23-Hz) dynamic strain at room temperature.

the rate-dependent nature of the reorientation processes of various submolecular constituents.

The advantage of using IR spectroscopy in dynamic studies of polymers is that such a measurement can be used to examine individual submolecular constituents or chemical functional groups by simply changing the wavelength of the IR probe. The variation of IR dichroism, depicted as a simple sinusoidal signal in Figure 1-4, can actually be measured as a function of not only time but also IR wavenumber. In other words, dynamic IR dichroism is obtained as a time-resolved spectrum representing the molecular level response of polymers, as shown in Figure 1-5.

For a given sinusoidal dynamic strain $\tilde{\varepsilon}(t)$ with a small amplitude $\hat{\varepsilon}$ and fre-

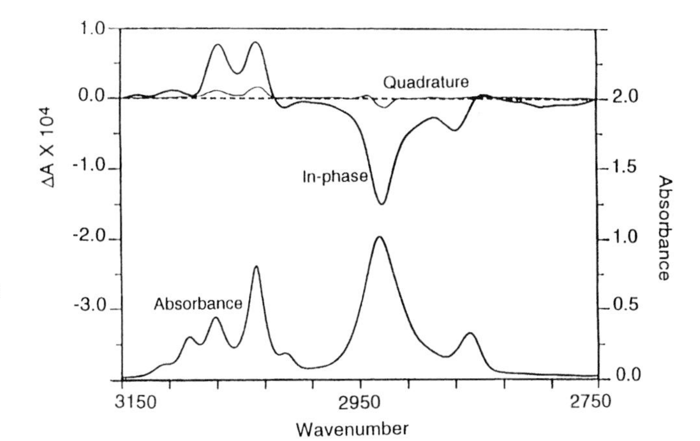

Figure 1-6. The *in-phase* and *quadrature* components of the DIRLD spectrum shown in Figure 1-5. A normal static absorbance spectrum of the sample is also provided for reference.

quency ω,

$$\tilde{\varepsilon}(t) = \hat{\varepsilon} \sin \omega t \tag{1-1}$$

a time-resolved DIRLD spectrum may be represented by

$$\Delta \tilde{A}(v, t) = \Delta \hat{A}(v) \sin[\omega t + \beta(v)] \tag{1-2}$$

where $\Delta \hat{A}(v)$ and $\beta(v)$ are, respectively, the *magnitude* and *phase (loss) angle* of the dynamic IR dichroism [23]. By using a simple trigonometric identity, the above expression can also be rewritten in the form

$$\Delta \tilde{A}(v, t) = \Delta A'(v) \sin \omega t + \Delta A''(v) \cos \omega t. \tag{1-3}$$

The wavenumber-dependent terms, $\Delta A'(v)$ and $\Delta A''(v)$, are known respectively as the *in-phase spectrum* and *quadrature spectrum* of the dynamic dichroism of the system. They represent the *real* (storage) and *imaginary* (loss) components of the time-dependent fluctuations of dichroism. Figure 1-6 shows an example of the in-phase and quadrature spectral pair extracted from the continuous time-resolved spectrum shown in Figure 1-5. These two ways of representing a DIRLD spectrum contain equivalent information about the reorientation dynamics of transition dipoles. However, the orthogonal representation of the time-resolved spectrum using the in-phase and quadrature spectra is obviously more compact and easier to interpret than the stacked-trace plot of the time-resolved spectrum.

1.2.3 IR Dichroism and Molecular Orientation

In a traditional characterization study of polymeric materials, the IR dichroism technique is most often employed for the determination of the degree of orientation

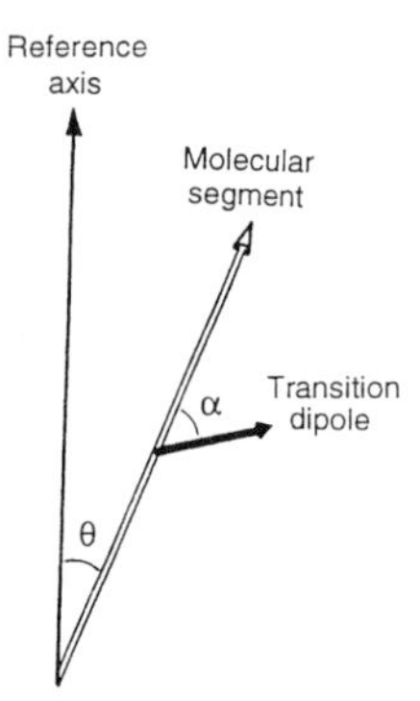

Figure 1-7. Uniaxial orientation of a polymer chain segment with respect to a reference axis and corresponding alignment of a transition dipole.

of molecular chain segments [10, 28–30]. For a uniaxially oriented polymer system (Figure 1-7), the *orientation factor* $P_2(\theta)$, i.e., the second moment of the orientation distribution function of polymer chain segments, is given by

$$P_2(\theta) = (3\langle\cos^2\theta\rangle - 1)/2 \tag{1-4}$$

where θ is the *orientation angle* between each polymer chain segment and the optical reference axis of the system. The notation $\langle\cos^2\theta\rangle$ indicates that the squared-cosine of orientation angles for individual segments are averaged over the entire space.

The orientation factor $P_2(\theta)$ is a convenient measure of the average degree of orientation of polymer chains in the system. Under an ideal orientation state where all polymer chains are aligned perfectly in the direction parallel to the reference axis, the value of $P_2(\theta)$ becomes unity. On the other hand, if polymer chain segments are all perpendicularly aligned, $P_2(\theta)$ becomes $-1/2$. An optically isotropic system gives the $P_2(\theta)$ value of zero.

According to the classical theory of the IR dichroism of polymers [10, 28, 29], the *dichroic ratio*, $D(v) \equiv A_{\|}(v)/A_{\perp}(v)$, of the system is directly related to the orientation of polymer chain segments by

$$P_2(\theta) = \frac{D(v) - 1}{D(v) + 2} \cdot \frac{D_\infty(v) + 2}{D_\infty(v) - 1} \tag{1-5}$$

The *ultimate dichroic ratio* $D_\infty(v)$ is the value of dichroic ratio if the polymer chain segments are all perfectly aligned in the direction of the reference axis, i.e., $P_2(\theta) = 1$. Given the local orientation angle α between the polymer chain segment and transition dipole probed at the IR wavenumber v (Figure 1-7), the ultimate dichroic ratio is given by

$$D_\infty(v) = 2\cot^2\alpha. \tag{1-6}$$

It can be easily shown that Eq. (1-5) simplifies to the form

$$P_2(\theta) = \Delta A(v)/\Delta A_\infty(v) \tag{1-7}$$

where $\Delta A_\infty(\nu)$ is the *ultimate dichroic difference* which is related to $D_\infty(\nu)$ by

$$\Delta A_\infty(\nu) = 3\,A_o(\nu)\frac{D_\infty(\nu) - 1}{D_\infty(\nu) + 2} \tag{1-8}$$

The *structural absorbance*, $A_o(\nu) \equiv [A_\parallel(\nu) + 2A_\perp(\nu)]/3$, is believed to be independent of the degree of molecular orientation as long as the system remains symmetric with respect to the optical axis.

If one accepts that the transition dipole orientation angle α is a fixed molecular constant unaffected by the conformation of polymer chains, then D_∞ and ΔA_∞ should also be independent of the state of chain orientation. The classical theory of IR dichroism for polymers thus makes an interesting assertion, according to Eq. (1-7), that IR dichroic difference $\Delta A(\nu)$ is *always* linearly proportional to the average orientation $P_2(\theta)$ of polymer chains regardless of the IR wavenumber ν. In other words, one should be able to determine the state of orientation of the entire polymer chain by simply observing the local orientation of a transition dipole associated with the molecular vibration of any arbitrarily chosen local functional group, as long as α is known.

1.2.4 Breakdown of the Classical Theory

An interesting observation often made during the DIRLD measurement of polymers is that the phase angle $\beta(\nu)$ between the applied strain and dynamic IR dichroic difference is strongly dependent on the IR wavenumber [3, 23–25]. Time-dependent dichroism intensities measured at certain wavenumbers change much faster than others. Individual dynamic IR dichroism signals thus become out of phase with each other as depicted in Figure 1-8. As already demonstrated in Figure 1-6, the shape of the in-phase spectrum also becomes quite different from that of the quadrature spectrum. The relative amount of dynamic dichroism signal appearing in the in-phase and quadrature components varies considerably as a function of wavenumber.

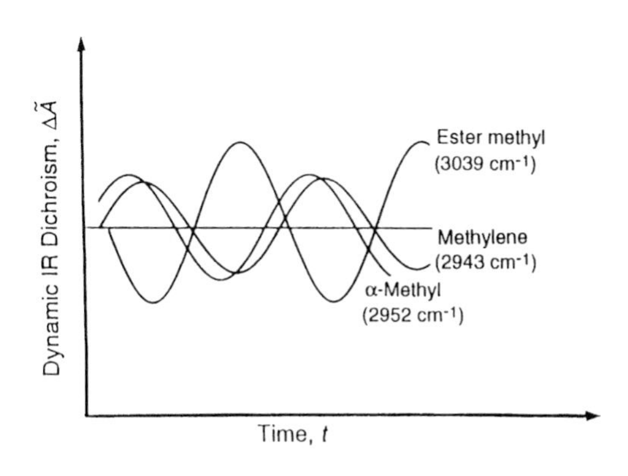

Figure 1-8. Time-dependent sinusoidal DIRLD signals from an atactic poly(methyl methacrylate) sample detected at different IR wavenumbers. Signals are out of phase with each other.

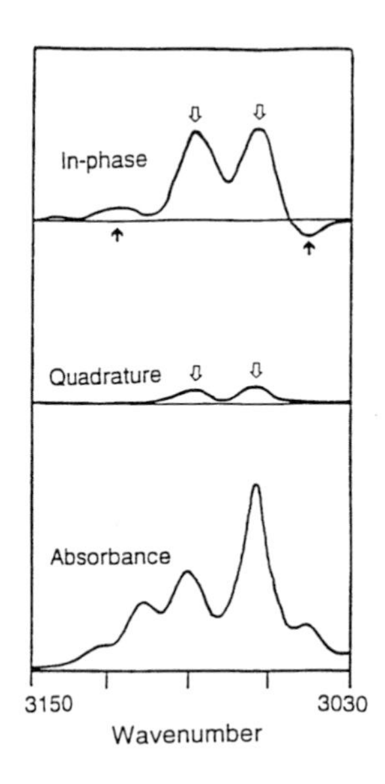

Figure 1-9. DIRLD spectra of atactic polystyrene in the phenyl CH-stretching region. Transition dipoles for IR bands marked with 🡅 are reorienting totally in phase with the applied dynamic strain, while those with ⇩ are moving at rates substantially different from the strain.

This surprising discovery reveals that a major discrepancy exists between the classical theory of IR dichroism and actual experimental observations made for reorientation dynamics of polymer chains. Figure 1-9, for example, indicates that the intensity of the quadrature spectrum is close to zero, i.e., signals are in phase with the applied strain for dichroism peaks marked with (🡅). If indeed IR dichroism signals at these wavenumbers reflect the time-dependent orientational state of the entire polymer chain, one would conclude that polymers are reorienting instantaneously under the dynamic deformation. On the other hand, if the dichroism signals are measured for other IR bands, for example, at the peaks marked with (⇩), the polymer chain seems to reorient at a rate substantially out of phase with the applied strain. This second statement clearly contradicts the conclusion made as a result of the previous observation. Thus, for a certain polymer system undergoing a dynamic deformation, the well-accepted classical view of IR dichroism (i.e., dichroic difference must always be linearly proportional to the average orientation of polymer chains, regardless of the wavenumber of IR probe, as described in Eq. (1-7)) is no longer valid.

The logical explanation for the experimentally observed, wavenumber-dependent behavior of the DIRLD phase angle is that the reorientation rates of individual transition dipoles in the system are not the same. For a given macroscopic perturbation such as dynamic strain, different submolecular constituents (e.g., backbone segments, side chains, and various functional groups comprising the polymer chain) may respond at different rates, more or less independently of each other. Consequently, transition dipoles associated with the molecular vibrations of different parts of a polymer chain have independent local reorientational responses not fully synchronized to the global motions of the entire polymer chain.

1.2.5 Two-Dimensional Correlation Analysis

The wavenumber-dependent orientation rates of individual transition dipoles observed for dynamically stimulated polymer systems poses several intriguing questions: What makes these transition dipoles reorient at different rates? Why do some of the transition dipoles seem to reorient at a rate similar to each other? Is there any underlying mechanism responsible for synchronization, or lack thereof, in the local reorientation processes of submolecular structures? To answer such questions, one must introduce an effective way of representing a measure of the similarity or difference of the reorientation rates of transition dipoles.

One convenient mathematical approach to comparing the behavior of a set of variables is a correlation analysis [2, 3, 8]. For a pair of dynamic dichroism signals, $\Delta\tilde{A}(\nu_1, t)$ and $\Delta\tilde{A}(\nu_2, t + \tau)$, which are observed at different instances separated by a fixed *correlation time* τ at two arbitrarily selected wavenumbers, ν_1 and ν_2, for a certain length of observation period T, a *two-dimensional cross-correlation function* $\mathrm{X}(\tau)$ is defined by

$$\mathrm{X}(\tau) = \lim_{T\to\infty} \frac{1}{T} \int_{-T/2}^{T/2} \Delta\tilde{A}(\nu_1, t) \cdot \Delta\tilde{A}(\nu_2, t + \tau)\, dt. \tag{1-9}$$

For a pair of sinusoidally varying signals with a fixed frequency ω, the cross-correlation function reduces to a simple form [3],

$$\mathrm{X}(\tau) = \Phi(\nu_1, \nu_2) \cos\omega\tau + \Psi(\nu_1, \nu_2) \sin\omega\tau. \tag{1-10}$$

The *real* and *imaginary* components, $\Phi(\nu_1, \nu_2)$ and $\Psi(\nu_1, \nu_2)$, of the cross-correlation function $\mathrm{X}(\tau)$ are referred to, respectively, as the *synchronous* and *asynchronous correlation intensity*. These quantities are related to the in-phase and quadrature spectra of dynamic dichroism by

$$\Phi(\nu_1, \nu_2) = [\Delta A'(\nu_1)\Delta A'(\nu_2) + \Delta A''(\nu_1)\Delta A''(\nu_2)]/2 \tag{1-11}$$

and

$$\Psi(\nu_1, \nu_2) = [\Delta A''(\nu_1)\Delta A'(\nu_2) - \Delta A'(\nu_1)\Delta A''(\nu_2)]/2. \tag{1-12}$$

The synchronous correlation intensity $\Phi(\nu_1, \nu_2)$ represents the *similarity* between the two dynamic dichroism signals measured at different wavenumbers. This quantity becomes significant if the two dynamic IR dichroism signals are changing at a similar rate but vanishes if the time dependency of the signals is very different. The asynchronous correlation intensity $\Psi(\nu_1, \nu_2)$, on the other hand, represents the *dissimilarity* between signals. It becomes significant only if the two IR dichroism signals are changing out of phase with each other but reduces to zero if they are changing together.

The two-dimensional nature of our correlation analysis arises from the fact that the correlation intensities are obtained by comparing the time dependence of

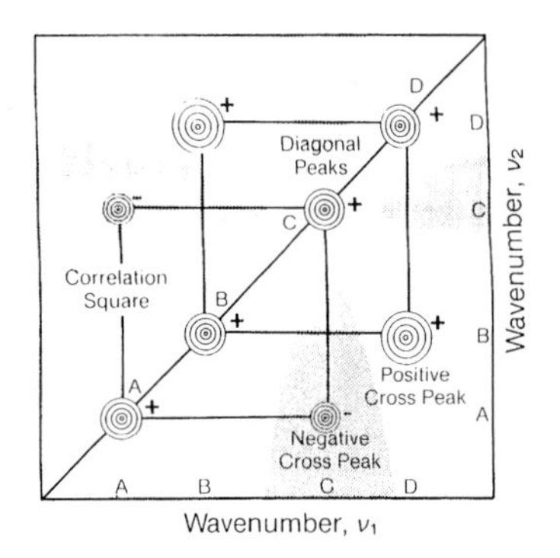

Figure 1-10. A schematic contour diagram of a synchronous 2D IR correlation spectrum [3]. Shaded areas represent negative correlation intensity.

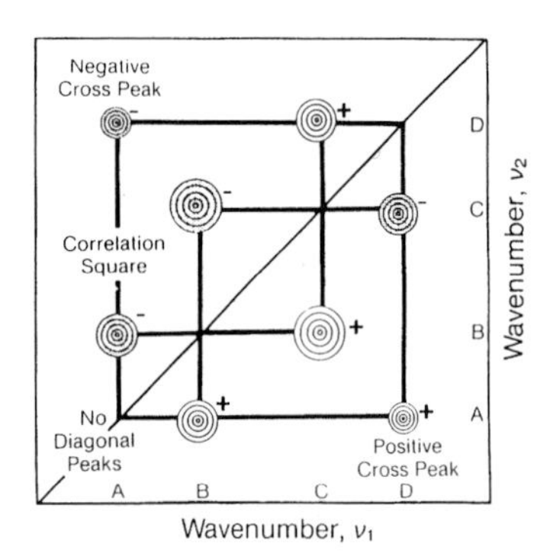

Figure 1-11. A schematic contour diagram of an asynchronous 2D IR correlation spectrum [3]. Shaded areas represent negative correlation intensity.

dynamic IR signals observed at two independently chosen wavenumbers. Plots of correlation intensities as functions of two wavenumber axes are referred to as *2D IR correlation spectra.* Such spectra reveal important and useful information not readily accessible from conventional one-dimensional spectra.

There are many different ways to plot a 2D IR correlation spectrum. A pseudo-three-dimensional representation (so-called fishnet plot) as shown in Figure 1-1 is well suited for providing the overall features of a 2D IR spectrum. Usually, however, 2D IR spectra are more conveniently displayed as contour maps to indicate clearly the location and intensity of peaks on a given spectral plane. In the following section, basic properties of and information extracted from 2D IR spectra are reviewed by using schematic contour maps (Figures 1-10 and 1-11).

1.3 Basic Properties of 2D IR Correlation Spectra

1.3.1 Synchronous 2D IR Spectrum

The correlation intensity at the diagonal position of a synchronous 2D spectrum (Figure 1-10) corresponds to the autocorrelation function of perturbation-induced dynamic fluctuations of the IR signals. Local intensity maxima along the diagonal are thus referred to as *autopeaks*. Since the magnitude of dynamic variations of IR dichroism represents the susceptibility of transition dipoles to reorient under a given external perturbation, autopeaks indirectly reflect the local mobility of chemical groups contributing to the molecular vibrations associated with the transition dipoles. The schematic example in Figure 1-10 indicates that functional groups contributing to the molecular vibrations with frequencies A, B, C, and D are undergoing reorientational motions induced by the applied dynamic strain.

Peaks located at off-diagonal positions of a 2D correlation spectrum are called *cross peaks*. They appear when the dynamic variations of IR signals at two different wavenumbers corresponding to the spectral coordinates (ν_1, ν_2) are correlated with each other. For a synchronous spectrum, this occurs when the two IR signals are fluctuating in phase (i.e., simultaneously) with each other. Synchronized variations of dichroism signals result from the simultaneous reorientation of transition dipoles associated with corresponding IR wavenumbers. Coordinated local motions of submolecular groups lead to such simultaneous reorientations. In turn, highly correlated local reorientations of chemical groups in response to a common external stimulus imply the possible existence of interactions or connectivity which restrict the independent motions of these submolecular structures.

Functional groups which are not strongly interacting, on the other hand, can move independently of each other. The transition dipoles associated with molecular vibrations of these groups may then reorient at different rates (or somewhat out of phase with each other), resulting in a much weaker synchronous correlation. Thus, as long as the normal modes of vibrations correspond to reasonably pure group frequencies, one can use the cross peaks in a synchronous 2D IR spectrum to map out the degree of intra- and intermolecular interactions of various functional groups. In Figure 1-10, functional groups contributing to the molecular vibrations with frequencies A and C may be interacting. Likewise, the pair B and D could be connected.

The signs of cross peaks indicate relative reorientation directions of transition dipoles and consequently their associated chemical groups. If the sign of a synchronous cross peak is positive, the corresponding pair of transition dipoles reorient in the same direction. If negative, on the other hand, the reorientation directions are perpendicular to each other. In Figure 1-10, mutually perpendicular reorientation of a pair of transition dipoles at wavenumbers A and C, as well as parallel reorientation for B and D, are observed.

1.3.2 Asynchronous 2D IR Spectrum

An asynchronous 2D IR spectrum (Figure 1-11) provides complementary information; cross peaks appear if IR signals are not synchronized to be completely in phase with each other. This feature is particularly useful, since IR bands arising from molecular vibrations of different functional groups, or even of similar groups in different local environments, may exhibit substantially different time-dependent intensity fluctuations. Thus, asynchronous 2D IR spectra can be used to differentiate highly overlapped IR bands, since asynchronous cross peaks should develop among these bands.

From the sign of asynchronous cross peaks, one can obtain temporal information about the perturbation-induced reorientation processes of transition dipoles and their corresponding chemical groups. A positive peak in an asynchronous spectrum indicates the transition dipole with the vibrational frequency ν_1 reorients *before* that for ν_2. If the sign is negative, ν_1 reorients *after* ν_2. However, this temporal relationship is reversed for perpendicularly reorienting pairs of transition dipoles, i.e., the synchronous correlation intensity at the same spectral coordinate becomes negative. In Figure 1-11, the reorientational motions of functional groups contributing to the molecular vibrations with frequencies B and D occur before those for A and C. More detailed discussions on the properties of 2D IR spectra are found elsewhere [3].

1.4 Instrumentation

Dynamic infrared spectra suitable for 2D correlation analysis can, in principle, be measured using any conventional IR spectrometer [7]. The spectrometer, however, must be equipped with the ability to stimulate samples by some physical means and measure the resulting time-dependent fluctuations of the IR signals. Both dispersive monochromators [23] and Fourier transform infrared (FT-IR) instrumentation [5] have been successfully used to measure dynamic 2D IR spectra. Although well-designed dispersive spectrometers can often achieve better signal-to-noise ratios (S/N) over small spectral regions, FT-IR measurements cover much broader spectral regions in less time and are, more importantly, readily available commercially. Conventional rapid-scan FT-IR spectrometers are not well suited for these types of measurements when the dynamic strain frequencies of interest are in the range between 0.1 Hz and 10 kHz. Using a step-scanning interferometer substantially simplifies dynamic 2D FT-IR measurements. Figure 1-12 shows an example of a

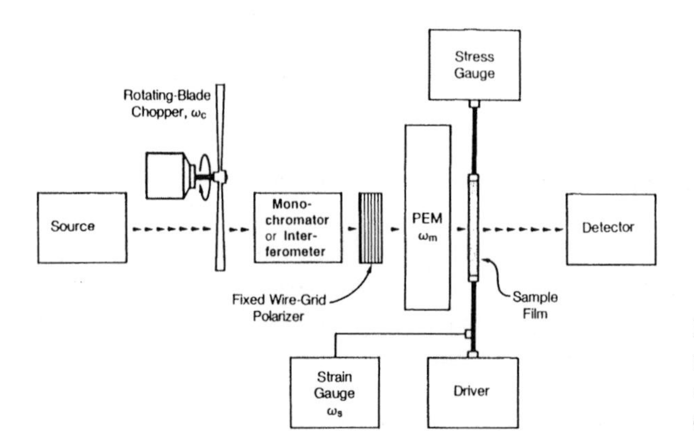

Figure 1-12. A schematic view of a DIRLD spectrometer based on a small-amplitude mechanical deformation [23].

2D IR spectrometer coupled with a dynamic rheometer capable of applying a small-amplitude mechanical perturbation [23].

In addition to the time-resolution constraints on dynamic infrared measurements, sensitive instrumentation is also required. The signals of interest, which are often 10^4 times smaller than the normal IR absorbance of the sample, exist only as a result of the small dynamic strain applied to the sample.

The type of perturbation applied to the sample is not restricted to simple mechanical strain. Electrical perturbations, for example, can be effectively used to induce the necessary fluctuations of IR signals to produce 2D IR spectra. The use of electrical stimuli has been especially successful in studies of nematic liquid crystalline systems [31–33], where selective orientation of liquid crystals under an alternating electric field was observed. Electrochemical experiments modulated with an alternating electrical current have also been used to generate dynamic IR spectra suitable for 2D correlation analysis [34, 35]. Recently, a two-dimensional photoacoustic spectroscopy (2D PAS) experiment was conducted with a step-scanning FT-IR spectrometer to obtain depth profiles of layered samples [36, 37]. The characteristic time dependence of PAS signals representing sample components located at different depths inside samples were successfully differentiated by 2D correlation analysis. Each of these applications is similar in that a dynamic perturbation is applied to the sample and the time-resolved response is measured using phase-sensitive detection [6–8].

1.4.1 Dispersive 2D IR Spectrometer

A high-optical-throughput dispersive monochromator turns out to be an excellent choice for dynamic 2D IR spectroscopy [3, 23]. As shown in Figure 1-12, the incident IR beam, originating from a high-intensity source, is modulated at three separate places by a chopper, photoelastic modulator (PEM), and dynamic mechanical analyzer (i.e., polymer stretcher). The chopper labels photons originating from the source so they can be distinguished from background IR emission. The PEM, which immediately follows a fixed linear polarizer in the beam path, enables the polarization direction of the beam to be switched back and forth rapidly ($\sim$100 kHz) between directions aligned parallel and perpendicular to the sample strain direction. Sample strain frequencies used are typically between 0.1 and 100 Hz. Each modulation frequency is thus separated from the other two modulation frequencies by at least one order of magnitude. This makes analysis of the individual signal components with lock-in amplifiers straightforward [23].

Figure 1-13 shows an example of a train of lock-in amplifiers used to demodulate DIRLD signals obtained with a spectrometer described in Figure 1-12. In order to obtain the time-dependent dynamic absorbance and DIRLD responses, quadrature lock-in amplifiers are used. These devices monitor signals both in phase and 90° out of phase (quadrature) with the sinusoidal strain reference signal. The monochromator is scanned one wavelength at a time through the spectrum. Data are collected on six separate channels (e.g., in-phase and quadrature dynamic dichroism, in-phase and quadrature dynamic absorbance, static dichroism, and normal IR absorbance)

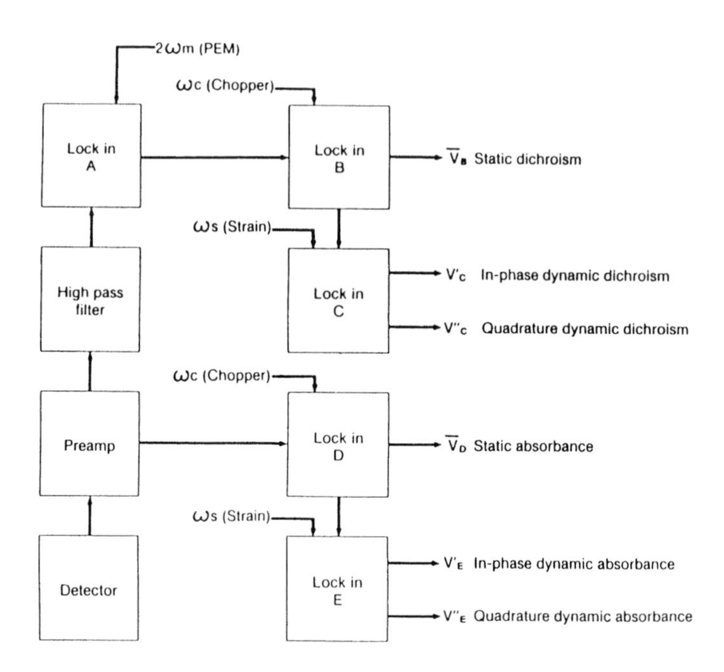

Figure 1-13. A demodulation process to extract DIRLD signals based on a series of lock-in amplifiers [23].

at each wavenumber position until an acceptable S/N is achieved. Multiple scans of the entire spectrum can be collected in order to further improve the S/N and to average out longer-term drift.

1.4.2 Step-Scanning 2D FT-IR Spectrometer

The dynamic 2D IR spectroscopy experiment performed on a step-scanning FT-IR spectrometer is in many ways similar to the way it is done on a dispersive instrument [5, 7]. For example, a step-scanning interferometer measures an interferogram one mirror retardation position at a time in the same way a monochromator scans through a spectrum one wavelength at a time. The mirror remains in a fixed retardation position while data are accumulated from the output channels of lock-in amplifiers tuned to the signals of interest (e.g., in-phase and quadrature dynamic absorbance, normal IR absorbance). As in the dispersive experiment, signals are time averaged long enough to achieve an acceptable S/N and multiple scans of the entire interferogram can be coadded to average out longer-term drift. When a scan is completed, the resulting interferogram does not depend on the moving mirror velocity, as in the case of an interferogram collected in rapid-scan mode. The Fourier frequencies in a step-scanning experiment are typically in the sub-Hz range where they are well separated from the polarization modulation and strain modulation frequencies. This is in contrast to a typical rapid-scan FT-IR experiment where each IR wavelength is modulated at a different acoustic-range frequency.

Unlike a dispersive spectrometer which uses a mechanical light chopper, the step-scanning interferometer often performs better when a technique called *phase modulation* is used to label the IR photons originating from the source. Phase modulation is achieved by dithering either the fixed or moving mirror of the interferometer a small amount (on the order of 1 μm) at a fixed frequency. This motion causes a modulation of the IR signal intensity due to a change in optical path difference between the two arms of the interferometer. The amplitude of the signal detected is thus the slope of the actual interferogram at the average optical path difference.

Most of the early measurements of DIRLD spectra were made using a photoelastic modulator (PEM) to modulate the polarization rapidly between parallel and perpendicular to the dynamic strain direction. Although polarization modulation does improve the S/N of dynamic IR spectroscopy, by about 50% over experiments using only a fixed polarizer aligned parallel to the sample strain direction, a PEM adds significant complexity to the experiment. More lock-in amplifiers and careful optical alignment are required when using polarization modulation. Indeed the first 2D IR spectra reported in the literature [2] were obtaining using unpolarized light. When a PEM is not in use, the step-scan FT-IR experiment will work equally well with either a room-temperature deuterated triglycine sulfate (DTGS) detector or a liquid-nitrogen-cooled mercury–cadmium–telluride (MCT) detector. Although MCT detectors are typically an order of magnitude more sensitive than DTGS detectors, the full throughput of a commercial FT-IR spectrometer is enough to saturate the most sensitive MCT detectors. Thus, the beam is normally attenuated by using smaller apertures, neutral-density filters, or optical filters when using an MCT detector.

Although dispersive dynamic IR experiments are presently more sensitive than FT experiments over limited spectral ranges, recent introductions of commercial FT-IR spectrometers with step-scanning capability should make dynamic 2D IR spectroscopy accessible to many more laboratories.

1.5 Applications

2D IR spectroscopy has been applied extensively to studies of polymeric materials. A recent review of 2D IR spectroscopy cites numerous applications in the study of polymers by this technique [6]. In this section, some representative examples of 2D IR analysis of polymers are presented. We will start our discussion with a simple homogeneous amorphous polymer then move to more complex multiphase systems, such as semicrystalline polymers. Alloys and blends consisting of more than one polymer components are of great scientific and technical importance. Both immiscible and miscible polymer blend systems may be studied by 2D IR spectroscopy. Analysis of microphase-separated block copolymers is also possible. Finally, the possible application of 2D IR spectroscopy to the studies of natural polymers of biological origin is explored.

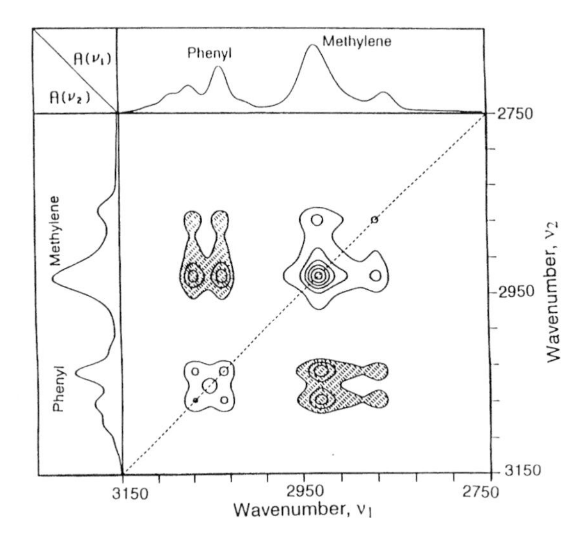

Figure 1-14. A synchronous 2D IR spectrum of a thin film of atactic polystyrene in the CH-stretching vibration region at room temperature. Regular one-dimensional spectra of the same system are provided at the top and left of the 2D spectrum for reference. Shaded areas represent negative correlation intensity.

1.5.1 Amorphous Polymers

A homogeneous one-phase system consisting of a single, noncrystallizing homopolymer, such as atactic polystyrene [3, 38–41] or poly(methyl methacrylate) [42–44], provides one of the least complicated testing grounds for the applicability of the 2D IR technique for the study of local dynamics of polymers. By using 2D IR, surprisingly rich information can be extracted even from a simple polymeric system [25, 39–41]. Figure 1-14 shows a contour-map representation of the synchronous 2D IR correlation spectrum of an atactic polystyrene film in the CH-stretching region. A three-dimensional fish-net plot of the same 2D IR spectrum corresponding to Figure 1-14 has already been shown in Figure 1-1.

DIRLD data used for the 2D correlation analysis was obtained by applying a 23-Hz dynamic strain with an amplitude of 0.1% to a thin solution-cast film of atactic polystyrene ($\overline{M}_w$ 150,000) at room temperature [39]. The resulting dynamic dichroism spectra have already been shown in Figures 1-5 and 1-6. The time-dependent reorientations of transition dipoles associated with the molecular vibrations of backbone and side groups are all observable in the original dynamic dichroism spectra. However, much more detailed features are recognized in the 2D correlation spectrum.

Autopeaks at the diagonal positions of Figure 1-14 represent transition dipoles that are susceptible to reorientational motions in the polystyrene sample. Peaks at 2855 cm^{-1} and 2925 cm^{-1} are, respectively, assignable to the symmetric and antisymmetric CH_2-stretching vibrations of the methylene groups in the backbone of the polystyrene chain; those peaks above 3000 cm^{-1} arise from the phenyl ring CH-stretching vibrations and indicate the reorientation of the side groups. The appearance of positive synchronous cross peaks at spectral coordinates corresponding to the two methylene bands indicates the transition dipoles associated with these two

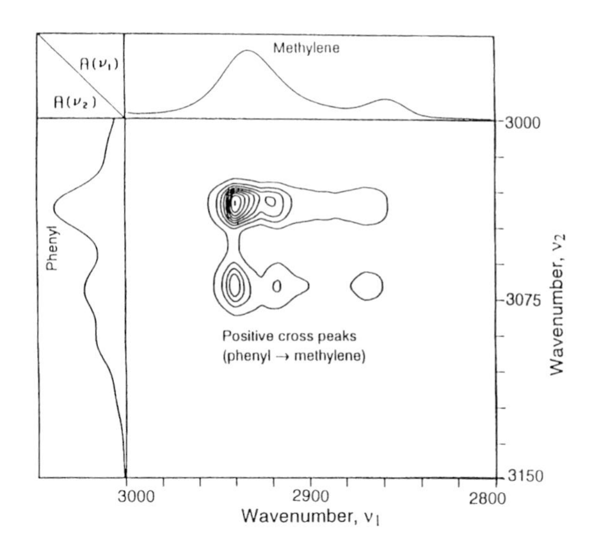

Figure 1-15. An asynchronous 2D IR spectrum of a atactic polystyrene film in the CH-stretching vibration region at room temperature.

wavenumbers reorient at a similar rate. Such synchronized responses are of course expected, since both IR bands originate from vibrations of the same submolecular structural unit.

The positive sign of the methylene cross peaks suggests the transition dipoles for the symmetric and antisymmetric methylene vibrations are both realigning in the same relative direction, i.e., perpendicular to the direction of applied strain. The molecular architecture dictates that these two transition dipoles are locally aligned more or less perpendicular to the backbone of the polymer segment. It is not difficult to conclude that the molecular chain segment of polystyrene must be realigning in the direction parallel to the applied dynamic tensile strain.

Negative (shaded) synchronous cross peaks between the methylene and phenyl bands indicate that some of the phenyl transition dipoles are orienting parallel to the strain direction. It is known [40, 41] that the signs of these particular cross peaks for atactic polystyrene become positive at temperatures above the glass-to-rubber transition temperature (ca. 100 °C). The inversion of the cross-peak sign, and consequently, the relative reorientation direction of the side group with respect to the main chain, indicates a dramatic change in the local mobility of the polymer segments and functional groups is occurring around the glass transition temperature.

Figure 1-15 shows the asynchronous 2D IR spectrum of the same sample comparing the reorientation dynamics of transition dipoles for backbone methylene and side-group phenyl. The spectral region corresponding to the asynchronous correlation of methylene stretching vibrations below 3000 cm^{-1} is not shown since no asynchronous cross peaks are observed. The appearance of strong asynchronous cross peaks between phenyl and methylene bands suggests the existence of rather complex reorientation dynamics involving the side groups. The rate of reorientational motion of the side groups under dynamic strain is apparently different from that of the backbone for this polymer. The signs of asynchronous cross peaks shown in this 2D spectrum are all positive. Coupled with the fact that the synchronous correlation intensities at corresponding coordinates (Figure 1-14) are all negative, one can conclude the side groups of this glassy polystyrene sample complete the realignment induced by the dynamic strain well before the main chain.

This observation cannot be explained if the average local orientation angle of the phenyl groups with respect to the chain axis of polystyrene is fixed. There apparently is substantial freedom of the side groups to temporarily realign ahead of the main chain [41]. This finding clearly demonstrates that it is impossible to monitor the main-chain dynamics of polystyrene by simply observing the reorientation of side groups. Furthermore, since phenyl groups contribute significantly toward the refractive index of polystyrene, the above observation also casts significant doubt over the validity of directly relating birefringence results of glassy polystyrene under a dynamic deformation to main-chain orientation dynamics.

The observation of highly localized reorientational motions of functional groups in glassy amorphous polymers is not limited to atactic polystyrene. Similar independent motions of different side groups are observed in many other systems, including atactic poly(methyl methacrylate) [42–44]. Different side groups attached to the same polymer chain, such as ester methyl and α-methyl groups of poly(methyl methacrylate), exhibited completely different reorientation rates (see Figure 1-8). 2D IR spectroscopy has been especially useful in elucidating submolecular-level realignment of functional groups controlling the mechanical properties of glassy amorphous polymers [4, 43].

The relationship between the local mobility of functional groups and glass-transition phenomenon has also been probed successfully. New insight into the dynamics of amorphous polymers going through glass-to-rubber transition processes was provided. Specifically, the local mobility of individual functional groups was found to play a very significant role in the glass-to-rubber transition of amorphous polymers [41]. In turn, it is now possible to determine if an amorphous polymer is in glassy or rubbery state by simply observing the characteristic local reorientation dynamics of functional groups using 2D IR spectroscopy.

1.5.2 Semicrystalline Polymers

Semicrystalline polymers, such as polyethylene [45–47] and polypropylene [5, 48], may also be studied by using the 2D IR technique. By taking advantage of the enhanced spectral resolution of 2D IR, overlapped IR bands assigned to the coexisting crystalline and amorphous phases of semicrystalline polymers can be easily differentiated. Such differentiation has become especially useful, for example, in the study of blends of high-density polyethylene and low-density polyethylene [47]. Here it was found that blends of polyethylenes are mixed at the molecular scale only in the amorphous phase, while each component crystallizes separately. In this section, an example of a 2D IR analysis applied to a film of linear low-density polyethylene is discussed [46].

Figure 1-16 shows the asynchronous 2D IR spectrum of a thin film of linear low-density polyethylene consisting mainly of ethylene repeat units with a small amount

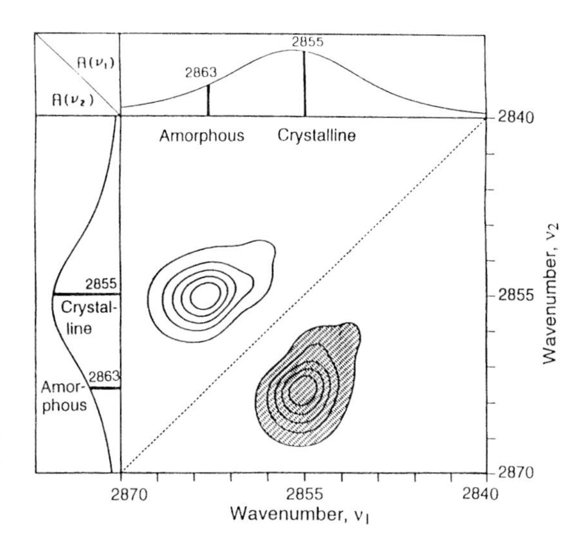

Figure 1-16. An asynchronous 2D IR spectrum for the main chain symmetric CH_2-stretching vibration of linear low-density polyethylene [46].

(ca. 3 mole%) of deuterium-substituted 1-octene comonomer units. The octene comonomer is incorporated into the polyethylene chain to produce short (six-carbon) side branches, which reduce the melt temperature and crystallinity of the polymer. By selectively labeling the octene units with deuterium, one can unambiguously differentiate the dynamics of main chain and short side branches. The spectral region shown in Figure 1-16 represents IR signals arising exclusively from the polyethylene chain, with little contribution from the octene side branches.

The presence of asynchronous peaks in the CH_2-stretching vibration of backbone methylene groups indicates there are two distinct bands in this region of the IR spectrum of the linear low density polyethylene sample. The band located near 2855 cm^{-1} can be assigned to the contribution from the crystalline phase of the sample, while the band near 2863 cm^{-1} is due to the amorphous component of this semicrystalline polymer. The above assignment can be easily verified by heating the sample above its melt temperature to remove the IR spectral contribution from the crystalline component of the specimen. Because of the significant difference in the reorientational responses of polymer chains located in the crystalline and amorphous phase domains to a given dynamic deformation, 2D IR spectroscopy can easily differentiate IR bands associated with the two-phase domains, even though these bands are highly overlapped.

Figure 1-17 shows an asynchronous 2D IR spectrum correlating the backbone of polyethylene segments located in either crystalline and amorphous domains and deuterium-substituted branches. Asynchronous cross peaks develop between the crystalline component of the symmetric CH_2-stretching band of the polyethylene segments and bands for octene side branches. Such asynchronicity suggests short side branches can move independently of the polyethylene chain located in the crystalline phase of the sample. No noticeable asynchronicity, however, is observed

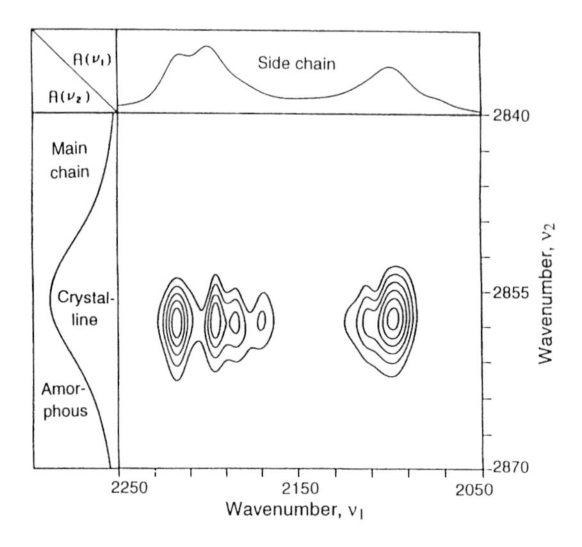

Figure 1-17. An asynchronous 2D IR spectrum of linear low-density polyethylene comparing the strain-induced local reorientational motions of the main chain and side chain [46].

between the reorientation dynamics of side branch and amorphous components of the system.

The above result reveals that the octene side branches are excluded from the crystalline lattice and are accumulating preferentially in the noncrystalline region of the polyethylene. This result agrees with the view that the depression of the melt temperature and crystallinity of linear low-density polyethylene is caused by the inability of the crystalline lattice to incorporate chain branches above a certain length. It is not possible, however, to determine conclusively from this 2D IR data if the short branches of linear low-density polyethylene are uniformly distributed within the amorphous phase domain, or preferentially accumulate near the interface between the amorphous and crystalline regions.

The high-resolution capability of 2D IR spectroscopy to differentiate highly overlapped crystalline and amorphous IR bands has been successfully employed for the characterization of many other semicrystalline polymers. Biodegradable poly(hydroxyalkanoate)s, for example, were studied by this technique to show that molecular defects created by the incorporation of comonomer units tend to accumulate in the amorphous regions [49] in a manner similar to the case of linear low-density polyethylene. Additives, such as plasticizers, which are miscible in the molten state of semicrystalline polymers also tend to be excluded from the crystal lattice and preferentially accumulate in the amorphous phase once the polymer is brought down to a temperature below the melt temperature.

1.5.3 Immiscible Polymer Blends

Mixtures of polymers are often immiscible and spontaneously phase separate because of the much reduced entropy contribution to the free energy of mixing [50].

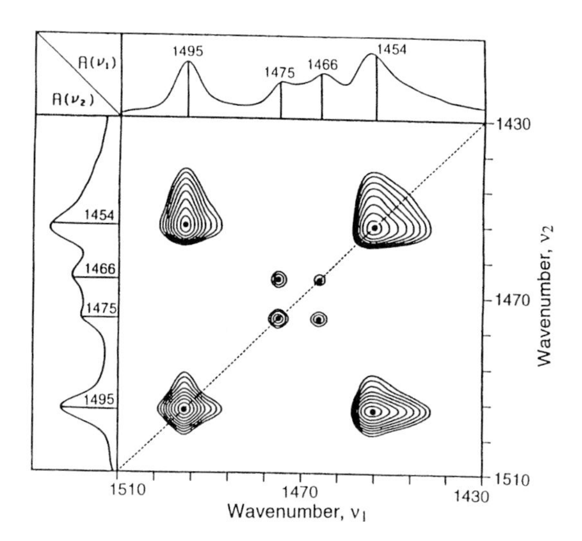

Figure 1-18. A synchronous 2D IR spectrum in the CH deformation and aromatic ring semicircle stretching vibration region of a blend of polystyrene and low-density polyethylene at room temperature [2].

Such phase-separated polymer mixtures serve as an excellent model system where molecular-level interactions between components are small. The system studied here is a film made of an immiscible binary mixture of atactic polystyrene and low-density polyethylene [2]. The dynamic IR measurement was carried out as before by mechanically stimulating the system at room temperature with a 23-Hz dynamic tensile strain with an amplitude of 0.1%. The time-dependent fluctuations of IR absorbance induced by the strain were recorded at a spectral resolution of 4 cm^{-1}.

Figure 1-18 shows the synchronous 2D IR spectrum of the blend. Autopeaks observed on the diagonal positions of the spectrum near 1454 and 1495 cm^{-1} represent the strain-induced local reorientation of polystyrene phenyl rings. The 1454-cm^{-1} band also contains a contribution from CH_2 deformation of the backbone of polystyrene [10, 11, 51]. A pair of intense cross peaks appear at the off-diagonal positions of the spectral plane near 1454 and 1495 cm^{-1}, indicating the existence of a strong synchronicity between the reorientation of transition dipoles associated with these two IR bands of polystyrene phenyl side groups.

Similarly, autopeaks corresponding to dynamic IR intensity variations for the CH_2 deformations of the polyethylene are observed near 1466 and 1475 cm^{-1}. These autopeaks arise from the reorientation of molecular chains in the amorphous and crystalline domains of polyethylene. A pair of cross peaks clearly correlate the two IR bands originating from the polyethylene component of this blend. It is important to note that there is little development of synchronous cross peaks correlating polystyrene bands to polyethylene bands.

The lack of synchronous cross peaks between polystyrene and polyethylene bands indicates these polymers are reorienting independently of each other. Cross peaks appearing in the asynchronous spectrum (Figure 1-19) also verify the above conclusion. For an immiscible blend of polyethylene and polystyrene, where molecular-level interactions between the phase-separated components are absent, the time-dependent behavior of IR intensity fluctuations of one component of the sample

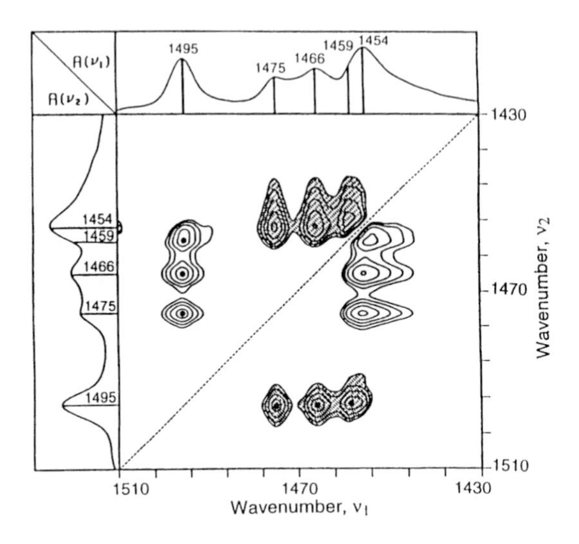

Figure 1-19. An asynchronous 2D IR spectrum of a blend of polystyrene and low-density polyethylene at room temperature [2].

may become substantially different from those of the other. Therefore, 2D IR correlation spectroscopy can easily differentiate individual spectral contributions from each of the immiscible components.

Another notable feature in Figure 1-19 is the development of asynchronous cross peaks correlating the 1459-cm^{-1} band attributed to the CH_2-deformation of the polystyrene backbone to the 1454-cm^{-1} and 1495-cm^{-1} bands arising from semicircle-stretching vibrations of the phenyl side groups. The asyncronicity among IR signals from these polystyrene bands again reflects the difference in mobilities of backbone and side-group functionalities, as already discussed in Section 1.5.1. The asynchronicity expected between the crystalline and amorphous bands of the polyethylene component are not apparent due to the intense cross peaks associated with polystyrene.

The assignments of the 1454-cm^{-1} and 1459-cm^{-1} bands can be independently verified by selectively substituting the hydrogen atoms of polystyrene backbone with deuterium atoms. The elimination of spectral contributions from the backbone of polystyrene leaves only the pure phenyl vibration spectrum in this region, as shown in Figure 1-20. The asynchronous 2D IR spectrum (Figure 1-19) differentiated the two IR bands even without deuterium substitution. This result clearly demonstrates the truly unique feature of 2D IR spectroscopy to enhance substantially the resolution of highly overlapped spectral regions by effectively spreading IR bands along the second dimension according to the characteristic reorientational response times of individual functional groups.

1.5.4 Miscible Polymer Blends

Certain classes of polymer blends are known to be truly miscible and form homogeneous, one-phase mixtures. A specific interaction at the submolecular level, e.g.,

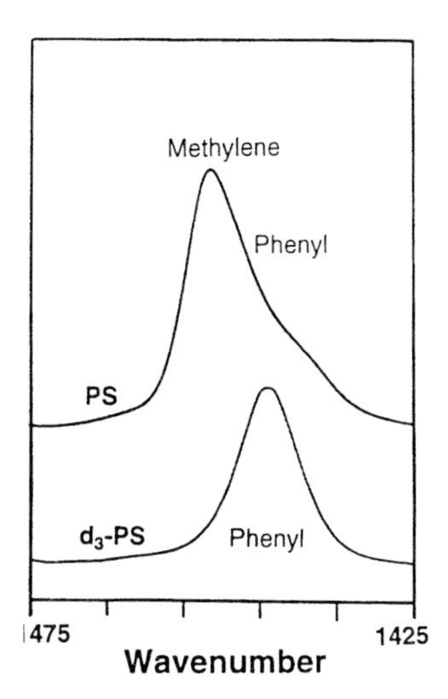

Figure 1-20. IR spectra of normal atactic polystyrene and backbone-deuterated polystyrene films.

hydrogen bonding, is often involved in such blends to overcome the tendency to phase segregate. In those cases, 2D IR spectroscopy is especially suited for the investigation of the molecular origin of such specific interactions. One of the well-known miscible systems, a blend of atactic polystyrene and poly(vinyl methyl ether) [19], was investigated using 2D IR to elucidate the molecular origin of the miscibility of these polymers [52, 53].

Interestingly, polystyrene and poly(vinyl methyl ether) are very different polymers. Polystyrene is a hard hydrophobic plastic resin, while poly(vinyl methyl ether) is a soft water-soluble polymer. It is quite surprising that such a pair of polymers could produce a molecularly mixed homogeneous one-phase system. The specific origin of the miscibility of this polymer pair is not well understood, although some level of interaction is speculated between the methoxyl groups of poly(vinyl methyl ether) and the phenyl groups of polystyrene [54].

To simplify the spectral assignment, polystyrene which had been selectively deuterium substituted to eliminate the backbone contribution (Figure 1-21) was blended with poly(vinyl methyl ether). Special attention was paid to the symmetric CH_2-stretching peak of the methoxyl group of the poly(vinyl methyl ether) component near 2820 cm^{-1}. Figure 1-22 shows the asynchronous 2D IR spectrum of poly(vinyl methyl ether) in this region. The presence of a pair of asynchronous cross peaks at the spectral coordinates at 2825 and 2813 cm^{-1} indicates there are two distinct populations of methoxyl groups belonging to the poly(vinyl methyl ether) component of this blend, each having a different IR absorption band.

Figure 1-23 shows that the methoxyl band of poly(vinyl methyl ether) at 2813 cm^{-1} is synchronously correlated with the IR bands located at 3055 cm^{-1} and 3024 cm^{-1} (assigned to the phenyl side groups of the polystyrene component of this blend). The presence of such synchronous correlation cross peaks strongly suggests the existence of a specific interaction synchronizing the local motions of the methoxyl groups of poly(vinyl methyl ether) with the polystyrene phenyl groups.

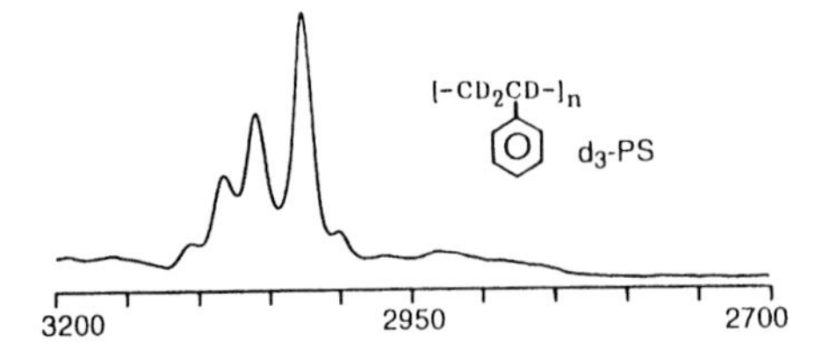

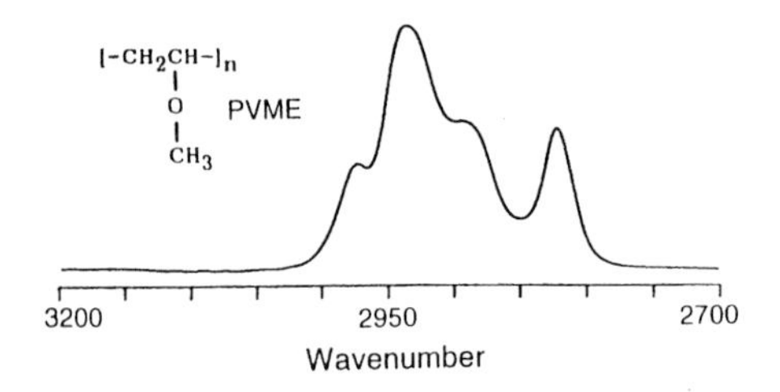

Figure 1-21. IR spectra of backbone-deuterated polystyrene and poly(vinyl methyl ether) in the CH-stretching vibration region.

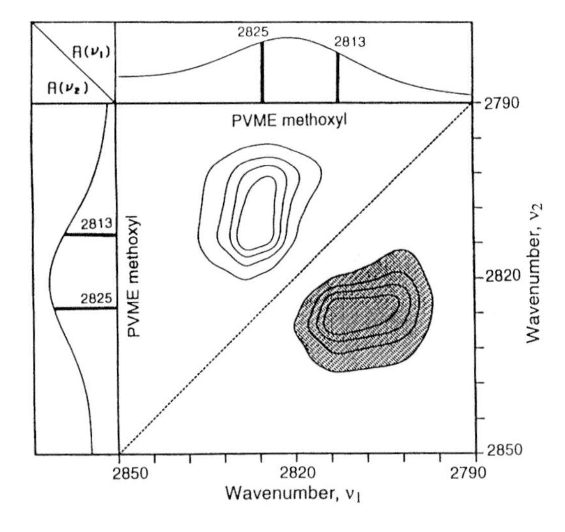

Figure 1-22. An asynchronous 2D IR spectrum for the methoxyl vibration of a blend of backbone-deuterated polystyrene and poly(vinyl methyl ether) [57].

Such a specific interaction most likely involves the lone-pair of electrons on the methoxyl oxygen atom of poly(vinyl methyl ether) and π-electrons of the polystyrene phenyl groups [54].

Figure 1-23 indicates only one methoxyl band is interacting with the polystyrene phenyl group, even though there are two distinct IR contributions of poly(vinyl methyl ether) methoxyl groups, as already shown in Figure 1-22. The other methoxyl band of poly(vinyl methyl ether) at 2824 cm^{-1} is asynchronously correlated with the phenyl bands of the polystyrene component (Figure 1-24). This result suggests that while one component of poly(vinyl methyl ether) methoxyl groups

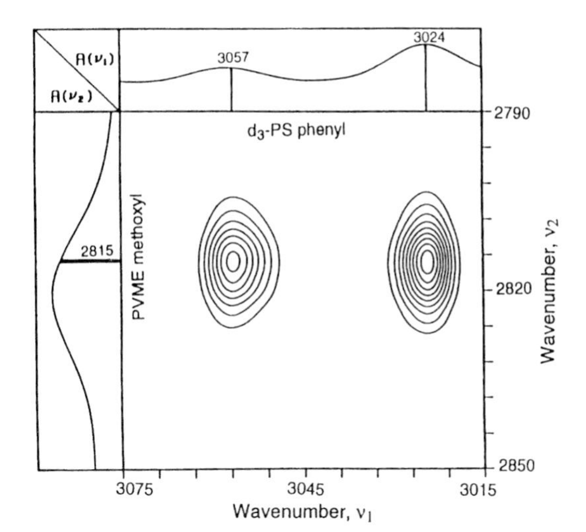

Figure 1-23. A synchronous 2D IR spectrum of a blend comparing the reorientational motions of transition dipoles associated with the polystyrene phenyl and poly(vinyl methyl ether) methoxyl groups [57].

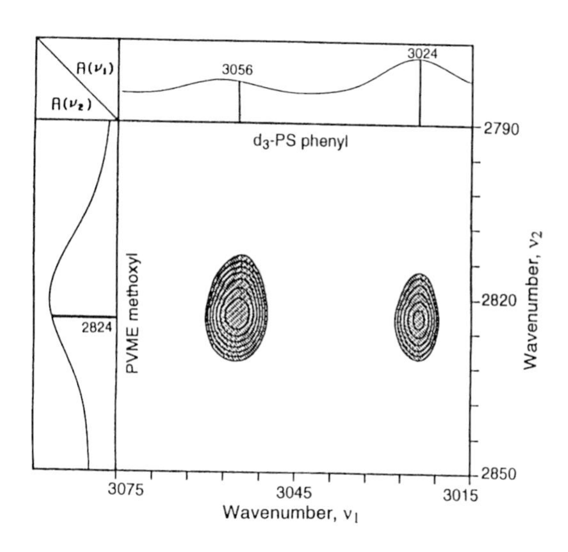

Figure 1-24. An asynchronous 2D IR spectrum of a blend comparing the reorientational motions of transition dipoles associated with the polystyrene phenyl and poly(vinyl methyl ether) methoxyl groups [57].

represented by the 2813-cm^{-1} band is strongly interacting with the polystyrene phenyl groups, the other component represented by the 2824 cm^{-1} band is not.

There are many other examples of miscible polymer blends arising from specific submolecular interactions of the components. The miscible polymer pair of polystyrene and poly(phenylene oxide), for example, was recently studied by the 2D FT-IR technique in an effort to understand the molecular origin of the miscibility of this system [55]. Specific interactions between polymers and various small molecules, such as additives and plasticizers, were also studied by 2D IR. The importance of localized submolecular level interactions among functional groups of mixture components was demonstrated.

1.5.5 Block Copolymers

Block copolymers made of dissimilar polymer segments joined by a covalent bond are interesting systems to study by 2D IR spectroscopy. Because of the repulsive interaction between the dissimilar segments, block copolymers often phase separate at the submolecular scale to form microdomains. It has been speculated for some time [56, 57] that the composition of block segments comprising such microdomains does not change sharply at the domain interface. Instead, a diffuse interlayer is thought to exist where substantial mixing of dissimilar block segments actually takes place. Analysis based on 2D IR was used to probe the existence of such an *interphase* region between microdomains [58, 59].

A thin solution-cast film of a diblock copolymer consisting of a polystyrene block ($\overline{M}_w$ 50,000) and a deuterium-substituted polyisoprene block ($\overline{M}_w$ 100,000) was stimulated at room temperature with a 0.1%-amplitude, 23-Hz dynamic tensile strain. In this composition range, diblock copolymers of styrene and isoprene produce a hexagonally packed regular array of rod-like polystyrene microdomains embedded in a continuous matrix of the polyisoprene block. The diameter of each polystyrene microdomain is well below 0.1 μm.

Figure 1-25 shows the synchronous 2D IR spectrum of this block copolymer in the CH-stretching region. Since the polyisoprene block of this polymer is fully deuterium substituted, the responses shown in Figure 1-25 represent exclusively the contribution from the polystyrene segment. One may expect the effect of the applied mechanical perturbation on the molecular reorientation of the block copolymer will be felt predominantly by the rubbery polyisoprene segments, since the dispersed polystyrene microdomains are in the glassy state at room temperature. It is, therefore, rather surprising to find autopeaks indicating substantial molecular mobility of polystyrene segments under dynamic strain at this temperature.

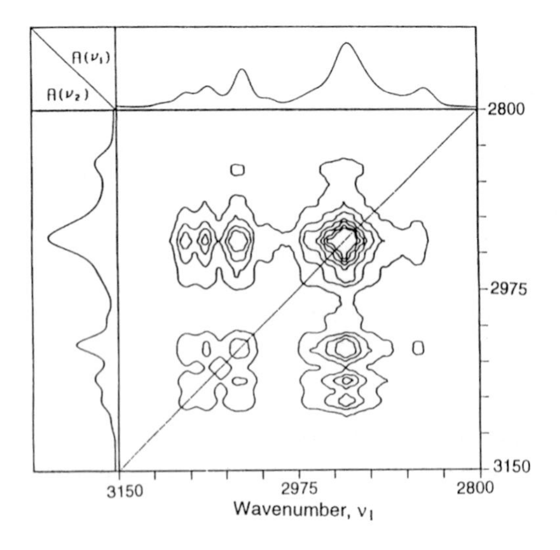

Figure 1-25. A synchronous 2D IR spectrum in the CH-stretching vibration region of a microphase-separated diblock copolymer consisting of polystyrene and perdeuterated polyisoprene segments.

Furthermore, the sign of the synchronous cross peaks correlating the motion of side-group phenyl and backbone methylene bands is positive. As discussed in Section 1.5.1, the positive cross peaks indicate the polystyrene segments giving rise to these dynamic IR signals are in the rubbery state, even though the measurement was made at room temperature. This result may be interpreted as the polystyrene segments showing rubber-like behavior being located in the diffuse interlayer region between the microdomains. The glass-transition temperature of the polystyrene segments in this region is substantially depressed due to mixing with the polyisoprene segments.

This hypothesis was later verified by selective deuterium substitution of part of the polystyrene segment [58, 59]. Near the junction between the two blocks (most likely to be located at the interphase region), polystyrene segments exhibited rubbery behavior even at room temperature. This part of the polystyrene segment reoriented extensively for a given dynamic deformation and contributed strongly to the dynamic IR dichroism signals. On the other hand, the tail end of the polystyrene segment (distributed more uniformly within the polystyrene microdomain) remained predominantly glassy. The local submolecular mobility of the tail end of polystyrene was noticeably limited.

1.5.6 Biopolymers

2D IR spectroscopy has also been used extensively to study not only synthetic polymers but also biopolymers. The exceptionally high spectral resolution of 2D IR has provided a major advantage over conventional IR analyses plagued by heavily overlapped bands of biopolymers. For example, amide I and II regions of IR spectra of various proteins are thought to be composed of numerous overlapping bands, each representing different protein conformation, e.g., helix, sheet, and random coils. However, it is usually very difficult to separate individual spectral contributions.

2D IR analysis of the protein component of human skin has been attempted in the past [3, 21]. Asynchronous 2D IR spectra suggest IR bands assignable to different protein conformations can be readily differentiated by the development of asynchronous cross peaks. Synchronous 2D IR spectra, on the other hand, can correlate different IR bands belonging to similar protein conformations. Thus, 2D IR spectroscopy may be used systematically to map out the IR band assignments of complex protein molecules.

Similar 2D IR studies have been carried out for human hair keratin [44, 60, 61]. Figure 1-26 shows the asynchronous 2D IR correlation spectrum of a protein film made from solubilized hair keratin [44]. In a normal one-dimensional IR spectrum, individual band components are highly overlapped and completely obscured. The enhanced spectral resolution is quite apparent in the 2D IR spectrum. The position of many of the asynchronous 2D IR cross peaks are in good agreement with other band assignments and normal-mode analyses [62].

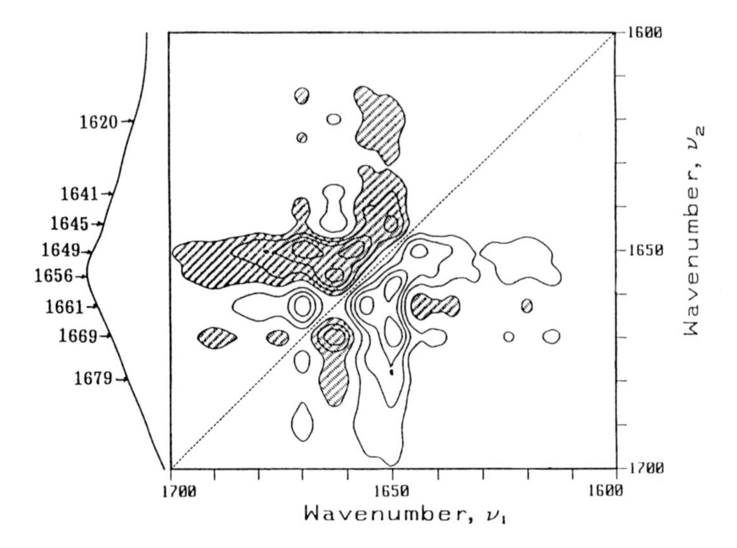

Figure 1-26. An asynchronous 2D IR spectrum in the amide I vibration region of a film made of solubilized human hair keratin [44].

1.6 Further Extension of 2D Correlation Technique

The basic concept of 2D IR spectroscopy based on the correlation analysis of perturbation-induced time-dependent fluctuations of IR intensities could be readily extended to other areas of polymers spectroscopy. The 2D correlation analysis has been successfully applied to the time-dependent variations of small angle X-ray scattering intensity measurement [4]. In this study, a small amplitude dynamic strain is applied a sheet of microphase-separated styrene–butadiene–styrene tri-block copolymer sample. Intensity variation of scattered X-ray beam due to the strain-induced changes in the interdomain Bragg distances coupled with the reorientation of microdomain structures is analyzed by using the 2D correlation map. Similarly, the formal approach of 2D correlation analysis to time-dependent spectral intensity fluctuations has been extended to UV, Raman [63], and near-IR spectroscopy [64]. There seems no intrinsic limitation to the application of this versatile technique in polymer spectroscopy.

The introduction of the generalized 2D correlation method [8] has made it possible to further extend the utility of 2D correlation analysis beyond the study of simple sinusoidally varying dynamic spectral signals as originally described in Eqs. (1-2) and (1-3). By utilizing the time-domain Fourier transform of spectral intensity variations, a formal definition of 2D correlation spectra resembling Eq. (1-9) can be derived in a straightforward manner. Using this generalized method, the 2D correlation analysis of any time-dependent spectral signals can be carried out. Highly nonlinear time-dependent reactions of crosslinking polymer systems, for example, are followed by the time-resolved IR measurement [65, 66]. The resulting sets of time-dependent IR spectra can be transformed into 2D correlation spectra to yield valuable information to probe details of curing reactions of polymers.

One of the very promising new developments in the recent 2D correlation analysis of polymers is the application of this versatile technique to the study of spectral changes recorded as functions of a general physical variable, which is no longer limited to time. For example, it is possible to apply the 2D correlation analysis to the near-IR spectral changes corresponding to the development of disorder in a hydrogen-bonded polyamide system induced by the rising temperature [67]. Of course, spectral changes of polymers induced by other variables, such as pressure, age, composition, and the like, may be analyzed in the same way.

1.7 Conclusions

2D IR spectroscopy based on the correlation analysis of the individual time-dependent behavior of localized characteristic reorientational motions of various submolecular moieties comprising a system has been shown to be very useful for a broad range of applications in the study of complex polymeric materials. In 2D IR, the spectral resolution is substantially enhanced by spreading the overlapped IR bands along the second dimension. The presence or lack of chemical interactions or connectivity among functional groups located in various parts of the polymer system are detected. The relative reorientation directions and the order of realignment sequence of submolecular units are also provided.

Notable discoveries made through the use of 2D IR spectroscopy include the fact that local dynamics of side groups can proceed independent of the polymer main chain, and there can be abrupt changes in the side group realignment mechanism above and below the glass transition temperature. It is possible to probe the microscopic spatial distribution of submolecular components of polymers, especially in phase-separated systems such as semicrystalline polymers and block copolymers, and the degree of interactions between various components can be estimated. Specific interactions between components in polymer mixtures are quite effectively identified. The application of 2D IR spectroscopy certainly is not limited to the characterization of traditional synthetic polymers. Complex macromolecules of biological origin can also be readily studied by this technique. The recent extension of the 2D correlation concept to (1) spectroscopic study of polymers using probes other than IR, (2) nonsinusoidally varying spectral changes, and (3) dependence on physical variables other than time has greatly expanded the scope of possible spectroscopic applications of this technique to the study of polymeric materials.

1.8 Symbols and Abbreviations

A	Absorbance
$A_{\parallel}$	Parallel absorbance
$A_{\perp}$	Perpendicular absorbance
A_o	Structural absorbance $\equiv A_{\parallel} + 2A_{\perp}$
P_2	Orientation factor
D	Dichroic ratio $\equiv A_{\parallel}/A_{\perp}$
D_{∞}	Ultimate dichroic ratio
DIRLD	Dynamic infrared linear dichroism
T	Observation period
t	Time
ΔA	Dichroic difference $\equiv A_{\parallel} - A_{\perp}$
ΔA_{∞}	Ultimate dichroic difference
$\Delta\tilde{A}$	Dynamic dichroic difference
$\Delta\hat{A}$	Amplitude of dynamic dichroic difference
$\Delta A'$	In-phase dynamic dichroic difference $\equiv \Delta\hat{A}\cos\beta$
$\Delta A''$	Quadrature dynamic dichroic difference $\equiv \Delta\hat{A}\sin\beta$
Φ	Synchronous correlation intensity
X	Two-dimensional cross-correlation function
Ψ	Asynchronous correlation intensity
α	Transition dipole orientation angle
β	Dynamic dichroism phase angle
$\tilde{\varepsilon}$	Dynamic strain
$\hat{\varepsilon}$	Dynamic strain amplitude
ν	Wavenumber
θ	Polymer chain segment orientation angle
τ	Correlation time
ω	Frequency of dynamic strain / dynamic dichroism
$\langle\rangle$	Spatial average

1.9 References

[1] Noda, I. *Bull. Am. Phys. Soc.* **1986**, *31*, 520.
[2] Noda, I. *J. Am. Chem. Soc.* **1989**, *111*, 8116.
[3] Noda, I. *Appl. Spectrosc.* **1990**, *44*, 550.
[4] Noda, I. *Chemtracts-Macromol. Ed.* **1990**, *1*, 89.
[5] Palmer, R. A.; Manning, C. J.; Chao, J. L.; Noda, I.; Dowrey, A. E.; Marcott, C. *Appl. Spectrosc.* **1991**, *45*, 12.
[6] Marcott, C.; Dowrey, A. E.; Noda, I. *Analyt. Chem.* **1994**, *66*, 1065A.
[7] Marcott, C.; Dowrey, A. E.; Noda, I. *Appl. Spectrosc.* **1993**, *47*, 1324.
[8] Noda, I. *Appl. Spectrosc.* **1993**, *47*, 1329.

[9] Colthup, N. B.; Daly, L. H.; Wilberley, S. E. *Introduction to Infrared and Raman Spectroscopy*; Academic: New York, 1975.
[10] Zbinden, R. *Infrared Spectroscopy of High Polymers;* Academic Press: New York, 1964.
[11] Painter, P. C.; Coleman, M. M.; Koenig, J. L. *The Theory of Vibrational Spectroscopy and Its Application to Polymeric Materials;* Wiley: New York, 1982.
[12] Koenig, J. L. *Spectroscopy of Polymers;* American Chemical Society: Washington, D.C., 1992.
[13] Aue, W. P.; Bartholdi, E.; Ernst, R. R.; *J. Chem. Phys.* **1976**, *64*, 2229.
[14] Nagayama, K.; Kumar, A.; Wüthrich, K.; Ernst, R. R. *J. Magn. Reson.* **1980**, *40*, 321.
[15] Ernst, R. R.; Bodenhausen, G.; Wokaun, A. *Principles of Nuclear Magnetic Resonance in One and Two Dimensions;* University Press: Oxford, 1987.
[16] Kessler, H.; Gehrke, M.; Griesinger, C. *Angew. Chem. Int. Ed. Engl.* **1988**, *27*, 490.
[17] Bovey, F. A. *Polym. Eng. Sci.* **1986**, *26*, 1419.
[18] Blümich, B.; Spiess, H. W. *Angew. Chem. Int. Ed. Engl.* **1988**, *27*, 1655.
[19] Mirau, P.; Tanaka, H.; Bovey, F. *Macromolecules* **1988**, *21*, 2929.
[20] Noda, I.; Dowrey, A. E.; Marcott, C. *J. Polym. Sci., Polym. Lett. Ed.* **1983**, *21*, 99.
[21] Palmer, R. A.; Manning, C. J.; Chao, J. L.; Noda, I.; Dowrey, A. E.; Marcott, C. *Appl. Spectrosc.* **1991**, *45*, 12.
[22] Meier, R. J. *Macromol. Symp.* **1997**, *119*, 25.
[23] Noda, I.; Dowrey, A. E.; Marcott, C. *Appl. Spectrosc.* **1988**, *42*, 203.
[24] Noda, I.; Dowrey, A. E.; Marcott, C. *J. Molec. Struct.* **1990**, *224*, 265.
[25] Noda, I.; Dowrey, A. E.; Marcott, C. *Polym. News* **1993**, *18*, 167.
[26] Ferry, J. D. *Viscoelastic Properties of Polymers;* 3rd ed.; Wiley: New York, 1980.
[27] McCrum, N. G.; Read, B. E.; Williams, G. *Anelastic and Dielectric Effects in Polymer Solids;* Wiley: New York, 1967.
[28] Fraser, R. D. B. *J. Chem. Phys.* **1953**, *21*, 1511.
[29] Fraser, R. D. B. *J. Chem. Phys.* **1958**, *28*, 1113.
[30] Stein, R. S. *J. Appl. Phys.* **1961**, *32*, 1280.
[31] Gregoriou, V. G.; Chao, J. L.; Toriumi, H.; Marcott, C.; Noda, I.; Palmer, R. A. *SPIE* **1991**, *1575*, 209.
[32] Gregoriou, V. G.; Chao, J. L.; Toriumi, H.; Palmer, R. A. *Chem. Phys. Lett.* **1991**, *179*, 491.
[33] Nakano, T.; Yokoyama, T.; Toriumi, H. *Appl. Spectrosc.* **1993**, *47*, 1354.
[34] Chazalviel, J.-N.; Dubin, V. M.; Mandal, K. C.; Ozaman, F. *Appl. Spectrosc.* **1993**, *47*, 1411.
[35] Budevska, B. O.; Griffiths, P. R.; *Analyt. Chem.* **1993**, *65*, 2963.
[36] Dittmar, R. M.; Chao, J. L.; Palmer, R. A., in: *Photoacoustic and Photothermal Phenomena III*: Bicanic, O. (Ed.). Springer: Berlin, 1992; pp. 492–496.
[37] Marcott, C.; Story, G. M.; Dowrey, A. E.; Reeder, R. C.; Noda, I., *Mikrochim. Acta [Suppl.]* **1997**, *14*, 157.
[38] Noda, I.; Dowrey, A. E.; Marcott, C. *Mikrochim. Acta [Wien]* **1988**, *I*, 101.
[39] Noda, I.; Dowrey, A. E.; Marcott, C.; *Polym. Prepr.* **1992**, *33*(1), 70.
[40] Noda, I.; Dowrey, A. E.; Marcott, C.; *Makromol. Chem., Macromol. Symp.* **1993**, *72*, 121.
[41] Noda, I.; Dowrey, A. E.; Marcott, C.; *J. Appl. Polym Sci.: Appl. Polym. Symp.* **1993**, *52*, 55.
[42] I. Noda, *Polym. Prepr. Japan, Engl. Ed.* **1990**, *39*(2), 1620.
[43] Noda, I.; Dowrey, A. E.; Marcott, C. *Polym. Prepr.* **1990**, *31*(1), 576.
[44] Noda, I.; Dowrey, A. E.; Marcott, C. in *Time-Resolved Vibrational Spectroscopy V*; Takahashi, H. Ed.; Springer-Verlag: 1992; pp. 331–334.
[45] Noda, I.; Dowrey, A. E.; Marcott, C. *J. Molec. Struct.* **1990**, *224*, 265.
[46] Stein, R. S.; Satkowski, M. M.; Noda, I. in *Polymer Blends, Solutions, and Interfaces,* Noda, I.; D. N. Rubingh, Eds.; Elsevier, New York: 1992; pp. 109–131.
[47] Gregoriou, V. G.; Noda, I.; Dowrey, A. E.; Marcott,C.; Chao, J. L.; Palmer, R. A.; J. Polym. Sci: Part B: Polym. Phys., **1993**, *31*, 1769.
[48] Palmer, R. A.; Manning, C. J.; Rzepicla, J. A.; Widder, J. M.; Thomas, P. J.; Chao, J. L.; Marcott, C.; Noda, I. *SPIE* **1989**, *1145*, 277.
[49] Marcott, C.; Noda, I.; Dowrey A. E. *Analytica Chim. Acta* **1991**, *250*, 131.
[50] Flory, P. J. *Principles of Polymer Chemistry;* Cornell University Press: Ithaca, 1953.
[51] Snyder, R. W.; Painter, P. C. *Polymer* **1981**, *22*, 1633.
[52] Noda, I.; Dowrey, A. E.; Marcott, C. *Oyo Buturi* **1991**, *60*, 1039.

[53] Satkowski, M. M.; Grothaus, J. T.; Smith, S. D.; Ashraf, A.; Marcott, C.; Dowrey, A. E.; Noda, I. in *Polymer Solutions, Blends, and Interfaces*, Noda, I.; Rubingh, D. N., Eds.; Elsevier: Amsterdam, 1992; pp. 89–108.
[54] Garcia, D. *J. Polym. Sci., Polym. Phys. Ed.* **1984**, *22*, 107.
[55] Palmer, R. A.; Gregoriou, V. G.; Chao, J. L. *Polym. Prepr.* **1992**, *33*(1), 1222.
[56] Learry, D. F.; Williams, M. C.; *J. Polym. Sci., Polym. Phys. Ed.* **1974**, *12*, 265.
[57] Hashimoto, T.; Fujimura, M.; Kawai, H. *Macromolecules* **1980**, *13*, 1660.
[58] Noda, I.; Smith, S. D.; Dowrey, A. E.; Grothaus, A. E.; Marcott, C. *Mat. Res. Soc. Symp. Proc.* **1990**, *171*, 117.
[59] Smith, S. D.; Noda, I.; Marcott, C.; Dowrey, A. E. in *Polymer Solutions, Blends, and Interfaces*, Noda, I.; Rubingh, D. N., Eds.; Elsevier: Amsterdam, 1992; pp. 43–64.
[60] Dowrey, A. E.; Hillebrand, G. G.; Noda, I. Marcott, C. *SPIE* **1989**, *1145*, 156.
[61] Noda, I.; Dowrey, A. E.; Marcott, C. *Polym. Prepr.* **1991**, *32*(3), 671.
[62] Byler, D. M.; Susi, H. *Biopolymers* **1986**, *25*, 469.
[63] Ebihara, K.; Takahashi, H.; Noda, I. *Appl. Spectrosc.* **1993**, *47*, 1343.
[64] Noda, I.; Story, G. M.; Dowrey, A. E.; Reeder, R. C.; Marcott, C. *Macromol. Symp.* **1997**, *119*, 1.
[65] Nakano, T.; Shimada, S.; Saitoh, R.; Noda, I. *Appl. Spectrosc.* **1993**, *47*, 1337.
[66] Shoda, K.; Katagiri, G.; Ishida, H.; Noda, I. *SPIE* **1994**, *2089*, 254.
[67] Ozaki, Y.; Liu, Y.; Noda, I. *Macromolecules* **1997**, *30*, 2391.

2 Segmental Mobility of Liquid Crystals and Liquid-Crystalline Polymers under External Fields: Characterization by Fourier-transform Infrared Polarization Spectroscopy

H. W. Siesler, I. Zebger, Ch. Kulinna, S. Okretic, S. Shilov and U. Hoffmann

2.1 Introduction

Liquid crystals and main-chain/side-chain liquid-crystalline polymers have gained scientific and technical importance due to their applications as display materials, their exceptional mechanical and thermal properties, and their prospective applications for optical information storage and nonlinear optics, respectively [1–13].

Among other aspects, the study of segmental mobility as a function of an external (electric, electromagnetic, or mechanical) perturbation is of basic interest for the understanding of the dynamics of molecular processes involved in technical applications of liquid-crystalline materials.

Fourier-transform infrared (FTIR) polarization spectroscopy has proved an extremely powerful technique to characterize the segmental mobility of specific functionalities of a polymer or a liquid-crystal(line polymer) under the influence of an external perturbation [14–23]. In a side-chain liquid-crystalline polymer (SCLCP) aligned due to the influence of an external field, for example, the orientation of the polymer backbone relative to the rigid mesogen is strongly influenced by the coupling via a more or less flexible spacer (Figure 2-1). Thus, the availability of characteristic absorption bands for these individual functionalities will allow a quantitative characterization of their relative alignment and contribute towards a better understanding of their structure–property correlation. Based on this knowledge, tailor-made systems can then be synthesized by a systematic variation of partial structures.

In what follows, different phenomena will be discussed for selected classes of liquid-crystalline compounds:

- structure-dependent alignment of side-chain liquid-crystalline polyacrylates on anisotropic surfaces
- electric field-induced orientation and relaxation of a liquid crystal, a side-chain liquid-crystalline polymer and a liquid-crystalline guest–host system
- reorientation dynamics of a ferroelectric side-chain liquid-crystalline polymer in a polarity-switched electric field
- alignment of side-chain liquid-crystalline polyesters under laser irradiation
- shear-induced orientation in chrial smectic liquid-crystalline copolysiloxanes

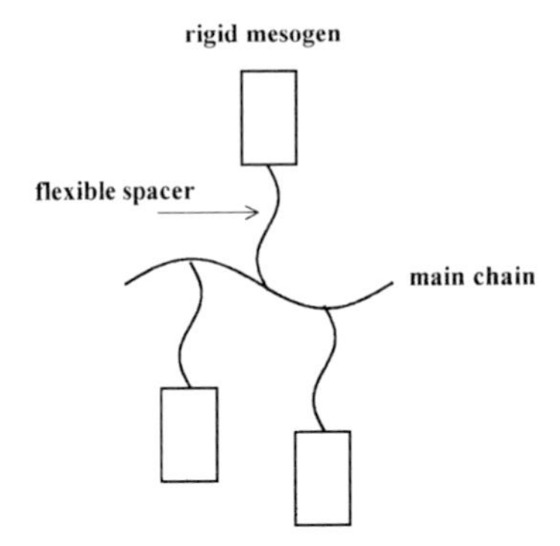

Figure 2-1. Schematic structure of a side-chain liquid-crystalline polymer.

- orientation mechanism of a liquid single crystal elastomer during cyclic deformation and recovery.

Static and time-resolved spectroscopic data acquired both, by the conventional rapid-scan and by the novel step-scan technique will be presented and interpreted in terms of their individual applications.

2.2 Measurement Techniques and Instrumentation

Since the mid-1970s, FTIR spectroscopy has been widely used for polymer analytical applications and for studies of time-dependent phenomena in chemical and physical processes [14]. With the current improvement in electronics and data acquisition procedures, the time resolution for such investigations with an acceptable signal/noise-ratio is presently about 20 ms. Time resolutions far beyond this value are not accessible by the conventional rapid-scan technique with continuous mirror movement without additional interferogram data manipulation. Thus, for time-resolved, rapid-scan measurements, generally, the retardation time of the moving mirror has to be considerably shorter than the time duration of the investigated event (Figure 2-2a). In search of new approaches, stroboscopic techniques have been proposed, where interferograms taken in the rapid-scan mode are subsequently subjected to a sampling of reordered interferogram segments and have thus led to an improvement in time resolution by a factor of about 500 [24, 25]. A further improvement in time resolution to the submicrosecond level has been achieved with the step-scan technique [16, 26, 27]. Here, the mirror is moved stepwise and the optical retardation is held constant during sampling of the individual interferogram elements. Thus, the time resolution is only limited by the response of

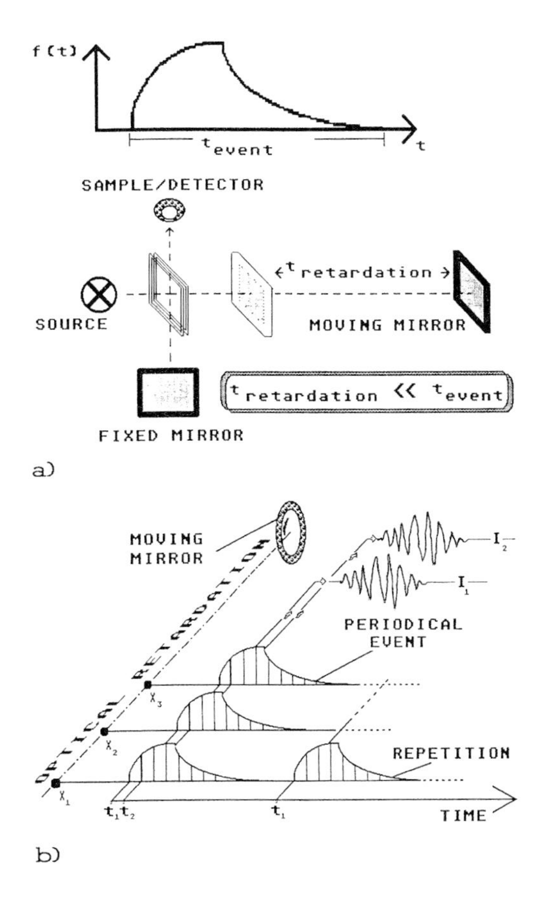

Figure 2-2. Principle of time-resolved rapid-scan (a) and step-scan (b) FTIR spectroscopy.

the IR-detector and the electronics. Finally, the interferogram elements are time-sorted and rearranged to obtain complete interferograms (Figure 2-2b). However, apart from the strict requirements on the stability of the scanner speed and the mirror positioning for the stroboscopic and step-scan techniques, respectively, it has to be pointed out clearly, that the achieved advantages in time resolution can be exploited only for the characterization of *reversible* processes [16].

The polarization measurements in the mid-infrared (IR) and near-infrared (NIR) were performed on a Bruker IFS 88 FTIR/FTNIR spectrometer utilizing a wire-grid polarizer on KRS5-substrate which could be rotated pneumatically parallel or perpendicular to the selected reference direction. For the experiments in the step-scan mode a mercury–cadmium–telluride (MCT) detector with a DC-coupled pre-amplifier was used. This allowed an absolute intensity at each mirror position to be recorded and the use of phase-modulation–demodulation techniques, which are often applied for the step-scan mode [28–30] to be avoided. Further specific instrumental details of the individual applications are given in the corresponding sections.

2.3 Theory

Throughout this chapter the general form of the order parameter S will be used [3, 31] to describe the long-range orientational order of the investigated liquid-crystalline materials. In contrast to the theory of Maier and Saupe [32, 33], which considers the uniaxial distribution of the mesogens in a nematic phase relative to the macroscopic preferential direction (director), this formalism also allows negative values. The relation of S to the inclination angle θ of the mesogenic/structural-unit axis and the reference direction (if not otherwise stated, the external field direction) is outlined for the case of an SCLCP in Figure 2-3. Also shown are the values of S for parallel and perpendicular orientation of the mesogens and their isotropic distribution. The derivation of S, which corresponds to the orientation factor f in general polymer orientation terminology [14, 17–20], is given in Figure 2-3, where R is the dichroic ratio determined from the experimental absorbance values measured with light polarized parallel and perpendicular, respectively, to the chosen reference direction. R_0 is the theoretical dichroic ratio for perfect alignment of the mesogen, and φ is the angle between the transition moment of the absorbing group and the local chain axis (in the present case the long axis of the mesogenic group) [14]. The structural absorbance A_0 has been chosen as parameter which represents the intensity of an absorption band exclusive of contributions from the orientation of the investigated polymer. It is given by [14]:

$$A_0 = \frac{A_{\parallel} + 2 \cdot A_{\perp}}{3} \tag{2-1}$$

Both, order parameter S and structural absorbance A_0 are based on the assumption of a uniaxial orientation symmetry [14, 18–20].

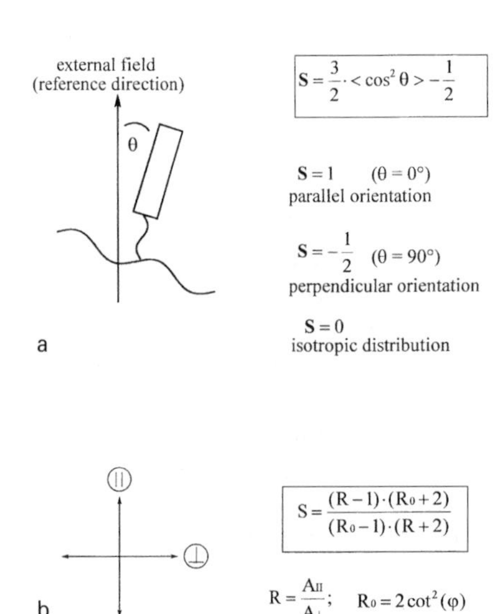

Figure 2-3. Geometrical considerations for the derivation of the order parameter S of a side-chain liquid-crystalline polymer under the influence of an external field (θ = inclination angle of the mesogen with the reference axis, φ = angle between the transition moment of the absorbing group and the local chain axis, R_0 = theoretical dichroic ratio for a perfectly aligned structural unit).

In the case of the laser irradiation experiments (see Section 2.6), the polarization direction of the incident laser beam was chosen as reference direction. Contrary to the former publications we have changed the reference direction for the calculation of the photo-induced anisotropy due to a better understanding of the underlying orientation mechanism. We assume that the mesogens are aligned preferentially perpendicular to the *polarization vector* [34, 35] and not to a *polarization plane* [15, 36] of the irradiating laser. Thus, we obtain negative S-values, whereas for the last-mentioned case the order parameter values would be positive, according to the Maier–Saupe theory.

2.4 Structure-Dependent Alignment of Side-Chain Liquid-Crystalline Polyacrylates on Anisotropic Surfaces

For the characterization of the anisotropic properties of liquid-crystalline (LC) materials it is necessary to induce a well-defined orientation of the mesogenic groups. Although spontaneous orientation phenomena can be regularly observed in limited domains, the necessary prerequisite for the application of LC-materials as optical construction elements, for example, is the preparation of uniform textures over larger areas. Such structures can be prepared by the alignment of LC-systems on anisotropic surfaces (e.g., glass, polymer films). The origin and theory of the interaction and the induced anisotropy has been treated in several books and articles [37–44]. Here, the attention is primarily directed towards the influence of the spacer length in a homologous series of nematic side-chain liquid-crystalline polyacrylates on their alignment on an anisotropic polyimide surface.

2.4.1 Materials and Experimental

The polymers had a molecular weight M_n (determined by gel permeation chromatography) of about 5000 [45] and their structural formula is shown alongside their nematic–isotropic transition temperatures (determined by polarization microscopy) [45] in Figure 2-4.

For the preparation of the substrate surface, KBr-plates have been spin-coated with a thin layer of polyimide (polyimide-kit ZLI 2650, Merck, Darmstadt, Germany). Upon drying and curing this coating to the final thickness of about 1 µm, the surface anisotropy was induced by unidirectional rubbing with a polymethacrylimide hardfoam-roll. The LC-polymers were sandwiched between the surface-treated KBr-plates above their clearing temperatures and finally annealed 3 °C below these temperatures for several hours. FTIR polarization spectra with a resolution of 4 cm^{-1} were recorded by accumulating 20 scans.

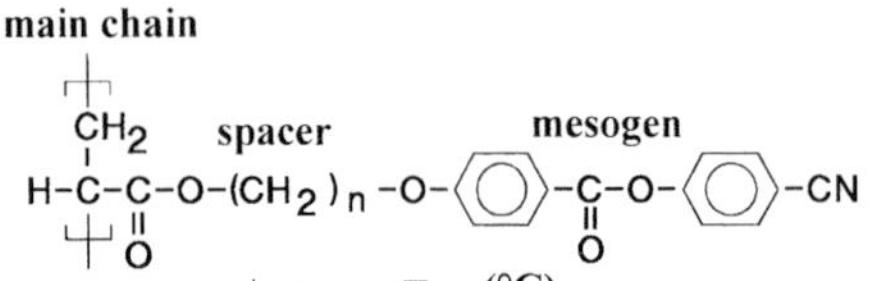

main chain

CH_2 spacer mesogen

$H-C-C-O-(CH_2)_n-O-$⟨◯⟩$-C(=O)-O-$⟨◯⟩$-CN$

	n	$T_{n,i}$ (°C)
P2CN	2	100
P4CN	4	95
P6CN	6	93

Figure 2-4. Structure and transition temperatures of the investigated side-chain liquid-crystalline polyacrylates.

2.4.2 Results

Despite identical sample pretreatment, gradually decreasing dichroic effects could be observed for the LC-polymers with decreasing spacer length. The FTIR polarization spectra of the P6CN- and P2CN-system, measured with radiation polarized parallel and perpendicular to the rubbing direction of the polyimide surface, respectively, are shown in Figure 2-5. From these data the coupling effect of the spacer between the main chain and the mesogen is clearly evident. Thus, the hexamethylene-spacer almost completely decouples the mesogen from the main chain and allows the induction of an extremely high orientation of the rigid mesogen ($S(\nu(C{\equiv}N)) = 0.62$) (see also insert in Figure 2-5). The short dimethylene-spacer, on the other hand, strongly links the mesogen to the polymer backbone and no orientation of the mesogen on the anisotropic surface can be observed. The spectra of the system with the tetramethylene-spacer (not shown here) display intermediate effects ($S(\nu(C{\equiv}N) = 0.46$). A detailed analysis of the dichroic effects in the polarization spectra of the whole series in terms of the relative orientation of the mesogen, spacer and main chain has been given in [36].

2.5 Electric Field-Induced Orientation and Relaxation of Liquid-Crystalline Systems

The reorientation of liquid crystals in an external electric field is the basis of their technical application in liquid-crystal displays (LCDs) and has been treated in numerous books and reviews [1–3, 12, 38, 46–50]. This section is intended to demonstrate the application of time-resolved FTIR spectroscopy for a better understanding of the electro-optical performance of liquid-crystalline systems. Due to the reversible character of the orientational and relaxational motions of liquid crystals under the perturbation of an electric field, the time-resolution of the FTIR measurements can be extended down to the microsecond range by the implementation of the step-scan technique (see Section 2.2). Thus, an insight on a molecular level

[9] Colthup, N. B.; Daly, L. H.; Wilberley, S. E. *Introduction to Infrared and Raman Spectroscopy*; Academic: New York, 1975.
[10] Zbinden, R. *Infrared Spectroscopy of High Polymers;* Academic Press: New York, 1964.
[11] Painter, P. C.; Coleman, M. M.; Koenig, J. L. *The Theory of Vibrational Spectroscopy and Its Application to Polymeric Materials;* Wiley: New York, 1982.
[12] Koenig, J. L. *Spectroscopy of Polymers;* American Chemical Society: Washington, D.C., 1992.
[13] Aue, W. P.; Bartholdi, E.; Ernst, R. R.; *J. Chem. Phys.* **1976**, *64*, 2229.
[14] Nagayama, K.; Kumar, A.; Wüthrich, K.; Ernst, R. R. *J. Magn. Reson.* **1980**, *40*, 321.
[15] Ernst, R. R.; Bodenhausen, G.; Wokaun, A. *Principles of Nuclear Magnetic Resonance in One and Two Dimensions;* University Press: Oxford, 1987.
[16] Kessler, H.; Gehrke, M.; Griesinger, C. *Angew. Chem. Int. Ed. Engl.* **1988**, *27*, 490.
[17] Bovey, F. A. *Polym. Eng. Sci.* **1986**, *26*, 1419.
[18] Blümich, B.; Spiess, H. W. *Angew. Chem. Int. Ed. Engl.* **1988**, *27*, 1655.
[19] Mirau, P.; Tanaka, H.; Bovey, F. *Macromolecules* **1988**, *21*, 2929.
[20] Noda, I.; Dowrey, A. E.; Marcott, C. *J. Polym. Sci., Polym. Lett. Ed.* **1983**, *21*, 99.
[21] Palmer, R. A.; Manning, C. J.; Chao, J. L.; Noda, I.; Dowrey, A. E.; Marcott, C. *Appl. Spectrosc.* **1991**, *45*, 12.
[22] Meier, R. J. *Macromol. Symp.* **1997**, *119*, 25.
[23] Noda, I.; Dowrey, A. E.; Marcott, C. *Appl. Spectrosc.* **1988**, *42*, 203.
[24] Noda, I.; Dowrey, A. E.; Marcott, C. *J. Molec. Struct.* **1990**, *224*, 265.
[25] Noda, I.; Dowrey, A. E.; Marcott, C. *Polym. News* **1993**, *18*, 167.
[26] Ferry, J. D. *Viscoelastic Properties of Polymers;* 3rd ed.; Wiley: New York, 1980.
[27] McCrum, N. G.; Read, B. E.; Williams, G. *Anelastic and Dielectric Effects in Polymer Solids;* Wiley: New York, 1967.
[28] Fraser, R. D. B. *J. Chem. Phys.* **1953**, *21*, 1511.
[29] Fraser, R. D. B. *J. Chem. Phys.* **1958**, *28*, 1113.
[30] Stein, R. S. *J. Appl. Phys.* **1961**, *32*, 1280.
[31] Gregoriou, V. G.; Chao, J. L.; Toriumi, H.; Marcott, C.; Noda, I.; Palmer, R. A. *SPIE* **1991**, *1575*, 209.
[32] Gregoriou, V. G.; Chao, J. L.; Toriumi, H.; Palmer, R. A. *Chem. Phys. Lett.* **1991**, *179*, 491.
[33] Nakano, T.; Yokoyama, T.; Toriumi, H. *Appl. Spectrosc.* **1993**, *47*, 1354.
[34] Chazalviel, J.-N.; Dubin, V. M.; Mandal, K. C.; Ozaman, F. *Appl. Spectrosc.* **1993**, *47*, 1411.
[35] Budevska, B. O.; Griffiths, P. R.; *Analyt. Chem.* **1993**, *65*, 2963.
[36] Dittmar, R. M.; Chao, J. L.; Palmer, R. A., in: *Photoacoustic and Photothermal Phenomena III*: Bicanic, O. (Ed.). Springer: Berlin, 1992; pp. 492–496.
[37] Marcott, C.; Story, G. M.; Dowrey, A. E.; Reeder, R. C.; Noda, I., *Mikrochim. Acta [Suppl.]* **1997**, *14*, 157.
[38] Noda, I.; Dowrey, A. E.; Marcott, C. *Mikrochim. Acta [Wien]* **1988**, *I*, 101.
[39] Noda, I.; Dowrey, A. E.; Marcott, C.; *Polym. Prepr.* **1992**, *33*(1), 70.
[40] Noda, I.; Dowrey, A. E.; Marcott, C.; *Makromol. Chem., Macromol. Symp.* **1993**, *72*, 121.
[41] Noda, I.; Dowrey, A. E.; Marcott, C.; *J. Appl. Polym Sci.: Appl. Polym. Symp.* **1993**, *52*, 55.
[42] I. Noda, *Polym. Prepr. Japan, Engl. Ed.* **1990**, *39*(2), 1620.
[43] Noda, I.; Dowrey, A. E.; Marcott, C. *Polym. Prepr.* **1990**, *31*(1), 576.
[44] Noda, I.; Dowrey, A. E.; Marcott, C. in *Time-Resolved Vibrational Spectroscopy V*; Takahashi, H. Ed.; Springer-Verlag: 1992; pp. 331–334.
[45] Noda, I.; Dowrey, A. E.; Marcott, C. *J. Molec. Struct.* **1990**, *224*, 265.
[46] Stein, R. S.; Satkowski, M. M.; Noda, I. in *Polymer Blends, Solutions, and Interfaces,* Noda, I.; D. N. Rubingh, Eds.; Elsevier, New York: 1992; pp. 109–131.
[47] Gregoriou, V. G.; Noda, I.; Dowrey, A. E.; Marcott,C.; Chao, J. L.; Palmer, R. A.; J. Polym. Sci: Part B: Polym. Phys., **1993**, *31*, 1769.
[48] Palmer, R. A.; Manning, C. J.; Rzepicla, J. A.; Widder, J. M.; Thomas, P. J.; Chao, J. L.; Marcott, C.; Noda, I. *SPIE* **1989**, *1145*, 277.
[49] Marcott, C.; Noda, I.; Dowrey A. E. *Analytica Chim. Acta* **1991**, *250*, 131.
[50] Flory, P. J. *Principles of Polymer Chemistry;* Cornell University Press: Ithaca, 1953.
[51] Snyder, R. W.; Painter, P. C. *Polymer* **1981**, *22*, 1633.
[52] Noda, I.; Dowrey, A. E.; Marcott, C. *Oyo Buturi* **1991**, *60*, 1039.

[53] Satkowski, M. M.; Grothaus, J. T.; Smith, S. D.; Ashraf, A.; Marcott, C.; Dowrey, A. E.; Noda, I. in *Polymer Solutions, Blends, and Interfaces*, Noda, I.; Rubingh, D. N., Eds.; Elsevier: Amsterdam, 1992; pp. 89–108.
[54] Garcia, D. *J. Polym. Sci., Polym. Phys. Ed.* **1984**, *22*, 107.
[55] Palmer, R. A.; Gregoriou, V. G.; Chao, J. L. *Polym. Prepr.* **1992**, *33*(1), 1222.
[56] Learry, D. F.; Williams, M. C.; *J. Polym. Sci., Polym. Phys. Ed.* **1974**, *12*, 265.
[57] Hashimoto, T.; Fujimura, M.; Kawai, H. *Macromolecules* **1980**, *13*, 1660.
[58] Noda, I.; Smith, S. D.; Dowrey, A. E.; Grothaus, A. E.; Marcott, C. *Mat. Res. Soc. Symp. Proc.* **1990**, *171*, 117.
[59] Smith, S. D.; Noda, I.; Marcott, C.; Dowrey, A. E. in *Polymer Solutions, Blends, and Interfaces*, Noda, I.; Rubingh, D. N., Eds.; Elsevier: Amsterdam, 1992; pp. 43–64.
[60] Dowrey, A. E.; Hillebrand, G. G.; Noda, I. Marcott, C. *SPIE* **1989**, *1145*, 156.
[61] Noda, I.; Dowrey, A. E.; Marcott, C. *Polym. Prepr.* **1991**, *32*(3), 671.
[62] Byler, D. M.; Susi, H. *Biopolymers* **1986**, *25*, 469.
[63] Ebihara, K.; Takahashi, H.; Noda, I. *Appl. Spectrosc.* **1993**, *47*, 1343.
[64] Noda, I.; Story, G. M.; Dowrey, A. E.; Reeder, R. C.; Marcott, C. *Macromol. Symp.* **1997**, *119*, 1.
[65] Nakano, T.; Shimada, S.; Saitoh, R.; Noda, I. *Appl. Spectrosc.* **1993**, *47*, 1337.
[66] Shoda, K.; Katagiri, G.; Ishida, H.; Noda, I. *SPIE* **1994**, *2089*, 254.
[67] Ozaki, Y.; Liu, Y.; Noda, I. *Macromolecules* **1997**, *30*, 2391.

2 Segmental Mobility of Liquid Crystals and Liquid-Crystalline Polymers under External Fields: Characterization by Fourier-transform Infrared Polarization Spectroscopy

H. W. Siesler, I. Zebger, Ch. Kulinna, S. Okretic, S. Shilov and U. Hoffmann

2.1 Introduction

Liquid crystals and main-chain/side-chain liquid-crystalline polymers have gained scientific and technical importance due to their applications as display materials, their exceptional mechanical and thermal properties, and their prospective applications for optical information storage and nonlinear optics, respectively [1–13].

Among other aspects, the study of segmental mobility as a function of an external (electric, electromagnetic, or mechanical) perturbation is of basic interest for the understanding of the dynamics of molecular processes involved in technical applications of liquid-crystalline materials.

Fourier-transform infrared (FTIR) polarization spectroscopy has proved an extremely powerful technique to characterize the segmental mobility of specific functionalities of a polymer or a liquid-crystal(line polymer) under the influence of an external perturbation [14–23]. In a side-chain liquid-crystalline polymer (SCLCP) aligned due to the influence of an external field, for example, the orientation of the polymer backbone relative to the rigid mesogen is strongly influenced by the coupling via a more or less flexible spacer (Figure 2-1). Thus, the availability of characteristic absorption bands for these individual functionalities will allow a quantitative characterization of their relative alignment and contribute towards a better understanding of their structure–property correlation. Based on this knowledge, tailor-made systems can then be synthesized by a systematic variation of partial structures.

In what follows, different phenomena will be discussed for selected classes of liquid-crystalline compounds:

- structure-dependent alignment of side-chain liquid-crystalline polyacrylates on anisotropic surfaces
- electric field-induced orientation and relaxation of a liquid crystal, a side-chain liquid-crystalline polymer and a liquid-crystalline guest–host system
- reorientation dynamics of a ferroelectric side-chain liquid-crystalline polymer in a polarity-switched electric field
- alignment of side-chain liquid-crystalline polyesters under laser irradiation
- shear-induced orientation in chrial smectic liquid-crystalline copolysiloxanes

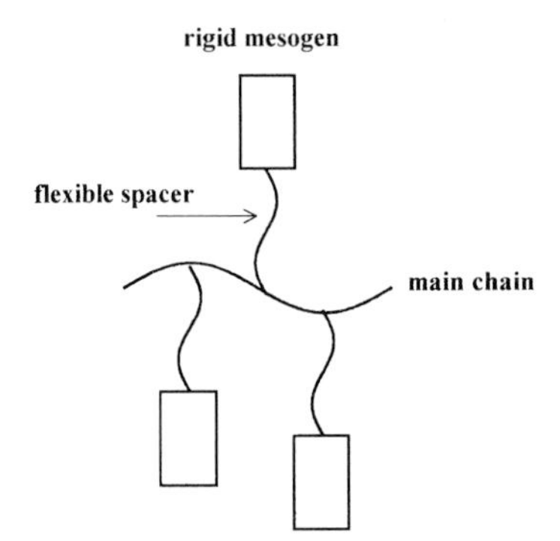

Figure 2-1. Schematic structure of a side-chain liquid-crystalline polymer.

- orientation mechanism of a liquid single crystal elastomer during cyclic deformation and recovery.

Static and time-resolved spectroscopic data acquired both, by the conventional rapid-scan and by the novel step-scan technique will be presented and interpreted in terms of their individual applications.

2.2 Measurement Techniques and Instrumentation

Since the mid-1970s, FTIR spectroscopy has been widely used for polymer analytical applications and for studies of time-dependent phenomena in chemical and physical processes [14]. With the current improvement in electronics and data acquisition procedures, the time resolution for such investigations with an acceptable signal/noise-ratio is presently about 20 ms. Time resolutions far beyond this value are not accessible by the conventional rapid-scan technique with continuous mirror movement without additional interferogram data manipulation. Thus, for time-resolved, rapid-scan measurements, generally, the retardation time of the moving mirror has to be considerably shorter than the time duration of the investigated event (Figure 2-2a). In search of new approaches, stroboscopic techniques have been proposed, where interferograms taken in the rapid-scan mode are subsequently subjected to a sampling of reordered interferogram segments and have thus led to an improvement in time resolution by a factor of about 500 [24, 25]. A further improvement in time resolution to the submicrosecond level has been achieved with the step-scan technique [16, 26, 27]. Here, the mirror is moved stepwise and the optical retardation is held constant during sampling of the individual interferogram elements. Thus, the time resolution is only limited by the response of

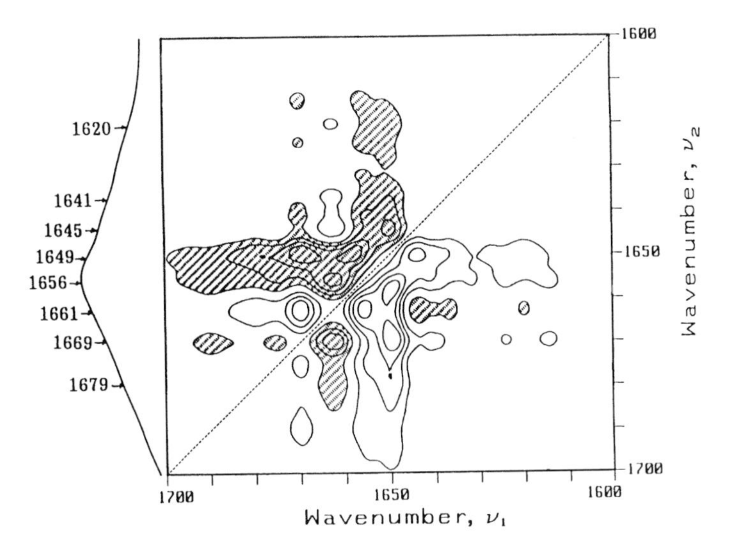

Figure 1-26. An asynchronous 2D IR spectrum in the amide I vibration region of a film made of solubilized human hair keratin [44].

1.6 Further Extension of 2D Correlation Technique

The basic concept of 2D IR spectroscopy based on the correlation analysis of perturbation-induced time-dependent fluctuations of IR intensities could be readily extended to other areas of polymers spectroscopy. The 2D correlation analysis has been successfully applied to the time-dependent variations of small angle X-ray scattering intensity measurement [4]. In this study, a small amplitude dynamic strain is applied a sheet of microphase-separated styrene–butadiene–styrene triblock copolymer sample. Intensity variation of scattered X-ray beam due to the strain-induced changes in the interdomain Bragg distances coupled with the reorientation of microdomain structures is analyzed by using the 2D correlation map. Similarly, the formal approach of 2D correlation analysis to time-dependent spectral intensity fluctuations has been extended to UV, Raman [63], and near-IR spectroscopy [64]. There seems no intrinsic limitation to the application of this versatile technique in polymer spectroscopy.

The introduction of the generalized 2D correlation method [8] has made it possible to further extend the utility of 2D correlation analysis beyond the study of simple sinusoidally varying dynamic spectral signals as originally described in Eqs. (1-2) and (1-3). By utilizing the time-domain Fourier transform of spectral intensity variations, a formal definition of 2D correlation spectra resembling Eq. (1-9) can be

Furthermore, the sign of the synchronous cross peaks correlating the motion of side-group phenyl and backbone methylene bands is positive. As discussed in Section 1.5.1, the positive cross peaks indicate the polystyrene segments giving rise to these dynamic IR signals are in the rubbery state, even though the measurement was made at room temperature. This result may be interpreted as the polystyrene segments showing rubber-like behavior being located in the diffuse interlayer region between the microdomains. The glass-transition temperature of the polystyrene segments in this region is substantially depressed due to mixing with the polyisoprene segments.

This hypothesis was later verified by selective deuterium substitution of part of the polystyrene segment [58, 59]. Near the junction between the two blocks (most likely to be located at the interphase region), polystyrene segments exhibited rubbery behavior even at room temperature. This part of the polystyrene segment reoriented extensively for a given dynamic deformation and contributed strongly to the dynamic IR dichroism signals. On the other hand, the tail end of the polystyrene segment (distributed more uniformly within the polystyrene microdomain) remained predominantly glassy. The local submolecular mobility of the tail end of polystyrene was noticeably limited.

1.5.6 Biopolymers

2D IR spectroscopy has also been used extensively to study not only synthetic polymers but also biopolymers. The exceptionally high spectral resolution of 2D IR has provided a major advantage over conventional IR analyses plagued by heavily overlapped bands of biopolymers. For example, amide I and II regions of IR spectra of various proteins are thought to be composed of numerous overlapping bands, each representing different protein conformation, e.g., helix, sheet, and random coils. However, it is usually very difficult to separate individual spectral contributions.

2D IR analysis of the protein component of human skin has been attempted in the past [3, 21]. Asynchronous 2D IR spectra suggest IR bands assignable to different protein conformations can be readily differentiated by the development of asynchronous cross peaks. Synchronous 2D IR spectra, on the other hand, can correlate different IR bands belonging to similar protein conformations. Thus, 2D IR spectroscopy may be used systematically to map out the IR band assignments of complex protein molecules.

Similar 2D IR studies have been carried out for human hair keratin [44, 60, 61]. Figure 1-26 shows the asynchronous 2D IR correlation spectrum of a protein film made from solubilized hair keratin [44]. In a normal one-dimensional IR spectrum, individual band components are highly overlapped and completely obscured. The enhanced spectral resolution is quite apparent in the 2D IR spectrum. The position of many of the asynchronous 2D IR cross peaks are in good agreement with other band assignments and normal-mode analyses [62].

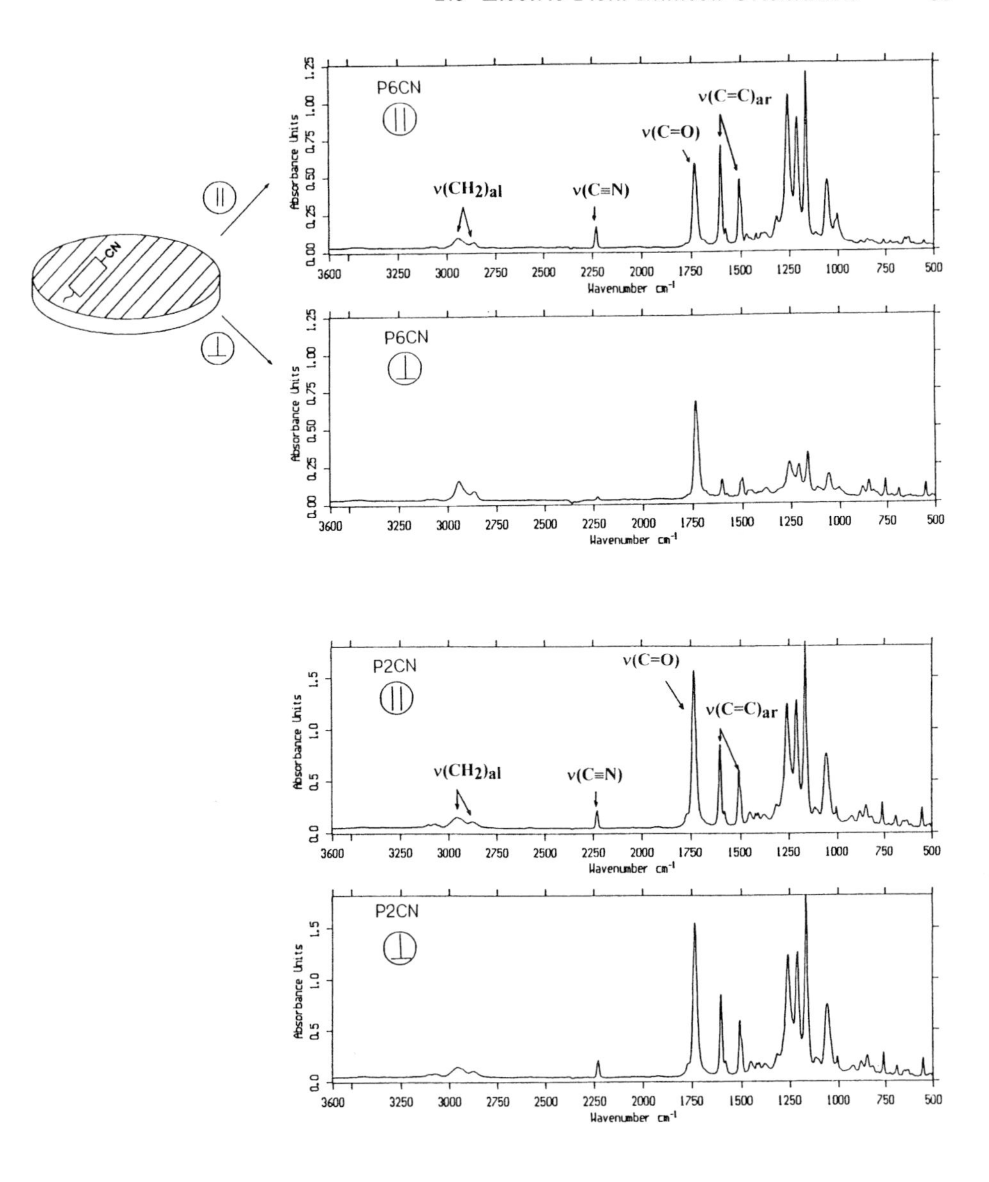

Figure 2-5. FTIR polarization spectra of the P6CN- and P2CN-system (the insert schematically shows the alignment of the mesogens on the anisotropic surface; parallel and perpendicular refers to the polarization of the IR radiation relative to the rubbing direction).

can be provided into the relevant mechanisms. In what follows, the application of vibrational spectroscopy to the investigation of different phenomena will be treated separately. On the one hand we will discuss the electric field-induced changes of nematic LC-systems (low-molecular weight compounds, side-chain liquid-crystalline

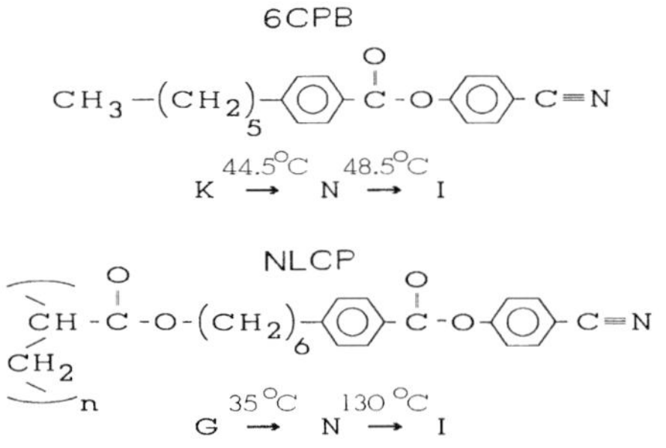

6CPB

$CH_3-(CH_2)_5-C_6H_4-CO-O-C_6H_4-C{\equiv}N$

K $\xrightarrow{44.5\,^\circ C}$ N $\xrightarrow{48.5\,^\circ C}$ I

NLCP

$[-CH_2-CH(-CO-O-(CH_2)_6-C_6H_4-CO-O-C_6H_4-C{\equiv}N)-]_n$

G $\xrightarrow{35\,^\circ C}$ N $\xrightarrow{130\,^\circ C}$ I

Figure 2-6. Structure and transition temperatures of the investigated low- and high-molecular weight LC systems (6CPB and NLCP).

polymers and liquid-crystalline guest–host system) in switch-on/switch-off experiments, and on the other hand we will describe the structural consequences incurred by a ferroelectric side-chain polymer in a poled electric field.

Nematic liquid-crystalline molecules, which consist of a rigid mesogenic unit with an attached flexible part (sequence of methylene units, for example) (Figure 2-6), are among the most characteristic thermotropic systems. FTIR spectroscopy has been applied to study the mechanism of orientation of the liquid crystals in an electric field; however, at present the experimental data are not always unequivocal, and conclusions obtained are not in complete agreement with each other. Thus, some authors did not detect any differences in orientational rates of the rigid and the flexible part of the LC molecule [16, 51–54], which is in accordance with the commonly accepted mechanism of the orientation of an LC as a cooperative motion of a molecular ensemble [1, 2, 38]. Other authors reported, that the motion of the flexible part precedes that of the mesogen in both orientation and relaxation [29, 55–57], and some investigators suggested that the motion of the mesogen precedes the motion of the flexible part [58, 59]. With respect to a more detailed exploitation of the spectroscopic data in terms of the synchronization of the segmental motions of different functionalities of the investigated liquid-crystalline systems, 2D-correlation analysis [16, 30, 60, 61] has proved an extremely valuable tool. In view of the detailed discussion of this mathematical technique in Chapter 1 of this book; however, this topic will not be treated here, although several results extracted from the presented spectroscopic data are based on the application of the 2D-correlation formalism.

2.5.1 Materials

To study the addressed phenomena, a commercial sample of p-cyanophenyl p-n-hexylbenzoate (hereafter abbreviated as 6CPB) (Figure 2-6) was obtained from Roche (Basel, Switzerland), and used without further purification [53, 54]. The transition temperatures of this LC are also presented in Figure 2-6. It should be mentioned, that the nematic phase will be retained during cooling down to 7 °C below the crystal–nematic transition point. To demonstrate the reduction of mobility in an external electric field when the mesogen is attached to a polymeric backbone, the nematic liquid-crystalline side-chain polymer (NLCP), whose side-chain is largely identical to the structure of the 6CPB (Figure 2-6), was investigated under analogous experimental conditions.

2.5.2 Experimental

The schematic construction of the measurement cell and the principles of operation are outlined in Figure 2-7. For the construction of this cell, two Ge-plates (approximately $10 \times 10 \times 2\ mm^3$) are used as IR-transparent window material and electrodes. The distance between the two Ge-plates is ensured by thin stripes of poly(ethyleneterephthalate) film (7 μm) acting as spacer. To avoid the occurrence of interference fringes in the spectra, the precise parallel alignment of the Ge-windows was intentionally disturbed by two supplementary spacers of polycarbonate film (2 μm) on one side of this assembly (Figure 2-7a). To fill the assembled cell, the LC-sample was heated above the nematic–isotropic transition temperature and then introduced by capillary forces. For the preparation of a prealigned monodomain-LC, the inner surface of each Ge-plate was covered with a thin layer of polyimide and then rubbed in one direction with polyamide cloth (see also Section 2.4.1). This pretreatment induced a homogeneous orientation of the investigated LC-molecules or mesogenic functionalities (in the case of the polymer) along the rubbing direction of the surface (Figure 2-7b). Upon application of an electrical voltage across the Ge-plates, the liquid crystals rotate with the long axes of their mesogens into the direction of the applied field (Figure 2-7c). Switching off the voltage leads to a relaxational motion back to the original state. The accompanying intensity changes were monitored with IR-radiation polarized parallel to the rubbing direction of the Ge-windows (Figure 2-7b). A home-built heating cell has been used to keep the investigated sample at a constant temperature within $\pm 0.1\,°C$ during the experiment (Figure 2-8).

The principle of the electronic set-up for the step-scan measurements, and the method of data acquisition relative to the experimental time-scale, are shown in Figures 2.9 and 2.10, respectively. When the scanning mirror has stepped to a definite retardation point, a control electronic unit coupled with a sine-wave generator sends a 6 Vpp (peak-to-peak height) pulse with a frequency of 10 kHz and a definite length (25 ms for the experiments presented here) to the sample cell to induce the reorientation of the LC. After this period the voltage is switched off during 225 ms to allow relaxation to the original state. Simultaneously to the application of the electric field, the same electronic unit sends a trigger pulse to the spectrometer to start the data collection. The data are sampled by the internal analog-to-digital converter (ADC) of the instrument during 50 ms with 0.05 ms time resolution. During this time interval 1000 data points are acquired for one fixed retardation (the applied data-acquisition scheme is imposed by the capacity of the computer). This cycle is repeated several times (five times for the presented results on 6CPB) and the data points corresponding to the same time in the period of the event are averaged to improve the signal-to-noise ratio. When the data collection at a given

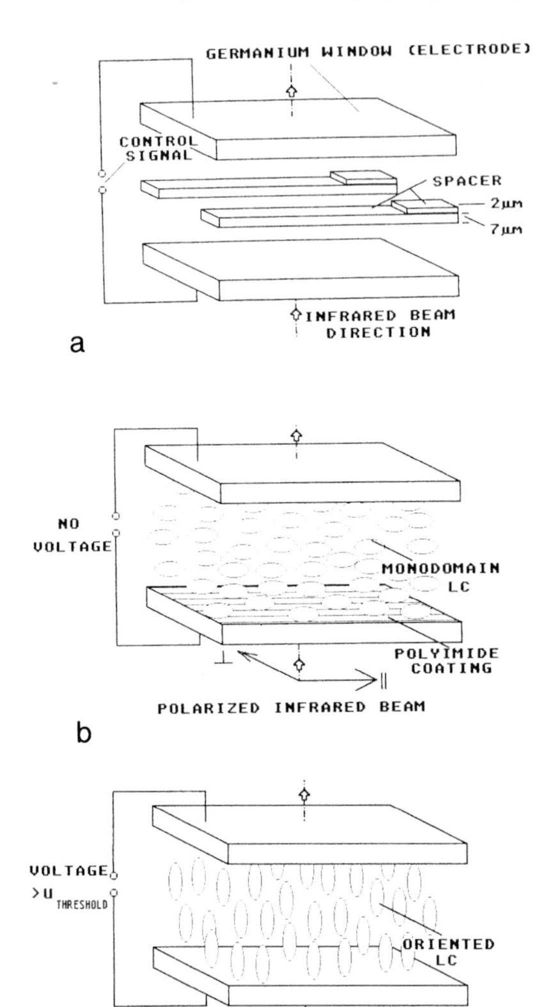

Figure 2-7. Electric field-induced orientation of a liquid crystal. (a) Construction of the measurement cell. (b) Homogeneous alignment of the LC between the anisotropic electrode surfaces before application of the electric field. (c) Homeotropic alignment of the LC during application of the electric field.

retardation is finished, the mirror is moved to the next retardation point (Figure 2-2b).

After a settling time of 40 ms, the experiment is repeated as described above. For the presented experiments all data were collected symmetrically around the centerburst of the interferogram. A total number of 2368 mirror positions were scanned, leading to a spectral resolution of 8 cm^{-1}. After completing the data collection at each retardation point, the data were resorted to interferograms on the time scale (Figure 2-2b) and transformed to the corresponding spectra. According to this

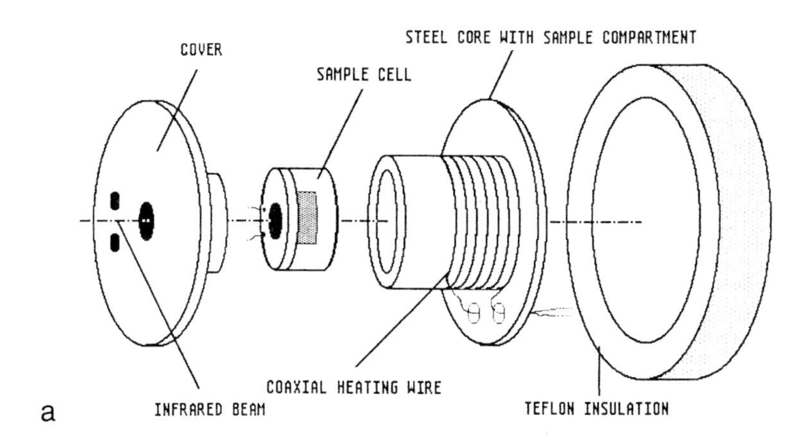

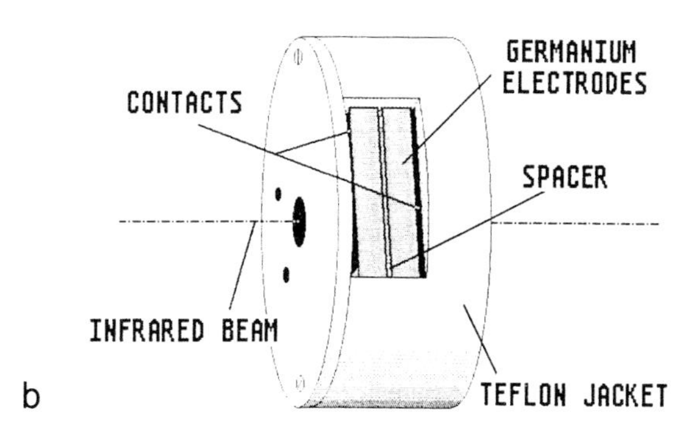

Figure 2-8. Variable-temperature cell utilized for the spectroscopic characterization of the electric field-induced alignment of liquid crystals. (a) Complete assembly of the accessory. (b) Expanded view of the sample cell.

procedure, 1000 spectra with a time resolution of 0.05 ms were processed. A further improvement of the signal-to-noise-ratio was obtained by averaging 10 spectra recorded in 0.5-ms intervals, starting with the time corresponding to the onset of applying the electric field. Thus, finally 100 spectra with a time resolution of 0.5 ms were obtained. The data collection for the whole experiment required approximately 50 minutes.

Due to the increase of viscosity in the polymeric NLCP system, the reorientation times varied between minutes and seconds, depending on the experimental temperature. Thus, the conventional rapid-scan technique (Figure 2-2a) was applied to follow the orientation and relaxation dynamics of the NLCP. In order to synchronize the application of the electric field with the data collection, a common switch was used to control the voltage [53].

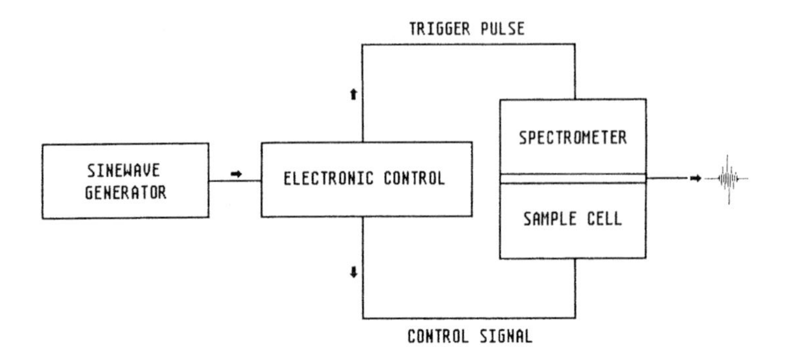

Figure 2-9. Block diagram of the electronic measurement set-up.

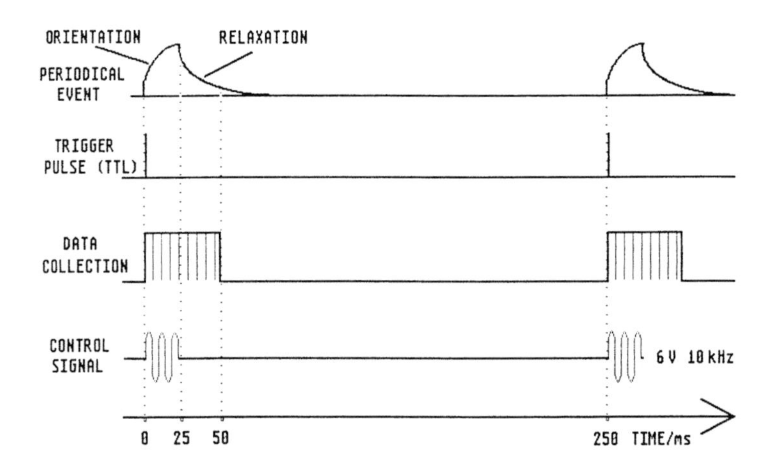

Figure 2-10. Time scale of the periodic event, control signal and data collection sequence for the step-scan measurements.

2.5.3 Results and Discussion

2.5.3.1 p-Cyanophenyl p-n-Hexylbenzoate (6CPB) in a Switching Experiment

The FTIR polarization spectra of a prealigned 6CPB in the nematic phase are presented in Figure 2-11. The band assignment of the most prominent absorptions based on the spectral analysis of similar compounds [52, 55] is given in Table 2-1. The $\nu(C{\equiv}N)$ band (2230 cm^{-1}) and the two $\nu(C{=}C)_{ar}$ stretching bands (1605 and 1503 cm^{-1}) are polarized along the long axis of the mesogenic group and show parallel dichroism with $A_{\parallel} > A_{\perp}$ ($A_{\parallel}$ and $A_{\perp}$ are the absorbance values measured

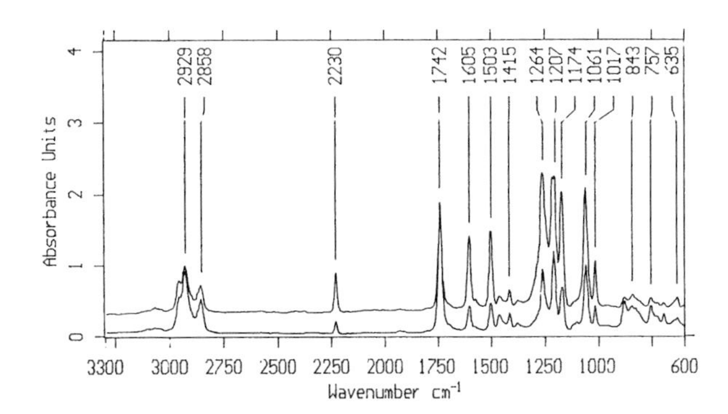

Figure 2-11. FTIR polarization spectra of 6CPB (41 °C) taken with radiation polarized parallel (upper spectrum) and perpendicular (lower spectrum) to the rubbing direction of the polyimide surface.

Table 2-1. Band assignment and wavenumber position of selected absorption bands for 6CPB (as = antisymmetric; s = symmetric; ar = aromatic; oop = out-of-plane).

Band assignment	Wavenumber (cm^{-1})
$\nu_{as}(CH_2)$	2929
$\nu_s(CH_2)$	2858
$\nu(C{\equiv}N)$	2230
$\nu(C{=}O)$	1742
$\nu(C{=}C)_{ar}$	1605
$\nu(C{=}C)_{ar}$	1503
$\nu(C{=}C)_{ar}$	1415
$\delta(CH_2)$	1464
$\nu(C{-}O{-}C)$	1264
$\nu(C{-}O) + \nu(C{-}O{-}C)$	1207–1017
$\gamma_{oop}(CH)_{ar}$	843
$\gamma_{oop}(CH)_{ar}$	757

with light polarized parallel and perpendicular to the rubbing direction of the polyimide layer, respectively). This means, that the long axis of the mesogenic unit is oriented predominantly parallel to the rubbing direction (Figure 2-7b). The 757 cm^{-1} band (ring CH out-of-plane deformation) exhibits a perpendicular dichroism with $A_\perp > A_\parallel$, since the transition moment for this band is perpendicular to the ring plane and hence, perpendicular to the long axis of the mesogen. These four bands are characteristic of the mesogenic units. For the characterization of the flexible part of the LC-molecule the 2929 cm^{-1} and 2858 cm^{-1} bands (antisymmetric and symmetric $\nu(CH_2)$-stretching vibrations) were evaluated. All the selected bands had absorbance values < 1 for the given cell thickness.

When a voltage higher than the threshold value is applied to the electrodes, the LC molecules undergo a transition from the homogeneous (parallel to the rubbing direction) to the homeotropic (parallel to the electric field) orientation. According to our data for 6CPB, the threshold voltage to induce such a reorientation is

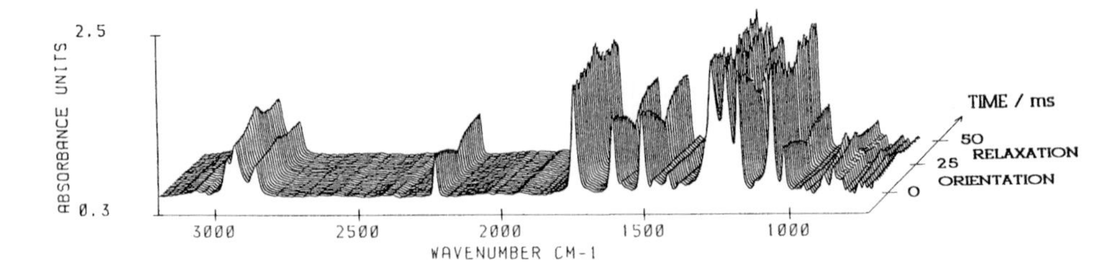

Figure 2-12. Stack-plot of spectra measured during the electric field-induced orientation and during part of the relaxation of the 6CPB molecules (41 °C).

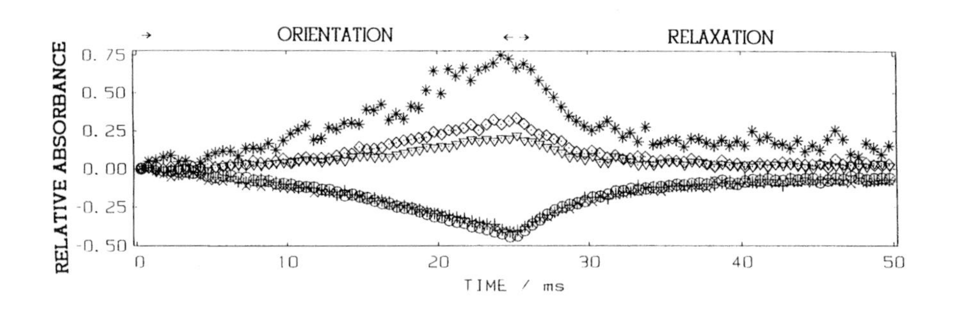

Figure 2-13. Relative absorbance/time-plot during orientation and relaxation of 6CPB (41 °C) for selected absorption bands: (◇) 2929 cm^{-1}, (∇) 2858 cm^{-1}, ($\circ$) 2230 cm^{-1}, (+) 1605 cm^{-1}, ($\times$) 1503 cm^{-1}, ($*$) 757 cm^{-1}.

approximately 1.5 Vpp. In the homeotropic type of orientation, the long axis of the LC molecule is normal to the electrode surface and parallel to the IR beam propagation. According to geometrical considerations [62] a reorientation to this homeotropic state thus leads to intensity changes of the $A_{\parallel}$-component only. This component asymptotically approaches the $A_{\perp}$ intensity value at sufficiently high fields. For this reason, all time-resolved measurements have been performed with light polarized parallel to the polyimide-layer rubbing direction.

A stack-plot of time-resolved spectra recorded during the homogeneous–homeotropic transition and part of the relaxation process is presented in Figure 2-12. A good signal-to-noise-ratio is observed for all spectral regions (except when the absorbance exceeds a value of 2). Relative absorbance changes A_r for selected absorption bands were calculated from:

$$A_r = (A_t - A_0)/A_0 \tag{2-2}$$

where A_t is the absorbance at time t and A_0 is the absorbance just before reorientation. These relative intensity changes of selected absorption bands are presented in Figure 2-13 and remarkable changes are observed only after a definite induction period of several milliseconds after application of the electric field. This phenomenon is well described in the literature [46]. The intensities of the 2230, 1605 and 1503 cm^{-1} bands decrease, whereas that for the 757 cm^{-1} band increases when the field is switched on. This indicates that the mesogens reorient with their long axes in the direction of the applied field (Figure 2-7c). The process of orientation at the given experimental conditions takes about 25 ms. After switching off the field, the relaxation to the initial homogeneous state (Figure 2-7b) takes place. The intensity changes of the 2929 and 2858 cm^{-1} bands reflect the orientation and relaxation process of the flexible part of the LC molecule. For an exact quantitative comparison of the orientational and relaxational rates of the rigid and flexible segments of the LC molecule, the order parameters of the different bands had to be calculated as a function of time. However, this is possible only for the case of a pure homogeneous orientation (i.e., the director of the LC monodomain and, hence, the axis of uniaxial orientation is perpendicular to the IR beam propagation) or for the case of a pure homeotropic orientation (where the axis of uniaxial orientation is parallel to the beam propagation). In intermediate states the orientation of the director is not constant. Thus, near the surface, due to strong interactions, the molecules are oriented perpendicular to the electric field whereas at the center of the cell they are oriented parallel to the electric-field direction [38]. For a qualitative comparison of the orientational rates of the rigid and the flexible parts of the investigated LC molecule a normalized intensity A_n was utilized which has been calculated from:

$$A_n = (A_t - A_0)/(A_m - A_0) \tag{2-3}$$

where A_t is the band peak absorbance at time t, A_0 is the value before the application of the electric field, and A_m is the corresponding absorbance value at $t = 25$ ms. The result for the 2929 and 2230 cm^{-1} bands are presented for a temperature of 41 °C in Figure 2-14a. According to these data there are no detectable differences in A_n both, during the orientation and relaxation process, for the rigid and the flexible segments of the LC molecule. This means that under the given experimental conditions and with the evaluation by the usual one-dimensional technique, the different segments of the LC molecule orient and relax at comparable rates. The same conclusion can be derived from Figure 2-14b where the corresponding A_n-values are presented for 45 °C. Due to the small temperature interval, the difference in orientational behavior is not significant. However, it is not possible to perform the experiment in a wider temperature range because of the narrow temperature interval of the investigated mesophase.

The results of similar rates for the reorientation of the flexible and rigid segments of the 6CPB molecule in an electric field are in accordance with the commonly accepted mechanism of orientation by a cooperative motion of the LC molecules [1, 2, 38]. However, in recent investigations it has been reported, that the response of the mesogens and the flexible part on the applied field may be different [29, 55–59]. If these observations are correct they could be the basis of a principally new description of LC orientation on a molecular level. Two explanations for the disagreement of our results with these views can be given:

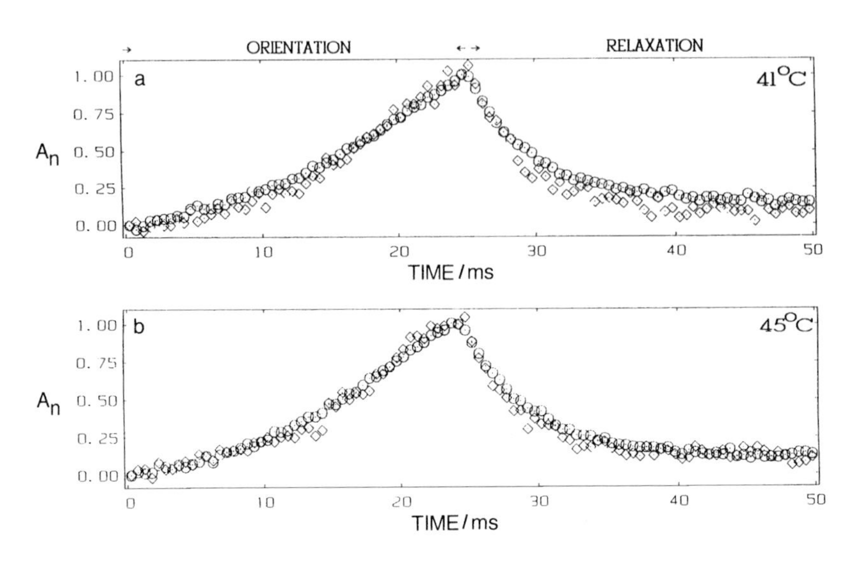

Figure 2-14. Normalized absorbance/time-plot during orientation and relaxation of 6CPB at two different temperatures for selected absorption bands: (◇) 2929 cm^{-1}, (○) 2230 cm^{-1}.

1. The effect of different response of the flexible and the rigid part of the LC molecule towards application of an electric field is not general. This is supported by results [56], where this phenomenon is detected under certain experimental conditions only. Moreover, up to now most experiments have been performed on one LC (4-n-pentyl-4′-cyanobiphenyl, 5CB) only, which has a different structure compared to the LC molecule investigated here.
2. The effect is too small to be detected under the given experimental conditions with conventional data analysis. To detect this effect, Urano et al. [55, 56] have applied time-resolved spectroscopy by using a modified dispersive instrument. With this instrument, absorbance changes down to 10^{-6} a.u. have been detected. However, the time requirement for the experiment in the step-scan mode depends on the period of the event (0.25 s in our case). To reach such a high signal-to-noise-ratio we had to average the data of many more events at the individual retardation points. For example, the data collection for 2368 retardation points over 100 events in each point (compared with our five events) would require approximately 17 h of experimental time. Due to uncontrolled changes in the environmental conditions and drifts of electronics during such a time period, the main requirement of the step-scan technique, i.e., the reproducibility of the periodical event, could be violated.

To extract more information, the data were processed with the newly developed method of two-dimensional spectroscopy [60, 61]. This technique is especially useful for the detection of small differences in the orientational rates of different molecular

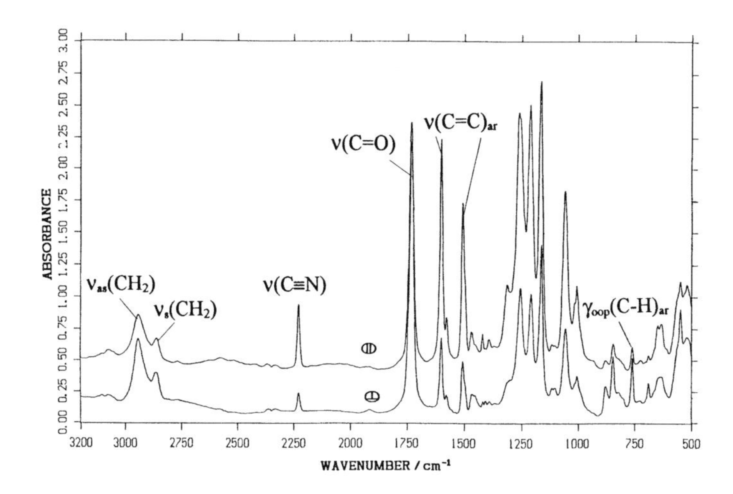

Figure 2-15. FTIR polarization spectra of NLCP (25 °C) taken with radiation polarized parallel and perpendicular to the rubbing direction.

groups. With the development of this formalism it is now possible to construct 2D correlation plots based on conventional time-resolved spectral data. The 2D analysis reveals [16], that there is a strong correlation in the synchronous spectrum and no appreciable asynchronicity in the asynchronous spectrum between the mesogen bands and the absorptions of the hydrocarbon chain (see Table 2-1). This means, that the mesogen and the flexible segment of the liquid crystal reorient synchronously. The strong correlation between the mesogen bands themselves indicates that the mesogen reorients as a rigid core. No differences in 2D correlation spectra between the orientation and relaxation processes were found.

2.5.3.2 Nematic Side-Chain Liquid-Crystalline Polymer in a Switching Experiment

The FTIR polarization spectra of the prealigned NLCP are presented in Figure 2-15. Due to the almost identical structure of the side-chain compared with the low-molecular weight liquid crystal (6CPB), most of the functionality-specific band assignments of Table 2-1 are also valid for the NLCP. The 1603 cm^{-1} and 1503 cm^{-1} (phenyl C=C-stretching vibrations) bands and the 2230 cm^{-1}(ν(C≡N)) band are polarized along the long axis of the mesogenic functionality and show parallel dichroism with $A_{\parallel} > A_{\perp}$. This means, that the long axis of the mesogenic unit is predominantly oriented parallel to the rubbing direction. The 757 cm^{-1} band (ring CH out-of-plane deformation) exhibits a perpendicular dichroism with $A_{\perp} > A_{\parallel}$, because the transition moment for this band is perpendicular to the ring plane and hence, perpendicular to the long axis of the mesogen. Since the transition moment

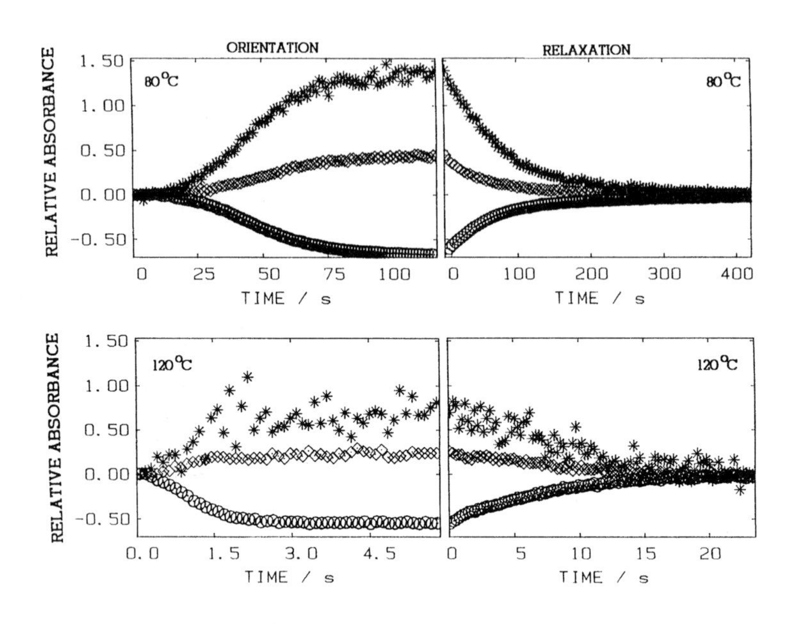

Figure 2-16. Relative absorbance/time-plot during orientation and relaxation of NLCP at different temperatures for selected absorption bands: (◇) 2926 cm^{-1}, (○) 2229 cm^{-1}, (∗) 757 cm^{-1}.

for this band is also nearly prependicular to the trans-polymethylene chain, this indicates that the spacer is oriented preferably parallel to the rubbing direction of the polyimide support and the mesogen. We used the 2230 and the 757 cm^{-1} bands for the characterization of the movement of the mesogen and the 2926 cm^{-1} band for that of the spacer. Unfortunately, the bands which characterize the backbone are heavily overlapped with absorptions of the spacer and they are much less intense than the corresponding spacer bands (a monomeric unit contains one methylene group in the main chain und six in the spacer). Therefore, in the case of this polymer, we cannot follow the movement of the backbone. Owing to the experimental geometry, the reorientation of NLCP led only to changes in intensity of light polarized parallel to the rubbing direction. For this reason, as before, we used only polarized radiation to monitor the molecular motion of the NLCP.

The relative intensity changes during orientation and relaxation at different temperatures for selected absorptions bands are shown in Figure 2-16. As can be seen from this figure, the orientation and relaxation rates increased more than 20-fold when the temperature increased from 80 to 120 °C. At temperatures close to the glass transition, the orientation times increase to a few tens of minutes. This is apparently related to the decrease of viscosity with increasing temperature. At each temperature the relaxation takes more time than the reoreination. The A_r values change less at higher temperatures than at lower temperatures (Figure 2-16). This is the consequense of the decrease in the order parameter of the NLCP with increasing temperature. The values of A_r for different absorption bands depend on the angle

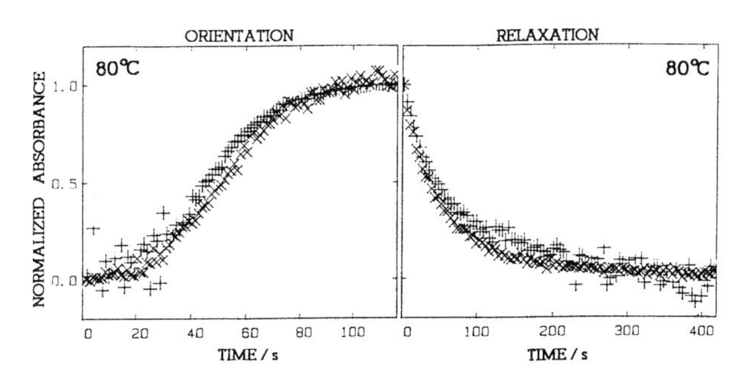

Figure 2-17. Normalized absorbance/time-plot during orientation and relaxation of NLCP at different temperatures for selected absorption bands: (×) 2926 cm^{-1}, (+) 2229 cm^{-1}.

between their transition moment and the molecular axis of the LC molecule. In order to compare the rates of molecular motion for the differents segments, the normalized absorbance A_n was used. The A_n values for bands characterizing the mesogen and the flexible part are shown in Figure 2-17. As is evident from this figure, no significant differences of their intensities can be detected during the orientation and relaxation process. Thus, it must be assumed, that the mesogen and the flexible part reorient as a rigid unit and the application of the electric field does not lead to conformational changes in the flexible spacer. This conclusion, based on the direct monitoring of the movement of different molecular segments, is in agreement with the results of [63] derived from the measurement of rotational viscosity of the same NLCP in a magnetic field and from theoretical modeling. However, it has been reported that the application of the electric field led to a decrease of trans conformers in a flexible spacer of an NLCP similar to that studied here. This conclusion was based on the analysis of the 1490–1410 cm^{-1} region which is sensitive to the conformation of the polymethylene sequence. As has been shown recently for a set of side-chain liquid crystalline polymers with different length of the polymethylene spacer, this region also contains absorption bands of the aromatic ring [64]. In our opinion, the conformational analysis based on the absorbance in this region is thus very complex and ambiguous.

If the main chain takes part in the reorientation, it is expected that its motion will be, at least partially, shifted in phase with respect to the side chain (spacer and mesogen). A detailed 2D analysis [30] in the region of the $\nu(CH_2)$ bands (2926 and 2863 cm^{-1}) does not reveal any significant asynchronicity. Hence, we may conclude that the main chain does not take part in the reorientation, or it moves only slightly.

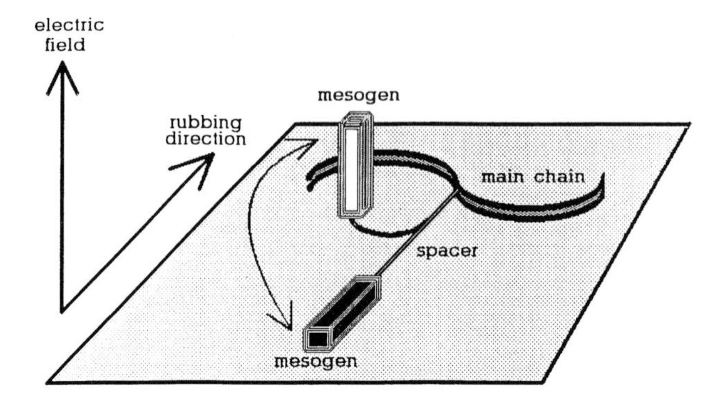

Figure 2-18. Schematic representation of the segmental motions of the NLCP during the switching process.

There is a strong correlation in the synchronous spectrum and no appreciable asynchronicity in the asynchronous spectrum between the mesogen band (2230 cm^{-1}) and the $\nu(CH_2)$ bands (2926 and 2863 cm^{-1}). The main part of the intensity changes in the region of 2850–2950 cm^{-1} is due to the spacer; thus, on the basis of the 2D results we may draw the conclusion that the spacer and the mesogen reorient simultaneously.

From a further detailed 2D analysis of selected absorption band intensities of the corresponding power spectrum [16, 60, 61] it was finally concluded that only part of the spacer of the NLCP takes part in the reorientation. This result is summarized graphically in Figure 2-18, where the orientational behavior of a NLCP-mesogen during the switching process is symbolized schematically relative to the entire polymeric structure, including the spacer and the main chain.

2.5.4 Nematic Liquid-Crystalline Guest–Host System in a Switching Experiment

Generally, the phenomenon of dissolving and aligning a molecule or a group of molecules, such as dyes or probes, for example, by a liquid crystal can be called the guest–host phenomenon [65, 66]. The host liquid crystal can be a single compound or a multicomponent mixture. Depending on its structural geometry (e.g., elongated or spherical) the guest molecule couples more or less significantly to the anisotropic intermolecular interaction field of the liquid-crystalline host. The liquid-crystalline solutions can be easily oriented by electric, magnetic, surface, or mechanical forces, resulting in highly oriented solvent and solute molecules. This phenomenon provides the basis for the application of liquid crystals as anisotropic solvents for spectroscopic investigations of anisotropic molecular properties [67, 68]. Actually, in a broad sense, most of the commercially available liquid-crystal mixtures may be regarded as being related to the guest–host effect, as they incorporate some non-mesomorphic molecules to generate the desired electro-optical effects in the mixtures [69–71]. In the present section we present an example of how a guest molecule is oriented anisotropically in a liquid-crystalline solution which is exposed to an

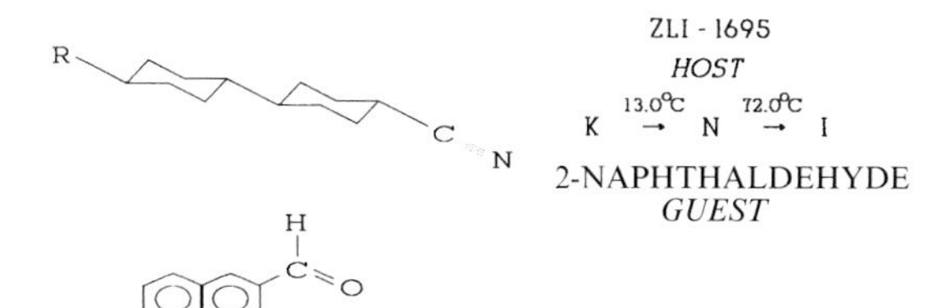

Figure 2-19. Chemical structures and transition temperatures of the investigated nematic liquid-crystalline guest–host system.

external electric field in a switching experiment. Time-resolved step-scan FTIR-spectroscopy has been applied to study how a solute follows the reorientation motion of the nematic solvent [72].

2.5.4.1 Materials

A 10% solution of 2-naphthaldehyde in the nematic liquid-crystal mixture 4-cyano-4′-alkylbicyclohexyls (ZLI-1695, Merck) (Figure 2-19) was prepared for the measurements. This nematic phase is very convenient as an anisotropic solvent because of its weak IR spectrum, allowing one to obtain polarization spectra in practically the whole mid-IR region.

2.5.4.2 Experimental

For the switching experiment, the cell described in Figures 2-7 and 2-8 was utilized (here the path length was 10 μm). The spectral resolution was 16 cm^{-1} and a function generator supplied the electrical signal (Figures 2-9 and 2-10) needed to orient the liquid-crystalline sample (16 V_{pp}, 1 kHz, 20 ms on, 260 ms off). The positive edge of this signal also synchronized the start of the data acqusition at each step during the complete step-scanning experiment. The time resolution of the experiment was 100 μs; each spectrum contains coadded data of two orientation/relaxation cycles.

The reorientation behavior of the nematic solvent was determined by the $\nu(C{\equiv}N)$ and the $\nu_s(CH_2)$ absorption bands at 2239 and 2857 cm^{-1}, respectively, whereas the $\nu(C{=}O)$ absorption at 1698 cm^{-1} was utilized to characterize the segmental motion of the solute (Figure 2-20).

2.5.4.3 Results and Discussion

The step-scan spectra of the studied solution, sampled at 0.0 and 19.5 ms, are presented in Figure 2-20. The observed intensity changes correspond to the reorientation of the solute and solvent molecules from their preferential alignment in the rubbing direction (0.0 ms) into the applied electric field axis (19.5 ms). The detailed relative intensity changes of the three references bands versus time after the electric field was switched off are shown in Figure 2-21. Here, the lower values for the rela-

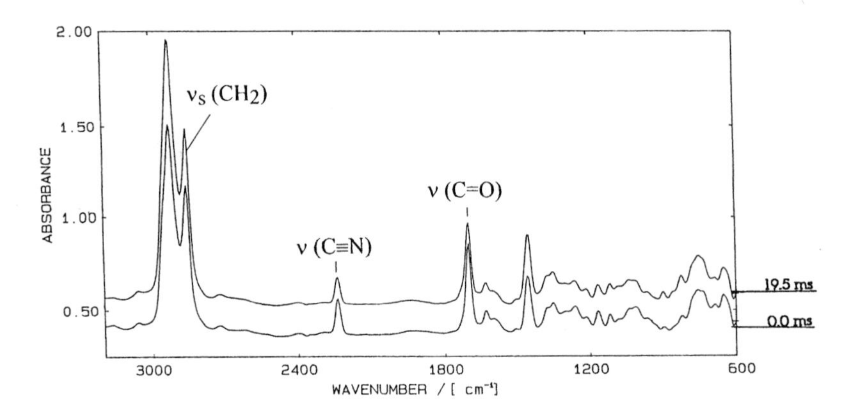

Figure 2-20. Step-scan FTIR spectra of the investigated nematic liquid-crystalline guest–host solution before (0.0 ms) and after (19.5 ms) reorientation in the electric field.

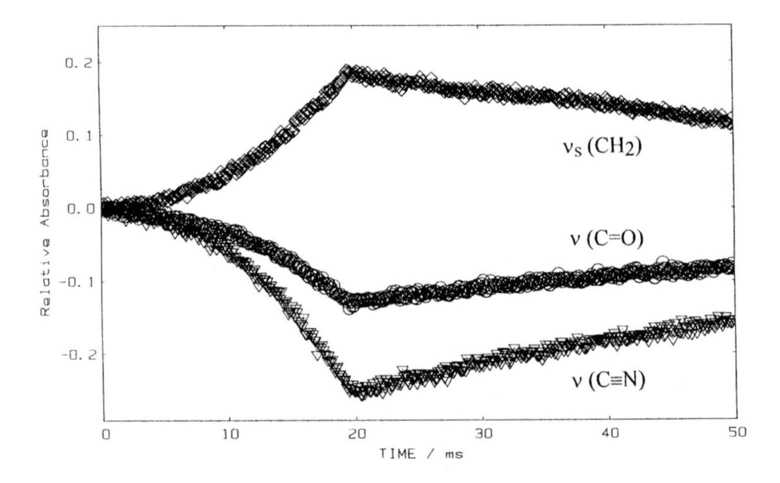

Figure 2-21. Relative absorbance changes versus time of the $\nu_s(CH_2)$ (◇) and $\nu(C{\equiv}N)$ (∇) absorption bands of the solvent and the solute-specific $\nu(C{=}O)$ (○) band during orientation and relaxation of the individual segments.

tive absorbance of the $\nu(C{\equiv}N)$ relative to the $\nu(C{=}O)$ band are due to two effects: (1) better initial orienation in the rubbing direction; and (2) the more efficient reorientation of the C≡N functionality into the electric field direction.

The positive values for the relative absorbance of the $\nu_s(CH_2)$ absorption reflect the preferential perpendicular alignment of the transition moment direction of this band relative to the long axis of the nematic solvent molecule. A recent study [29]

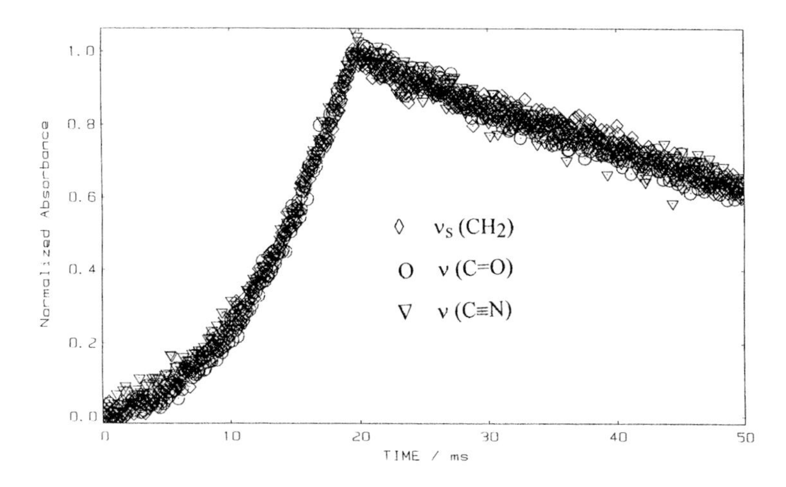

Figure 2-22. Normalized absorbance versus time during orientation and relaxation of the solvent- and solute-specific functionalities (see Figure 2-21).

on the nematic liquid-crystal 4-cyano-4′-pentylbiphenyl (often denoted as 5CB) by means of time-resolved FTIR spectroscopy under the conditions of dynamic reorientation give evidence for different reorientation times of its non-ridgidly bonded fragments.

Although Figure 2-21 shows that neither the orientation nor the relaxation of the solute and solvent molecules is complete under the experimental conditions, we expected that the delay phenomena between the solvent and solute molecules would be clearly expressed in a normalized absorbance versus time plot. This plot, however, as shown in Figure 2-22, does not provide any indications of such a behavior. In other words, the reorientation rates are equal for the solvent and the solute. Without drawing any general conclusions from a single example, the present results show that one may expect similar behavior in other nematic solutions.

2.5.5 Reorientation Dynamics of a Ferroelectric Side-Chain Liquid-Crystalline Polymer in a Polarity-Switched Electric Field

The surface-stabilized ferroelectric liquid crystals in the smectic C* (SmC*) phase are among the most interesting types of liquid-crystalline systems because of their potential applications in high-resolution flat panel displays and fast electro-optical devices [73–76]. Within this class of compounds, ferroelectric liquid-crystalline polymers (FLCPs) have gained theoretical and practical interest as systems which combine the properties of polymers and ferroelectric liquid crystals. This combination is achieved by attaching the ferroelectric mesogen to a main chain via a flexible spacer

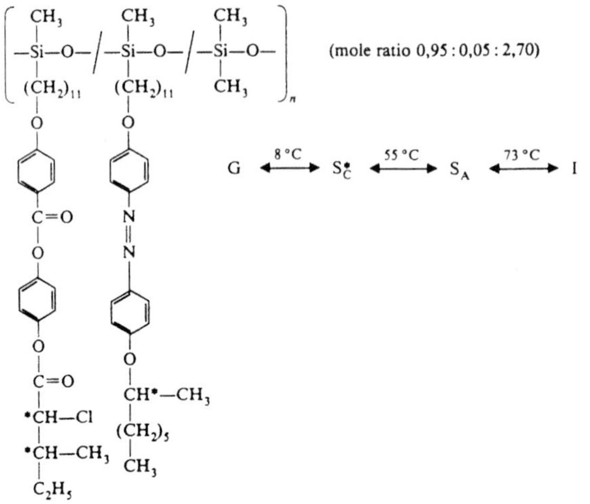

$$\left[\begin{array}{c}CH_3\\ |\\ -Si-O-\\ |\\ (CH_2)_{11}\end{array} / \begin{array}{c}CH_3\\ |\\ -Si-O-\\ |\\ (CH_2)_{11}\end{array} / \begin{array}{c}CH_3\\ |\\ -Si-O-\\ |\\ CH_3\end{array}\right]_n$$

(mole ratio 0,95 : 0,05 : 2,70)

$G \overset{8\,°C}{\longleftrightarrow} S_C^* \overset{55\,°C}{\longleftrightarrow} S_A \overset{73\,°C}{\longleftrightarrow} I$

Side chain 1: O–C₆H₄–C=O–O–C₆H₄–O–C=O–•CH–Cl–•CH–CH₃–C₂H₅

Side chain 2: O–C₆H₄–N=N–C₆H₄–O–CH•–CH₃–(CH₂)₅–CH₃

Figure 2-23. Chemical structure, composition and transition temperatures of the side-chain FLCP under investigation.

[77, 78]. Since the spacer is effectively decoupling the movement of the mesogen from the polymer backbone, a different behavior of mesogen, spacer, and main chain can be expected in a reorientation process. So far, most of the research on the mobility of FLCPs has been limited to the determination of relaxation times by using dielectric spectroscopy [79] or response times by using electro-optical switching studies [80] without differentiation of the motions of individual functionalities in the macromolecule and only few time-resolved vibrational spectroscopic measurements are available on this subject [81–83]. However, the movements of the individual segments control the reorientation process and hence, the study of their dynamics is a key to understand the behavior of an FLCP in an electric field. Based on the recent advances in data-acquisition techniques [16] we have attempted in the present work to follow the reorientation dynamics of the different segments of an FLCP from one stable state to another on the submillisecond time scale during poling of an electric field [83].

2.5.5.1 Materials

The chemical structure, composition and the transition temperatures of the investigated side-chain FLCP are shown in Figure 2-23. The molecular weight of this polymer is approximately 18800 and the synthesis and characterization have been described in detail elsewhere [84].

2.5.5.2 Experimental

The variable-temperature cell was the same as described in Section 2.5.2, the only difference being that a spacer with a uniform thickness of 5 μm was used between the electrodes. The empty cell and the FLCP were heated to 90 °C to introduce the

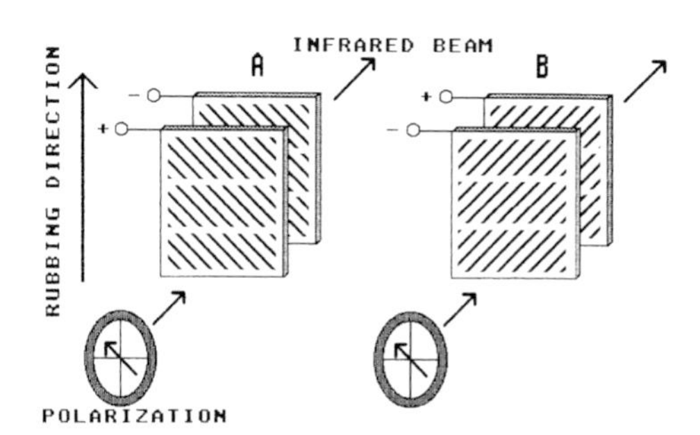

Figure 2-24. Polarization geometry and mesogen alignment of the investigated FLCP in the polarity-switched electric field (A: polarity +, B: polarity −).

polymer into the cell through the gap between the electrodes by capillary forces. After filling the cell, it was slowly cooled (0.5 °C/min) and an AC-voltage of 30 Vpp, 10 Hz, was applied during cooling with the aim to reduce defects in orientation. The FLCP was investigated in the SmC*-phase where two surface-stabilized orientational states exist, depending on whether the mesogens incline to the right or to the left relative to the rubbing direction of the polyimide support (Figure 2-24). The switching between these two states is possible by changing the polarity of the applied field. The propagation direction of the polarized IR beam was perpendicular to the plane of the Ge-windows. The largest absorption intensity changes during the switching process were obtained by fixing the polarizer at 45° relative to the rubbing direction of the polyimide coating. All spectra were collected with a spectral resolution of 8 cm^{-1} and a time resolution of 0.1 ms. For a more detailed characterization of the reorientation mechanism, the experiments were carried out at different temperatures (28 °C, 42 °C and 47 °C ($\pm$0.1 °C)). At the lower temperature only rapid-scan measurements were required during orientation at 20 V (DC), whereas at the two higher temperatures the step-scan technique had to be applied due to the much faster segmental movement. Here, for the switching between the two orientational states a bipolar rectangular signal with a voltage of 20 Vpp and a frequency of 20 Hz and 100 Hz was applied at 42 °C and 47 °C, respectively (Figure 2-25). The spectral data collection was synchronized with the negative edge of the electric signal.

2.5.5.3 Results and Discussion

Polarization spectra of the preoriented FLCP are shown in Figure 2-26. For the characterization of the mesogen and the flexible spacer, the 1607 cm^{-1} ($\nu(C{=}C)_{ar}$) and 2926 cm^{-1} ($\nu_{as}(CH_2)$) absorptions, respectively, were chosen. Unfortunately, no bands free of overlap can be attributed to the backbone. The band at 1020 cm^{-1} (ν(Si–O–Si)) is strongly overlapped by the ν_s(C–O–C) absorption of the ester group and the band near 807 cm^{-1} is also superimposed in a rather complex manner. For the characterization of the methyl substituents of the siloxane-backbone, the $\delta(Si{-}CH_3)$ vibration at 1260 cm^{-1} was chosen, although it contains a contribution from the ν_{as}(C–O–C) vibration related to the mesogen. The 1607 and 1260 cm^{-1} bands have parallel dichroism and the 2926 cm^{-1} band has perpendicular dichro-

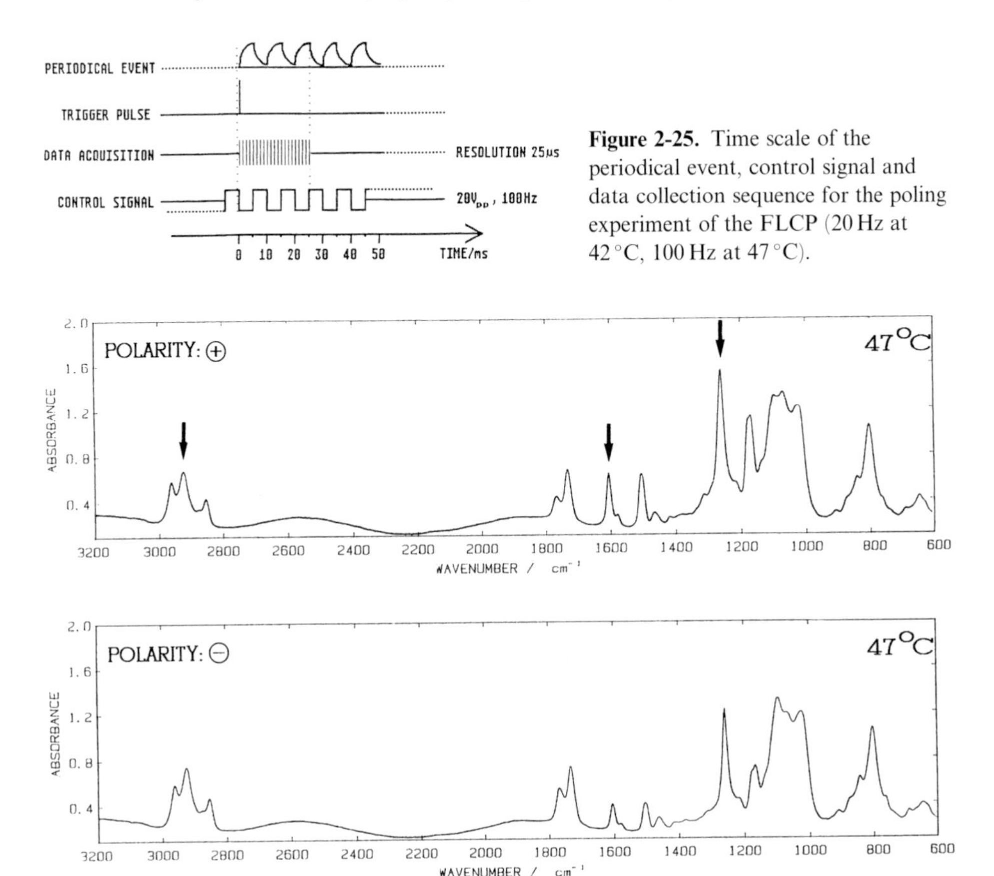

Figure 2-25. Time scale of the periodical event, control signal and data collection sequence for the poling experiment of the FLCP (20 Hz at 42 °C, 100 Hz at 47 °C).

Figure 2-26. FTIR-polarization spectra of the investigated FLCP prealigned at 47 °C and different polarity of the electric field (radiation polarized at 45° to the rubbing direction according to the geometry of Figure 2-24).

ism. Due to the perpendicular transition moment of the 1260 cm^{-1} band relative to the local chain axis it can be concluded that in the FLCP the mesogen and the spacer are aligned preferentially parallel to each other but are perpendicular to the backbone [16, 53, 83]. The sinusoidal changes observed in the baseline of the spectra are a consequence of the interference of the IR-radiation between the Ge-windows, but do not disturb the evaluation of the absorption bands in the present case. A stack-plot of time-resolved spectra recorded during several cycles of polarity changes is shown in Figure 2-27. A good signal-to-noise-ratio is observed for all spectral regions, except where the intensity exceeds an absorbance of 1.7.

Normalized absorbance/time-plots of the selected absorption bands are plotted for experiments at different temperatures in Figure 2-28. As can be seen from this figure, the reorientation at 47 °C takes only approximately 5 ms compared with

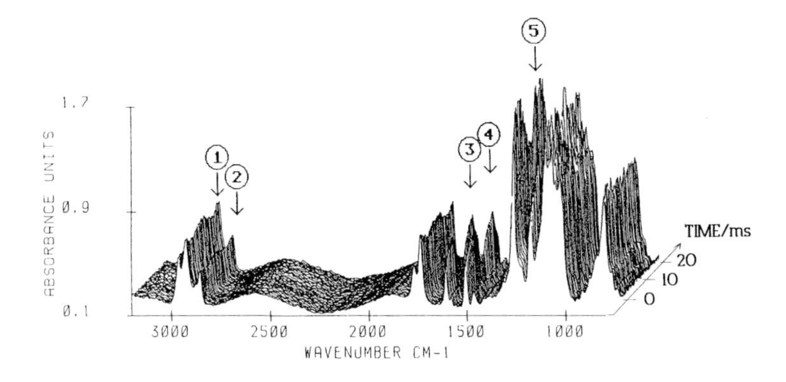

Figure 2-27. Spectroscopic response of the FLCP upon changing the polarity of the applied electric field (1: 2926 cm^{-1}, 2: 2854 cm^{-1}, 3: 1607 cm^{-1}, 4: 1507 cm^{-1}, 5: 1260 cm^{-1}).

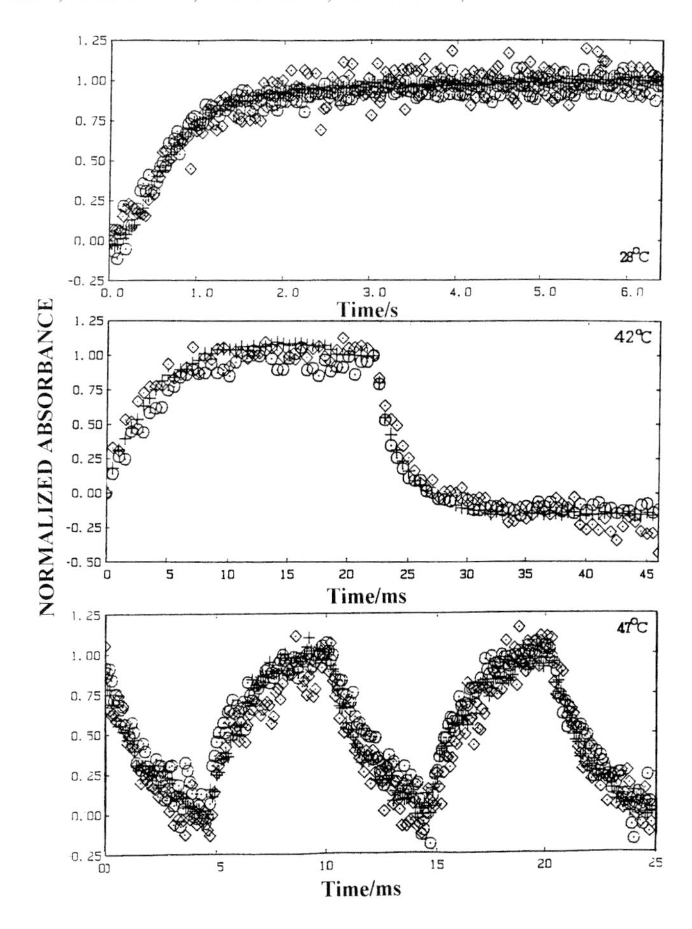

Figure 2-28. Normalized absorbance/time-plots for poling experiments of the FLCP at different temperatures. (a) 28 °C, rapid-scan measurement, 20 V (DC); (b) 42 °C, step-scan measurement, 20 V_{pp}, 20 Hz; (c) 47 °C, step-scan measurement, 20 V_{pp}, 100 Hz, data acquisition for 2.5 periods. (1260 cm^{-1} (○), 1607 cm^{-1} (+), 2926 cm^{-1} (◇)).

about 1.5 s for 28 °C. The shape of the response curves are similar and can be described as exponential. The reorientation starts without time delay when the polarity of the applied field is changed. This means, that there is no retardation in the response of different structural segments. No distinct differences have been found in the orientation rates of the flexible and the rigid part of the molecules of ferroelectric nonpolymeric LCs [52, 58]. However, it has been shown that the molecular motion cannot be described by a single exponential process but consists primarily of a fast and a subsequent slow process [52]. This motion also includes conformational changes in the flexible part of the molecule. In the polymeric ferroelectric system under investigation, such conformational changes may take place in the spacer and the main chain. Due to the lack of conformationally sensitive absorptions in the FLCP, we were not able to detect or differentiate such changes. Taking into account the assignment of the absorption bands under consideration and their dichroic changes during poling the electric field, we can conclude from Figure 2-28 [83] that not only the mesogen but also the spacer and the main chain (at least some molecular groups which are in the neighborhood of the spacer attachment to the backbone) take part in the reorientation process during poling the electric field. This picture, however, is not supported by the 2D-correlation analysis, which shows, that there is a strong correlation between the ring-stretching bands (1506, 1607 cm^{-1}) and the two carbonyl bands around 1750 cm^{-1} [16]. This is evidence that the mesogen reorients as a rigid unit. The strong correlation in the synchronous spectrum and no appreciable asynchronicity in the asynchronous spectrum between the mesogen bands and the $\nu(CH_2)$ bands (2926 and 2855 cm^{-1}), which are mainly due to the spacer, indicates that the mesogen and the spacer reorient synchronously. Figure 2-29 compares the power spectrum of the FLCP to the reference spectrum (average spectrum of a reorientation process). One can easily notice, that the intensities of the spacer bands in the power spectrum are considerably reduced relative to the mesogen bands. Based on these data, we have to conclude that only a small part of the spacer, directly attached to the mesogen, takes part in the reorientation. The 2D analysis does not supply any evidence that the main chain of the FLCP participates in the reorientation. Based on these data, the picture shown in Figure 2-30 can be derived for the response of the FLCP to a poled electric field.

2.6 Alignment of Side-Chain Liquid-Crystalline Polyesters Under Laser Irradiation

It has been demonstrated that the photo-induced orientation of dye-containing SCLCPs can be used for reversible optical data storage [85–90]. FTIR polarization spectroscopy has been applied to monitor the orientation of the main chain and the mesogenic side-chains which are of critical importance in these storage processes [15, 91–94]. Here, some results obtained by laser irradiation of a homologous series

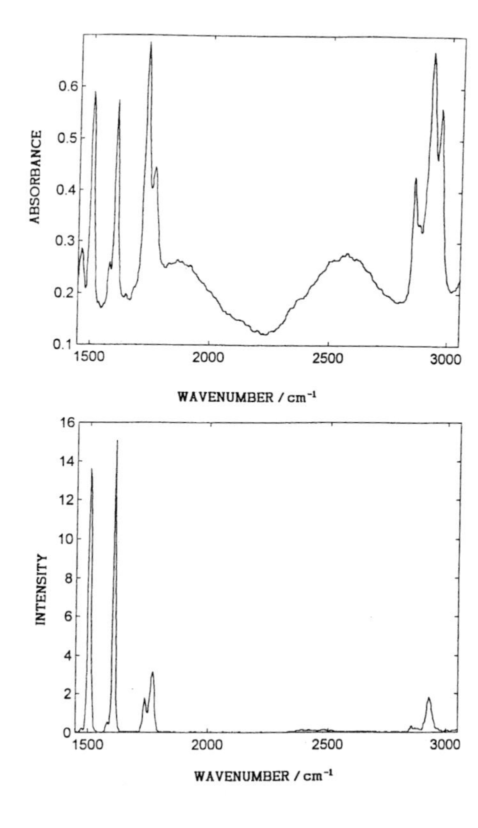

Figure 2-29. Comparison of the reference (top) and power (bottom) spectra intensities of the FLCP for the reorientation process during poling the external electric field.

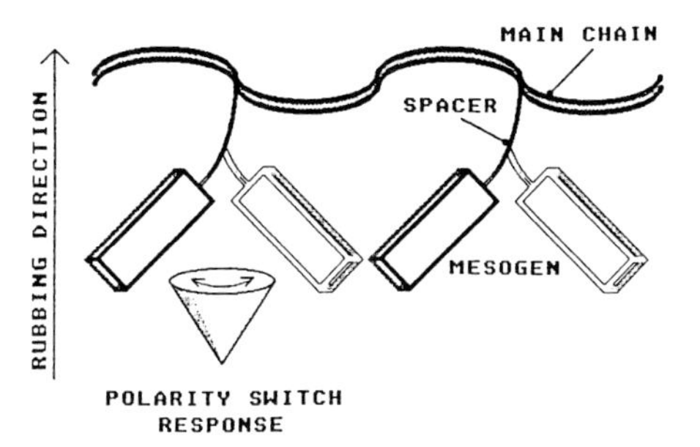

Figure 2-30. Schematic representation of the segmental motions of the FLCP during poling the electric field.

of novel SCLC-polyesters [15] are presented. In unoriented polyester films with a dodecamethylene spacing of the ester groups in the main chain, holograms with a resolution of 6000 lines/mm can be recorded by the interference of an Argon-ion laser beam [95]. The thermal erasure and rewriting of the laser-induced alignment in the SCLC polyesters is the basis for such storage processes. In order to investigate the mechanism of the involved processes in more detail, a variable-temperature, computer-controlled sampling system for on-line polarization FTIR measurements during laser irradiation as well as thermal erasure has been designed and constructed.

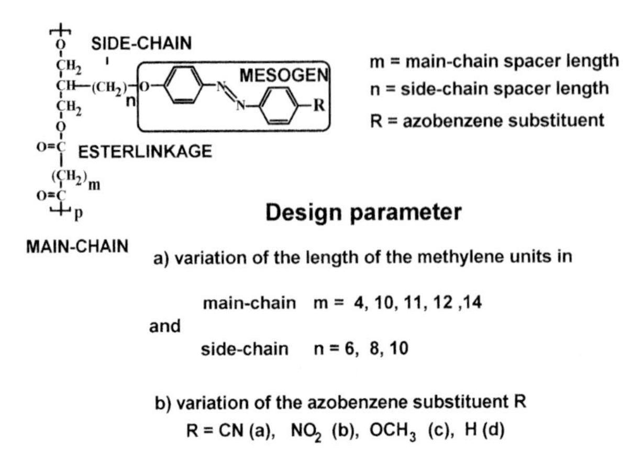

Figure 2-31. General structure and nomenclature (*PnRm*) of the investigated SCLC polyesters used for reversible optical data storage (e.g., P6a12: polyester with 12 CH_2 groups in the acid part of the main chain and six CH_2-groups in the side-chain spacer and a CN-substituent on the mesogen).

Table 2-2. Phase behavior of P8a12 [96].

Structure	4	3	2	1
Transition temperature range	30–40 °C	46–54 °C	59–63 °C	65–71 °C

2.6.1 Materials

The variations in chemical structure of the investigated SCLC polyesters and their nomenclature is outlined in Figure 2-31. Differential-scanning-calorimetry (DSC), polarization-optical-microscopy (POM) and X-ray investigations [95–97] reveal a rather complex phase behavior of this homologous polyester series, which strongly depends on the length (m) of the main-chain. For the polyadipates ($m = 4$), for example, an enantiotropic liquid-crystalline behavior is observed, involving disordered smectic mesophases in the range from T_g at about 10 °C to the smectic ↔ isotropic transition (from about 55–70 °C, varying with the side chain spacer length $n = 6, 8, 10$). The longer homologs, with a main chain length of 10–14 methylene units could be classified as monotropic LCPs, with polymorphic character. Some of them exhibit up to four first-order transitions, which partially overlap, involving many different organized phases. An exact phase assignment is very difficult. Candidates with the highest potential for optical data storage are the polytetradecanedioates ($m = 12$) with 6 and 8 methylene groups in the spacer [95, 98]. DSC and X-ray investigations reveal for both polyesters two low-order mesomorphic phases (structures 3 and 4) and two high-order mesomorphic or crystalline phases (structures 1 and 2). The glass-transition temperature of both systems is about 25 °C. Representative of the discussed SCLC polyesters, the phase behavior of P8a12 is detailed in Table 2-2.

The influence of azobenzene substituents with varying dipole moments has also

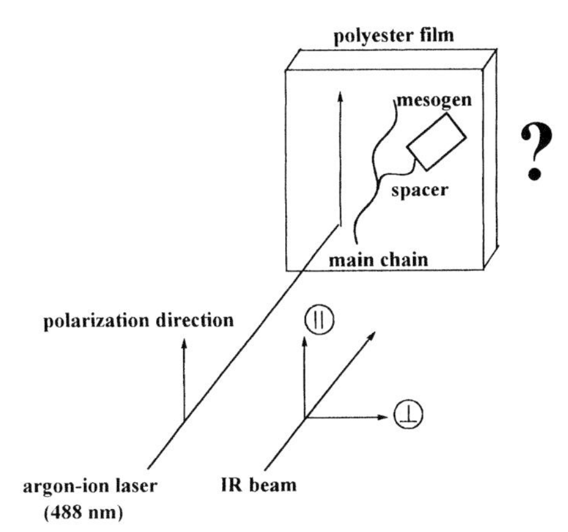

Figure 2-32. Simplified picture of the experimental set-up for laser-irradiation and symbolic representation of the motivation for the FTIR-spectroscopic determination of the order parameters of the polyester functionalities.

been studied, but no obvious improvement of the data storage capability [99] could be observed.

Thin films of the polyesters were obtained by casting a chloroform solution (3 mg/200 μl) of the polymer onto KBr plates with a $15 \times 10\,\text{mm}^2$ area. After evaporation of the solvent, the films were dried at room temperature under vacuum for 2 h. The film thickness obtained by this preparation procedure was approximately 5 μm.

2.6.2 Experimental

Figure 2-32 provides a simplified picture of a laser irradiation/FTIR measurement experiment symbolizing the inherent problem to determine the laser-induced orientation of the different polymer segments. The detailed instrumental design of the apparatus used for these on-line measurements is given in Figure 2-33 and shows that the following experimental parameters can be varied:

- Based on an internally calibrated photo-diode, the laser intensity can be reduced from 800 mW/cm^2 in a controlled fashion with a neutral-glass filter wheel.
- The irradiation time is adjusted by the shutter.
- The temperature of the sample is controlled by a Peltier element.

The acquisition of the FTIR polarization spectra is performed by using the dedicated computer of the spectrometer, which is synchronized with the different devices. The laser beam passes through the shutter and the filter wheel and is directed via two mirrors onto the film sample, which is mounted on the Peltier heater/cooler arrangement. This set-up was used to follow the photo-induced anisotropy during and after irradiation as a function of time and temperature, as well as its thermal stability during erasing cycles. Time-resolved spectra with a resolution of 4 cm^{-1} were taken with radiation polarized alternately parallel and

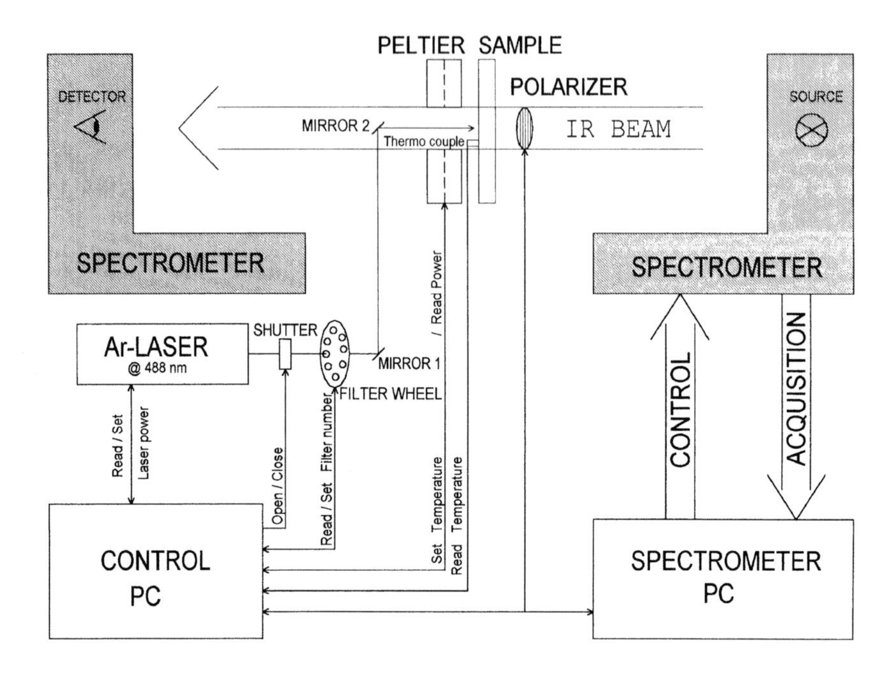

Figure 2-33. Instrumental details of the apparatus used for on-line FTIR measurements during laser irradiation.

perpendicular to the argon-ion laser beam polarization direction, respectively. The irradiated sample area was about 3 mm^2.

2.6.3 Results and Discussion

In the FTIR polarization spectra of the films prepared by casting from solution no dichroism could be detected before laser irradiation. Figure 2-34 shows the polarization spectra of a P8a12 polyester after irradiation with the 488 nm argon-ion laser line (laser power 800 mW/cm^2, irradiation time 1000 s). A detailed assignment of the individual absorptions bands to specific structural units is given in Table 2-3. The observable dichroic effects clearly demonstrate the significant orientation induced in the different functionalities of the polymer. Thus, the mesogenic side groups are preferentially oriented perpendicular to the polarization direction of the laser beam. This picture is based on the σ-dichroism of the $\nu(C{\equiv}N)$ band and the aromatic ring modes (see also Table 2-3). Additionally, the π-dichroism of the $\nu(CH_2)$ absorptions indicates a perpendicular orientation of the side and main chains relative to the laser beam polarization direction, whereas the smaller π-dichroism of the ester linkage indicates a somewhat lower orientation. As the electromagnetic radiation influences only the orientation of the mesogenic groups with

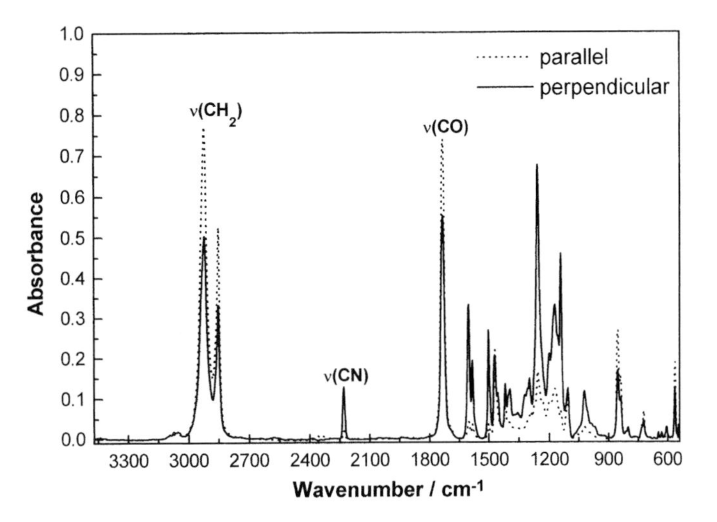

Figure 2-34. FTIR polarization spectra of an irradiated P8a12 polyester (after 1000 s of irradiation with the 488 nm line of an argon-ion laser at an intensity of 800 mW/cm^2). The IR beam was polarized parallel and perpendicular to the polarization direction of the laser beam.

Table 2-3. Band assignment, wavenumber position and structural origin of selected absorptions of the polyester P8a12 (as = antisymmetric; s = symmetric; ar = aromatic; oop = out-of-plane).

Band assignment	Wavenumber (cm^{-1})	Structural origin
$\nu_{as}(CH_2)$	2920	main chain / spacer
$\nu_s(CH_2)$	2851	main chain / spacer
$\nu(C{\equiv}N)$	2229	mesogen
$\nu(C{=}O)$	1733	main chain / (ester linkage)
$\nu(C{=}C)_{ar}$	1603 / 1583	mesogen
$\nu(C{=}C)_{ar}$	1501	mesogen
$\omega(CH_2)$	1403–1355	main chain / spacer
$\nu_{as}(C{-}O{-}Ar)$	1257	mesogen / spacer
Various modes of coupled COOR group vibrations	1201 ... 1105	main chain / (ester linkage)
$\nu(N{-}Ar)$	1141	mesogen
$\nu_s(C{-}O{-}Ar)$	1022	mesogen / spacer
$\gamma_{oop}(CH)_{ar}$	839	mesogen
$\gamma_r(CH_2)$	723	main chain / spacer

the azobenzene moieties, the alignment of the spacer and the main chain can only be induced by coupling effects. Investigations of polyesters with specifically deuterated aliphatic sites in the spacer and main chain demonstrate, that the perpendicular alignment of the acid-specific main chain relative to the laser polarization direction is even better than the alignment of the side-chain spacer units [100]. Thus, Figure 2-35a shows the region of the $\nu(CH_2)$ and $\nu(CD_2)$ polarization spectra of a P6a12 polyester with a completely deuterated main chain (except the CH-group

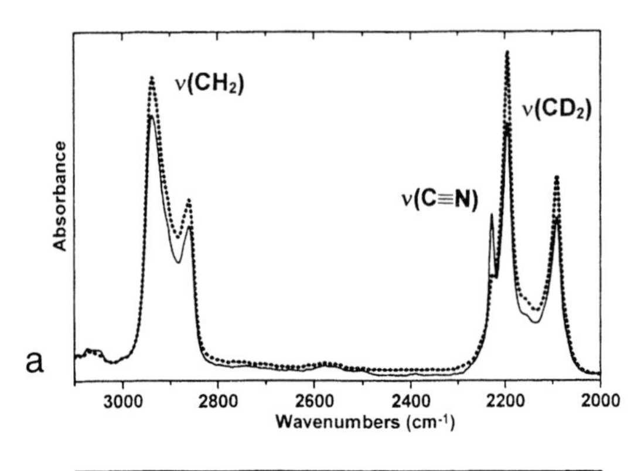

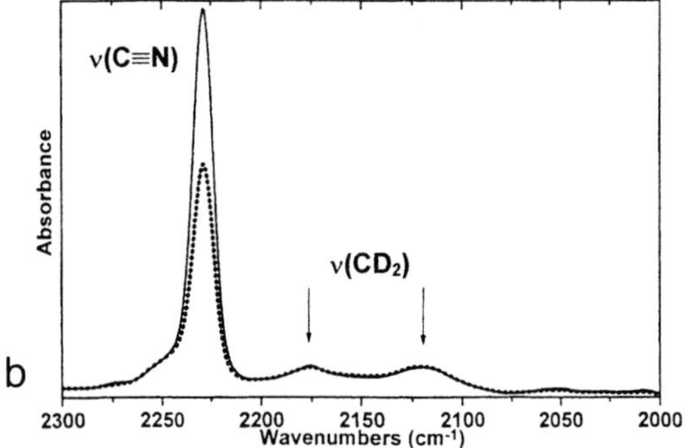

Figure 2-35. (a) FTIR polarization spectra of the $\nu(CH_2)$, $\nu(C{\equiv}N)$ and $\nu(CD_2)$ absorptions of a specifically deuterated P6a12 (see text) after laser irradiation. (b) FTIR polarization spectra of the $\nu(CD_2)$ and $\nu(C{\equiv}N)$ absorptions of a specifically deuterated P6a12 (see text) after laser irradiation. (IR radiation polarized parallel (···) and perpendicular (—) to the laser polarization).

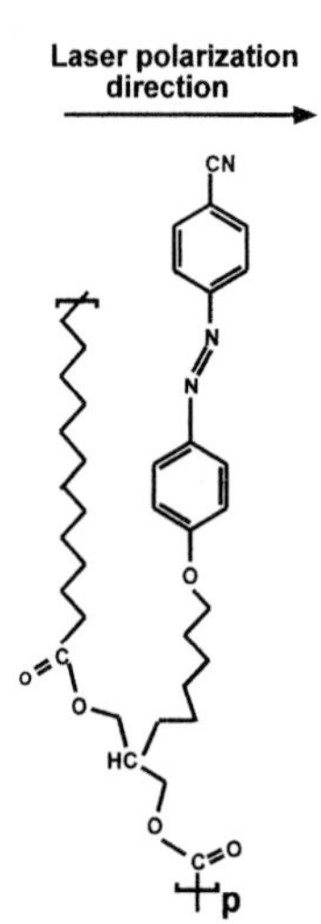

Figure 2-36. Schematic model of the irradiation-induced segmental alignment in the polyesters relative to the polarization direction of the argon-ion laser beam.

where the spacer merges into the main chain) and Figure 2-35b demonstrates the almost negligible dichroism of the $\nu(CD_2)$-region of a polyester which has been deuterated in the alcoholic part of the main chain [100]. On the basis of these experimental results, the model shown in Figure 2-36 can be derived for the laser-induced alignment of the investigated polyester and the following scenario is suggested for optical storage: the film obtained by casting from solution consists of macroscopically isotropic micro-domains. The irradiation with linearly polarized laser light leads to a melting of these domains, caused by direct heating due to absorption or a deposition of heat through *trans–cis–trans* isomerization cycles of the azobenzene moieties [95]. The origin of the orientation effect is attributed to *trans–cis* photoisomerization of the azo-group, followed by a *cis–trans* thermal or photoisomerization. In the *cis*-form, this functionality is able to undergo some reorientation with respect to its original alignment. The absorption of the polarized 'writing' beam will continue as long as a component of the electric dipole vector of the azo-group lies in the direction of the laser polarization. The *trans–cis* reorientation process will therefore continue until all the azo-groups have been oriented perpendicular to the polarization vector of the laser beam [86, 101]. In the polyesters investigated here, this orientation remains after the 'writing' beam is turned off and a long-term stable, macroscopic orientation is established.

Apart from this long-term stability, the possibility to erase and reinduce this anisotropy in the polymer is a prerequisite for its application as a reversible optical data-storage medium. To demonstrate the loss of mesogen-alignment during heating and the perfect reversibility of a 'write–erase–write' cycle, Figure 2-37a shows a stack plot of the $\nu(C{\equiv}N)$-band region in the polarization spectra of a P10a12 film during heating from 32 °C to 100 °C (isotropization temperature, 68 °C) and in Figure 2-37b the dichroism of the $\nu(C{\equiv}N)$-band of the same polymer is shown (from left to right) after irradiation, upon heating the irradiated sample to 100 °C and after reirradiation. The almost perfect coincidence of repeated writing cycles is demonstrated in more detail in Figure 2-38a, which displays the simultaneous photo-induced alignment of the different segments during laser irradiation in an order parameter/time-plot. Figure 2-38b exhibits the intermediate thermal erasure of the first orientation as a function of temperature. The loss of orientation for the different polymeric segments is also coupled and takes place in a narrow temperature interval. A comparison of Figure 2-38b with the transition temperature ranges of the individual phases in Table 2-2 shows that, during erasure, the anisotropy is mainly retained in the less ordered mesophases 3 and 4. Beyond 54 °C, only a small, residual anisotropy is observed due to recrystallization effects [96], which completely disappears at the clearing point of phase 2.

2.7 Orientation of Liquid Crystals Under Mechanical Forces

Apart from the orientation mechanisms discussed in the previous sections, a preferential alignment of liquid crystals or liquid-crystalline polymers may also be induced by the application of mechanical forces. Depending on the investigated system, this can be either done by shearing between plates or in stress–strain mea-

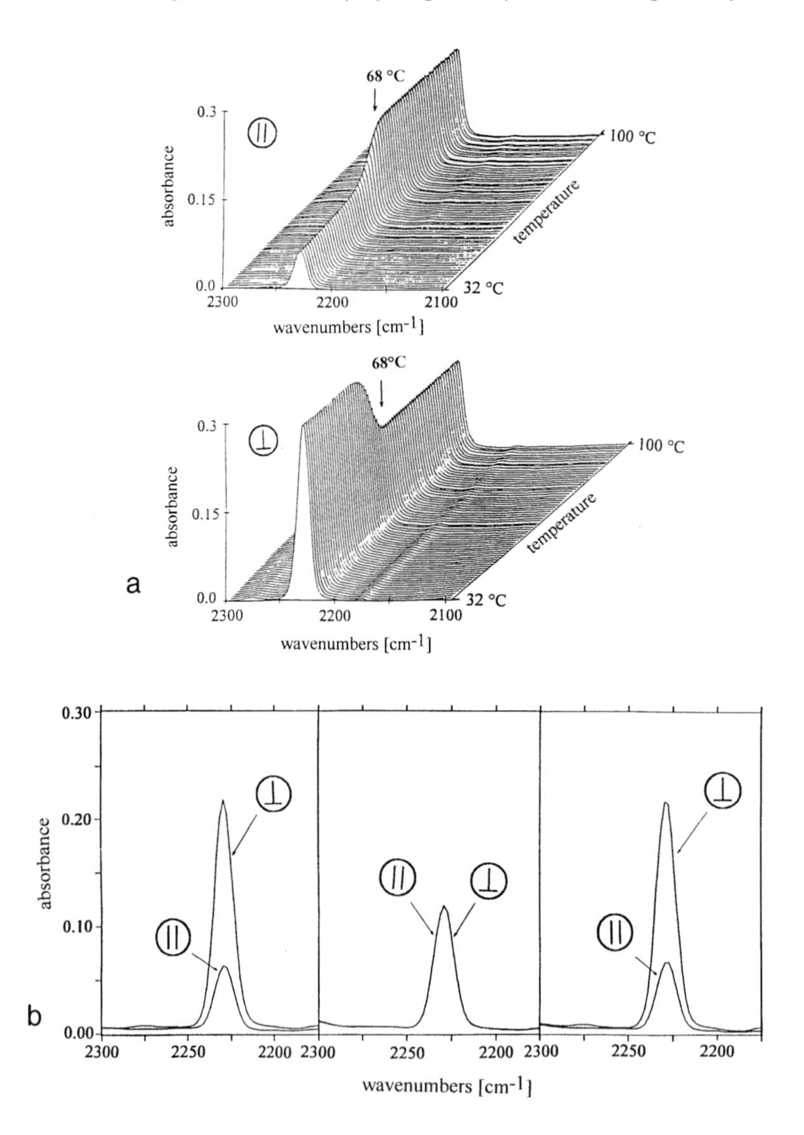

Figure 2-37. Thermal erasure of a laser-induced alignment in a P10a12 polyester demonstrated by the disappearance of the $\nu(C{\equiv}N)$ dichroism at the isotropization point (68 °C) (a) and the effect of irradiation, thermal erasure and reirradiation on the $\nu(C{\equiv}N)$ dichroism of the same polymer (b).

surements of bulk samples [36, 102–104]. Spectroscopic data on the orientation induced by these pretreatments may then be utilized for a better understanding of the different responses of specific functionalities to the external force field, or they can establish the basis for a correlation of the mechanical properties and the structure of the liquid-crystalline system.

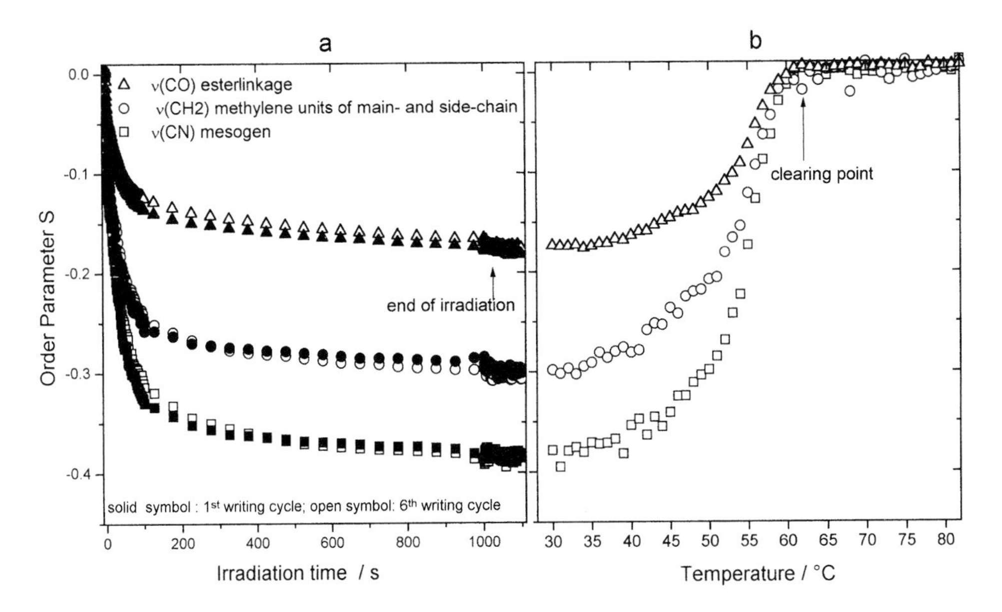

Figure 2-38. Reversibility of repeated write/erase-cycles demonstrated in time-resolved FTIR polarization measurements by the order parameter for the different segments of P8a12: (a) As a function of irradiation time (488 nm line of an argon-ion laser, at 30 °C, 800 mW/cm^2) in the 1st and 6th cycle; and (b) Versus temperature during thermal erasure with a heating rate of 30 °C/min.

2.7.1 Shear-Induced Orientation in Side-Chain Liquid-Crystalline Copolysiloxanes

As pointed out already in Section 2.5.5, low-molecular weight ferroelectric liquid crystals (FLCs) and FLCPs are attracting a lot of interest because of their potential for electro-optical applications. The polymers offer new possibilities, e.g., as elastomers for piezoelectric elements or by copolymerization [77, 78, 105] due to the formation of intrinsic mixtures between SmC* mesogenic units and other comonomers. This leads to FLCPs combining several material properties which might be utilized for colored displays in the case of comonomers containing chromophores. For the differentiated evaluation of such copolymers with reference to the possible exploitation of nonlinear optical (NLO) properties, the interplay of the different orientation tendencies of the side-chain functionalities is of crucial importance [36, 106].

2.7.1.1 Materials

The copolymers presented here are copolysiloxanes whose liquid-crystalline phase behavior is characterized by the formation of a SmC* phase. By the additional

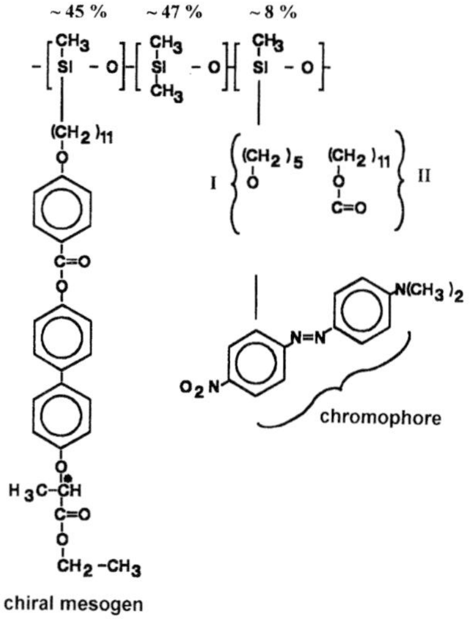

		phase transition SmC* - isotropic / °C
	main chain + mesogen	130
copolymer	+ chromphore I (~ 8%)	98
	+ chromphore II (~ 8%)	118

Figure 2-39. Chemical structure, composition and phase-transition temperatures of the different investigated copolymers.

introduction of chromophores with extended π-conjugation it was attempted to combine the ferroelectric properties of the mesomorphic phase with the NLO-properties of the chromophore. The synthesis has been described in detail [106], and Figure 2-39 shows the structure and composition of the investigated homopolymer (containing no chromophore) and the modified copolymers alongside their corresponding phase-transition temperatures.

2.7.1.2 Experimental

Oriented polymeric films were obtained in a mechanical force field by applying a shear stress. The NaCl plates used as sample supports were coated with a thin polyimide layer to improve the surface properties. The shearing process along a defined direction was first performed in the isotropic state above the clearing temperature and then at the isotropic/SmC*-transition. Afterwards, the temperature was gradually decreased to room temperature. The mechanical process was simultaneously analyzed using polarization optical microscopy. The experimental set-up is shown in Figure 2-40. For the subsequent dichroic measurements the samples were mounted in the spectrometer so that the shearing direction and the polarization direction of the incoming radiation were initially parallel. By rotating the

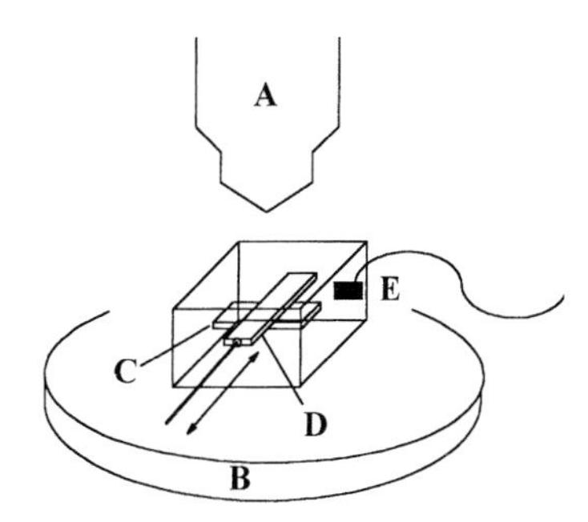

Figure 2-40. Experimental set-up for the shearing experiment combined with POM measurements (A: microscope, B: goniometer, C: fixed NaCl crystal, D: shearing NaCl crystal, E: temperature control).

polarizer for 90° the spectrum with perpendicular polarization relative to the shearing direction was recorded.

2.7.1.3 Results and Discussion

The question to be clarified by the FTIR spectroscopic polarization studies was whether:

- the interaction between the polymer chain and the side-on linked mesogenic dye functionalities dominates and the chromophores orient parallel to the polymer chain [107, 108] and thus perpendicular to the non-chromophoric mesogens (Figure 2-41a); or
- the interaction between the different mesogens (non-chromophoric and chromophoric) is dominating and leads to their parallel alignment (perpendicular to the chain direction) (Figure 2-41b).

In order to determine the orientational behavior of the main chain, the spacer and the mesogenic and chromophoric side groups relative to the shearing direction, detailed band assignments of the monomers, homopolymer and copolymers preceded the dichroic measurements of the sheared polymers [36, 106]. Based on these investigations, the polarization spectra of the sheared copolymers I and II (Figure 2-42) yielded the surprising qualitative result, that the orientation of the side-on fixed chromophores is dominated by the interaction with the mesogenic groups and leads to the structure represented in Figure 2-41b. Thus, the dichroism of specific absorptions indicates only low orientation effects for the main chain parallel to the shearing direction, whereas the mesogens and the spacers clearly orient preferentially perpendicular to the direction of shear. The σ-dichroism of the $\nu_s(NO_2)$-absorption (Figure 2-42) supports the perpendicular alignment of the chromophore long axis relative to the shearing direction, and thus the validity of the model shown in Figure 2-41b [36]. These dye-containing copolymers may therefore not be useful for nonlinear optics [106].

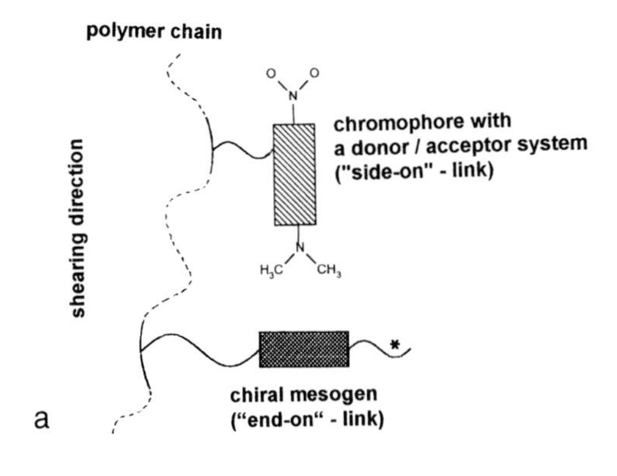

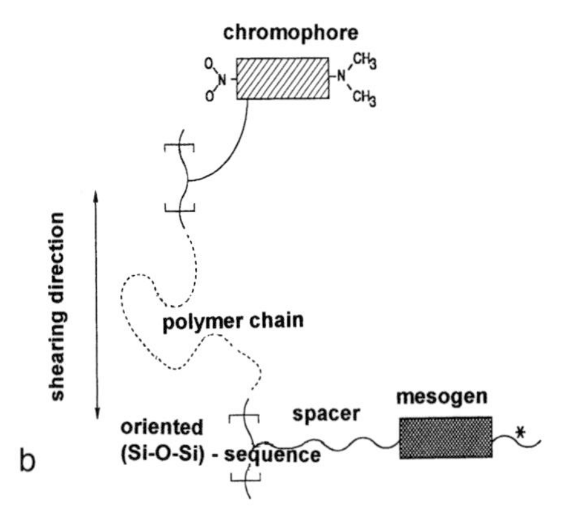

Figure 2-41. Possible models of mesogen and chromophore alignment in the sheared FLC-copolymers. (a) The chromophore orients parallel to the local polymer chain axis and the shearing direction, whereas the mesogen aligns perpendicular to the chromophore. (b) The interaction between the chromophores and the mesogens dominates and induces a parallel alignment to each other and a perpendicular orientation to the shearing direction.

2.7.2 Orientation Mechanism of Liquid Single Crystal Elastomers During Cyclic Deformation and Recovery

The application of the previously discussed techniques to induce monodomain structures in side-chain liquid-crystalline polymers by the application of electric or electromagnetic fields, by shearing or on anisotropic surfaces, frequently leads to comparatively low, macroscopically uniform orientation. Additionally, the methods are limited to a sample thickness of about 100 μm. Liquid-crystalline side-chain elastomers do not have this restriction, because a high macroscopic orientation can be induced in polymeric networks by mechanical deformation up to a sample thickness of about a centimeter [103, 109]. The synthesis of such systems can be performed by crosslinking linear, side-chain liquid-crystalline polymers to networks [110]. The inherent combination of rubber elasticity and liquid-crystalline phase behavior, may then be exploited for the induction of a macroscopic mesogen orientation by mechanical deformation.

Liquid single crystal elastomers (LSCEs) are characterized by a high macroscopically uniform orientation in the liquid-crystalline phase without the require-

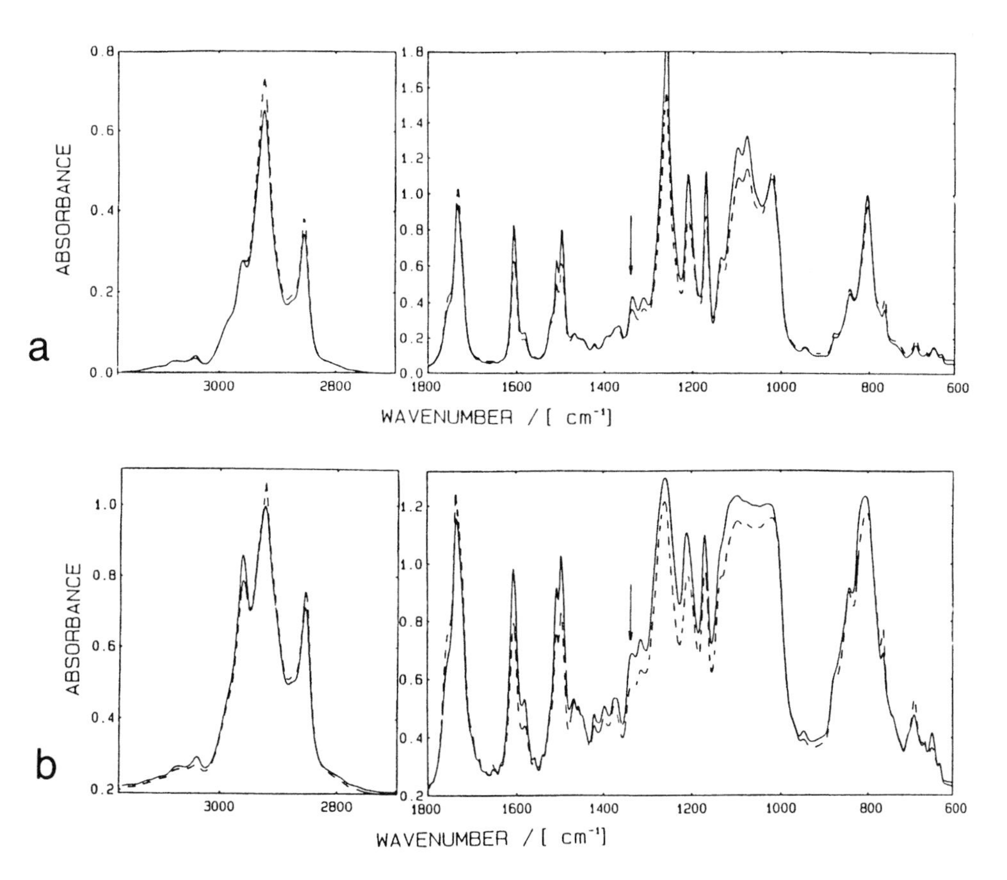

Figure 2-42. FTIR polarization spectra of the copolymers with chromophore I (a) and II (b) after shearing (IR-radiation polarized parallel (- - -) and perpendicular (—) to the shearing direction) (the arrow indicates the $\nu_s(NO_2)$ absorption of the chromophore).

ment of an external stabilizing (mechanical) field. A characteristic feature of this state of order is the optical transparency of the material. The optical behavior of such elastomers is comparable with organic or inorganic single crystals. Küpfer and Finkelmann [111] have described the synthetic route for LSCEs with a permanent mesogen orientation.

2.7.2.1 Materials

The LSCE system investigated in the present work is based on a methylsiloxane elastomer which has been crosslinked in a two-step procedure (Figure 2-43). The mesogenic side groups have been attached via a vinyl end-group (compound 2) and the crosslinking agents are a hydrochinonedialkylether with vinyl end-groups

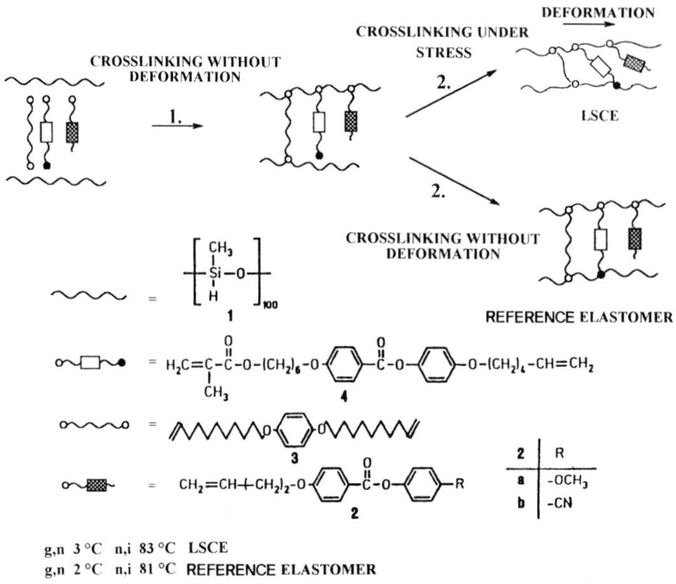

CROSSLINKING WITHOUT DEFORMATION

1.

CROSSLINKING UNDER STRESS

2.

DEFORMATION

LSCE

2.

CROSSLINKING WITHOUT DEFORMATION

REFERENCE ELASTOMER

2	R
a	$-OCH_3$
b	$-CN$

g,n 3 °C n,i 83 °C LSCE

g,n 2 °C n,i 81 °C REFERENCE ELASTOMER

Figure 2-43. Chemical structure, transition temperatures and schematic presentation of the cross-linking processes involved in the synthesis of the investigated LSCE and reference elastomer [111].

(compound 3) and a vinyl- and methacryloyl-substituted derivative (compound 4), which has also liquid-crystalline properties. The key point of the crosslinking reactions is the fact that, under the experimental conditions, the addition of the vinyl group is faster by approximately two orders of magnitude [111]. Thus, in the first step, a network with a low crosslink-density is synthesized because predominantly compound 3 becomes active. In the second reaction step, the mesogens of this network are macroscopically oriented by application of a stress at elevated temperature and the induced anisotropy is fixed in the system by crosslinking via the methacryloyl groups. The crosslinking reactions were performed in the nematic phase of the system. Part of the methoxy functionalities in compound 2 have been replaced by nitrile groups to provide an IR-active probe for the orientation measurements. Additionally to the LSCE system, a structurally identical reference elastomer was investigated, which had not been subjected to a mechanical deformation during the second crosslinking step (Figure 2-43). In the reference polymer the mesogens do not exhibit a macroscopic orientation and, contrary to the LSCE, this material is opaque.

2.7.2.2 Rheo-Optical FTIR Spectroscopy

The increased need for more detailed data and a better understanding of the mechanisms involved in polymer deformation and relaxation has led to the search

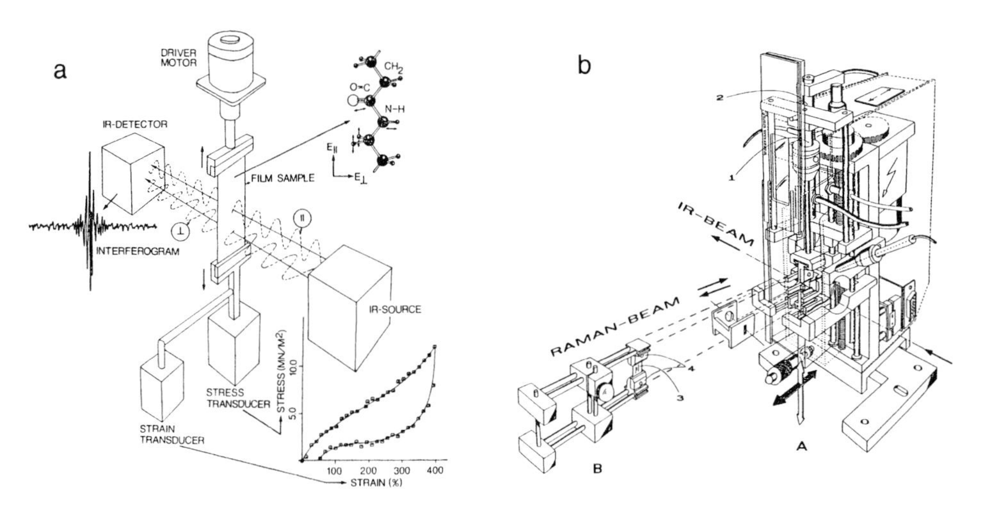

Figure 2-44. (a) Principle of rheo-optical FTIR spectroscopy of polymer films. (b) Variable-temperature stretching machine for rheo-optical FTIR and FT-Raman spectroscopy. (A) stretching machine; (B) sample transfer mechanism; (1) stress transducer; (2) strain transducer; (3) film sample; (4) sample clamps [17, 18].

for new experimental techniques to characterize transient structural changes during mechanical processes. With the advent of rapid-scanning FTIR spectroscopy, rheo-optical measurements have been applied to obtain data on the orientation, conformation, and crystallization during mechanical treatment of a wide variety of polymers [14, 17–20, 112]. The experimental principle of rheo-optical FTIR spectroscopy is based on the simultaneous acquisition of polarization spectra and stress–strain diagrams during deformation, recovery or stress–relaxation of the polymer under investigation in a miniaturized, computer-controlled stretching machine which fits into the sample compartment of the spectrometer (Figure 2-44). The implementation of a heating cell as a closed, nitrogen-purged system offers the additional possibility to study the above mechanical phenomena under controlled temperature conditions (± 0.5 °C) up to 200 °C. The electromechanical apparatus, the experimental procedure and the evaluation of the relevant parameters from the polarization spectra series have been described in detail in several publications [17–20, 112].

2.7.2.3 Near-Infrared Spectroscopy

For samples exceeding the thickness accessible for transmission measurements by mid-infrared spectroscopy, near-infrared (NIR) spectroscopy may be used as an alternative technique [113]. Adjacent to the conventional mid-infrared region (4000–400 cm^{-1}) the NIR region covers the interval from about 12500–4000 cm^{-1}.

The absorption bands observed in this region can be assigned to overtones and combinations of characteristic fundamental frequencies. Apart from analytical applications, NIR spectroscopy has been demonstrated to be of value also to rheo-optical applications [17, 114, 115]. Thus, although NIR spectra are generally less informative than their mid-infrared analogs due to overlap of absorption bands, the details of the changes in crystallinity, conformation, and orientation are also contained in the overtone and combination vibrations [17, 113].

2.7.2.4 Experimental

The rheo-optical investigations were performed in the stretching machine shown in Figure 2-44b. For this purpose, film samples of the LSCE and the reference polymer (sample thickness about 300–350 μm, original length 8 mm, width 5 mm) were subjected to an elongation–recovery cycle up to about 45% strain, with a strain rate of 11%/min and 20-scan interferograms with a spectral resolution of 5 cm^{-1} were collected simultaneously with radiation alternatingly polarized parallel and perpendicular to the drawing direction. After completion of the experiment the interferograms were transformed to the corresponding spectra and evaluated in terms of the order parameter (orientation function) [14, 17–20, 112] of selected absorption bands. The experiments discussed here were performed at 27 °C. In order to study the reorientation mechanism of the preoriented LSCE [116–121] and the isotropic reference elastomer during the mechanical treatment of the rheo-optical experiment, the samples were mounted in the clamps of the stretching machine after a rotation of 90° from the deformation direction in the second crosslinking step (Figure 2-45). Thus, in the LSCE the mesogens are originally oriented perpendicular to the subsequent drawing direction. This is reflected in the

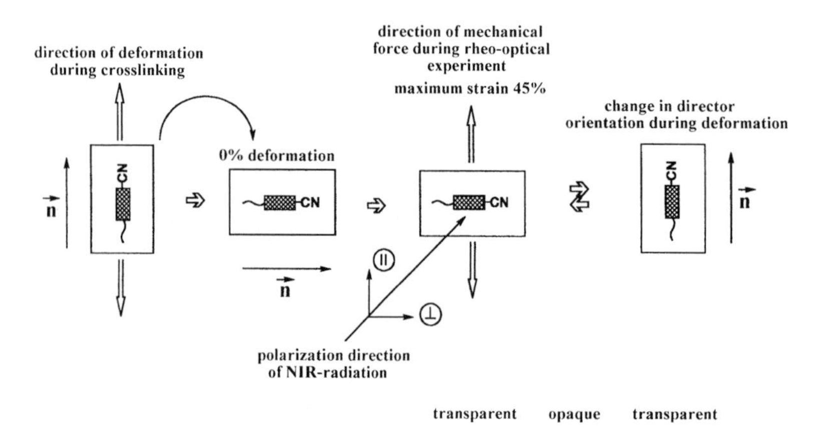

Figure 2-45. Experimental sequence for the observation of the director reorientation during the rheo-optical elongation–recovery cycle.

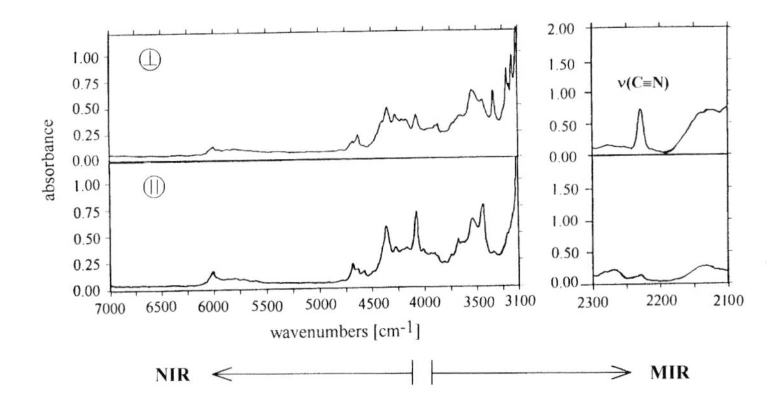

Figure 2-46. FTIR/FTNIR polarization spectra of the undrawn LSCE system (the radiation was polarized parallel and perpendicular to the subsequent drawing direction of the rheo-optical experiment).

dichroic effects of the polarization spectra taken before the elongation procedure of (Figure 2-46). In contrast, no dichroism can be observed in the undrawn reference polymer (see below).

Due to the large thickness of the samples, the spectrometer had been modified to collect NIR spectra extending into the mid-infrared region (10000–2000 cm^{-1}) (tungsten–halogen source, CaF_2-beamsplitter, InSb-detector). Despite an extensive band-overlap in the NIR-region, several absorptions in the spectrum of Figure 2-46 could be assigned by spectral comparison with low-molecular weight reference compounds [36]. Here, however, only the data derived from the $\nu(C{\equiv}N)$ absorption will be discussed in some detail.

2.7.2.5 Results and Discussion

In Figure 2-47 the stress–strain diagram of an elongation–recovery cycle up to 43% strain at 27 °C is shown for the reference elastomer. Further elongation–recovery cycles do not reflect significant differences. The change in order parameter of the $\nu(C{\equiv}N)$ absorption band during the mechanical treatment is shown in Figure 2-48. Throughout the whole experiment the drawing direction was the polarization reference direction. Due to the crosslinking conditions, no orientation ($S = 0$) can be observed for the original sample and uniaxial elongation leads to a linear increase of the order parameter up to $S = 0.45$ at the maximum strain of 43%. During recovery, an almost linear decrease of S can be observed with a small residual orientation of the mesogens upon return to the original starting position. This is in agreement with the hysteresis effect and the permanent elongation observed in the stress–strain diagram (Figure 2-47). When the structural absorbance A_0 of the $\nu(C{\equiv}N)$ absorption is plotted as a function of elongation [36], an almost linear decrease is observed which is in good agreement with the theoretically calculated

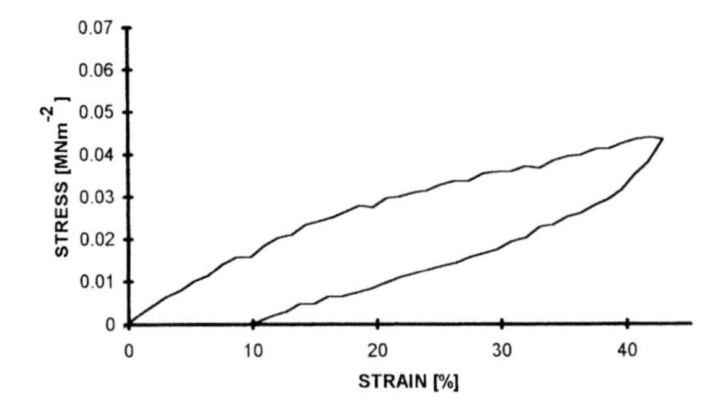

Figure 2-47. Stress–strain diagram of an elongation–recovery cycle of the reference elastomer at 27 °C.

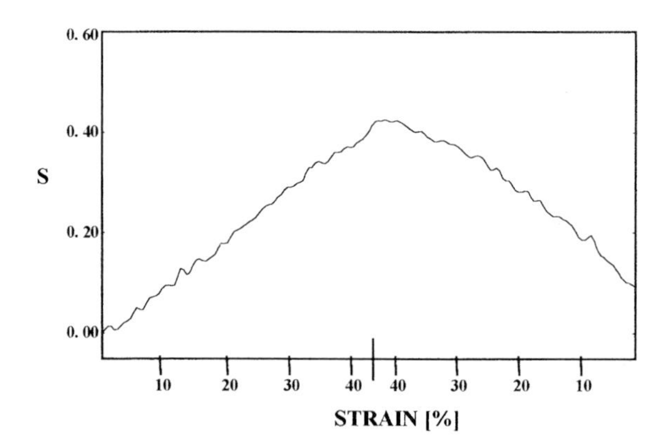

Figure 2-48. Order parameter/strain-plot for the $\nu(C{\equiv}N)$ absorption corresponding to the elongation–recovery cycle shown in Figure 2-47.

decrease in sample thickness based on the assumption of a constant sample volume during elongation [20, 36]. A more complex behavior is observed during the elongation–recovery cycle of the LSCE system to a maximum strain of 45% at 27 °C. The corresponding stress–strain curve is shown in Figure 2-49 and has been subdivided into different strain intervals. The order parameter of the $\nu(C{\equiv}N)$ absorption derived from the spectra acquired during this mechanical treatment is presented in Figure 2-50. Here too, the analogous strain intervals are indicated to assist the correlation and interpretation of the mechanical and spectroscopic changes. Due to the anisotropy of the starting material and the reorientation of the mesogens during the mechanical treatment, specific assumptions in terms of the reference direction were necessary. Thus, for the derivation of the order parameter during elongation between 0 and 32% strain (intervals I and II in Figure 2-50), the deformation direction during the second crosslinking step (which is perpendicular to the applied deformation in the rheo-optical experiment) has been chosen as polarization reference direction. Beyond the inversion point of S = 0, the machine direction of the rheo-optical experiment is applied as polarization reference (interval III). During recovery this procedure is reversed. Therefore, the order parameter always has a positive value throughout the whole experiment. From Figure 2-50 the following picture can be derived for the orientational changes taking place during the elongation–recovery cycle:

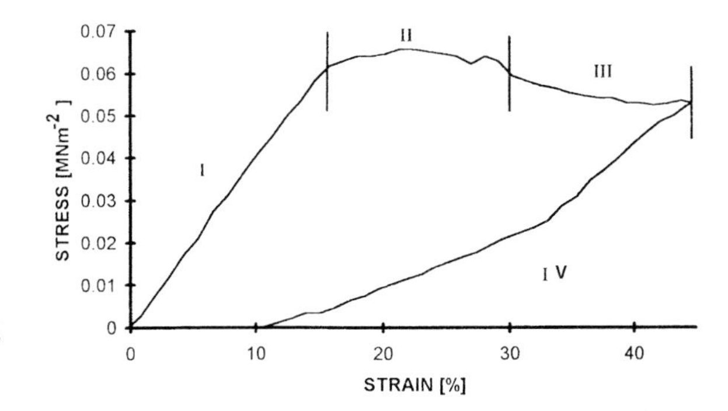

Figure 2-49. Stress–strain diagram of an elongation–recovery cycle of the LSCE at 27 °C.

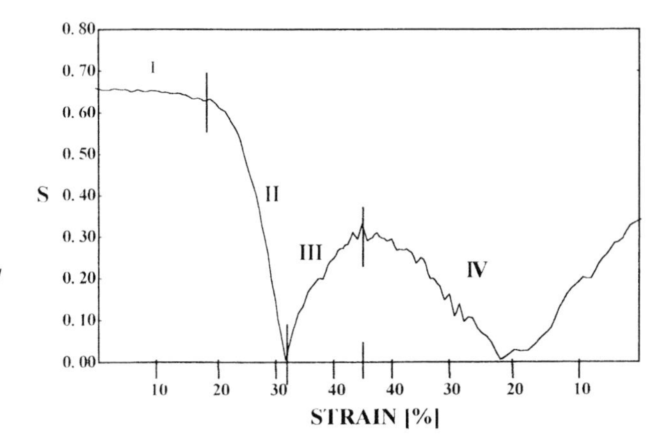

Figure 2-50. Order parameter/strain-plot for the $\nu(C{\equiv}N)$ absorption corresponding to the elongation–recovery cycle shown in Figure 2-49 (see text).

- During interval I the stress increases linearly with strain and no significant changes in the mesogen orientation can be observed.
- When the external stress (σ_e) reaches the internal stress (σ_i) of the LSCE system (induced by the second crosslinking procedure) [111] at about 15% strain, the stress–strain curve levels off and simultaneously the order parameter decreases rapidly towards zero (interval II) (Figures 2-49 and 2-50).
- During interval III, the reorientation of the mesogens in the direction of the external field takes place until maximum elongation. This procedure is accompanied by a slight decrease in stress due to the cooperativity of the reorientational motion after exceeding the internal stress of the LSCE.
- Depending on the experimental conditions (temperature, maximum strain), the described phenomena are completely reversible upon recovery (Figure 2-51).

Due to the intrinsic anisotropy in the original LSCE network, the spectroscopically monitored inversion point of the order parameter (S = 0) could principally be interpreted by the three different mesogen alignments shown schematically in Figure 2-52 [36]. For the elucidation of this question the structural absorbance A_0

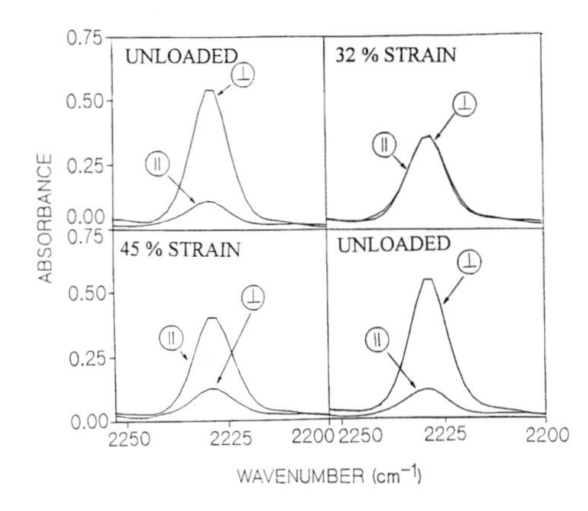

Figure 2-51. Reversible dichroic behavior of the ν(C≡N) absorption of the LSCE during an elongation–recovery cycle to 45% strain at 27 °C.

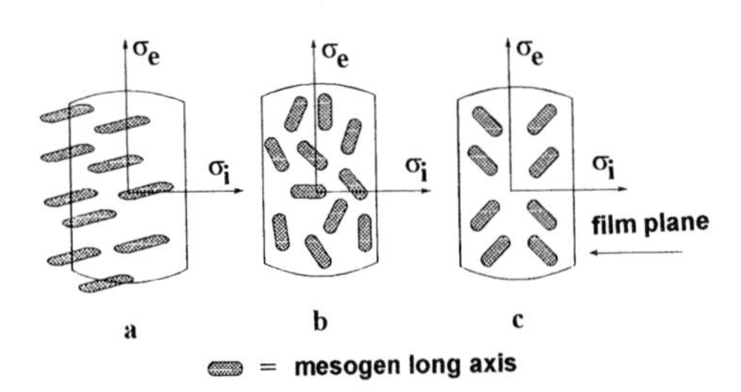

Figure 2-52. Possible mesogen orientations during the reorientation process at S=0 (σ_i: direction of internal stress induced during the second crosslinking step, σ_e: direction of external stress applied during the rheo-optical experiment). (a) Homeotropic orientation perpendicular to the film plane; (b) isotropic orientation; (c) planar biaxial orientation.

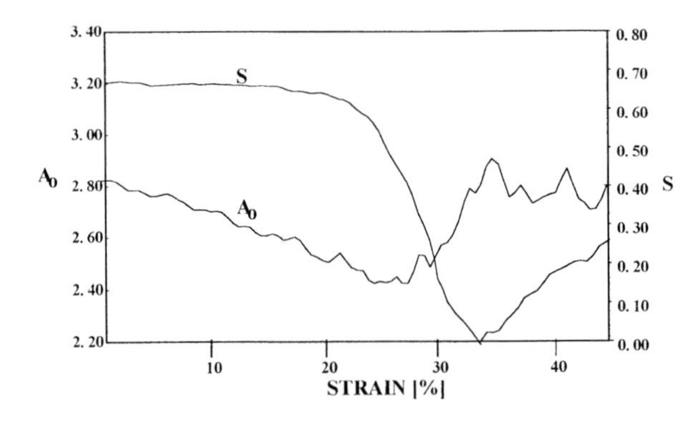

Figure 2-53. Structural absorbance and order parameter/strain-plots for the ν(C≡N) absorption corresponding to the elongation of the LSCE to 45% strain at 27 °C.

of the ν(C≡N) absorption has been plotted alongside the order parameter for the LSCE elongation step in Figure 2-53. The reference direction for the evaluation of the structural absorbance has been chosen in accordance with the assumption made for the order parameter (see Figure 2-50). As can be seen from Figure 2-53 the structural absorbance decreases linearly up to about 25% strain. This is in agreement with the reference polymer and can be attributed to the decrease in sample thickness. Between 25% and 35% strain, however, the structural absorbance shows an abrupt increase followed by only a slight decrease up to maximum elongational strain. As A_0 is an experimentally determined parameter based on a uniaxial orientation model whose value should represent the intensity of an absorption band exclusive of orientation effects, the anomalous increase of the structural absorbance during elongation–where the sample thickness actually decreases–points towards an invalidity of the uniaxial symmetry. Thus, the geometries a and b of Figure 2-52 can be excluded on the basis that both would lead to a decrease of A_0 during elongation. The occurrence of the structural model c at the inversion point of the order parameter has first been proposed on the basis of small-angle X-ray measurements by Finkelmann and Küpfer [117]. Thus, the appearance of four diffraction maxima have been explained in terms of a biaxial, planar orientation of the mesogens in the film plane (Figure 2-52c). Depending on the inclination angle of the mesogens with the stretching direction, such a transition state would lead to an order parameter close to zero and could also explain the anomalous behavior of the structural absorbance [36]. In conclusion, it could be shown that, during elongation of the anisotropic LSCE network, the network structure changes from a uniaxial symmetry perpendicular to the applied mechanical field (parallel to the orientation induced in the second crosslinking step) via a planar–biaxial model (inversion of the order parameter) to a system aligned in the stretching direction of the rheo-optical experiment [36]. Simultaneously, the originally transparent polymer becomes opaque at the inversion point of the order parameter and turns transparent again upon reorientation of the mesogens into the machine direction.

2.8 Conclusions

The aim of this contribution was to demonstrate the potential of FTIR/FTNIR polarization spectroscopy at variable temperatures for the characterization of segmental mobility in liquid crystals and liquid-crystalline polymers under the influence of external fields. Selected examples have been discussed, where either time-resolved measurements or static experiments have been utilized to monitor the effects of electric, electromagnetic, or mechanical perturbations on the liquid-crystalline system under investigation. For the extremely fast, reversible switching processes of low-molecular weight liquid crystals and side-chain liquid-crystalline polymers in electric fields, the novel step-scan FTIR technique has been applied. The selective character of vibrational spectroscopy generally allowed us to separate the orientational effects in different functionalities of the liquid-crystalline compound. Thus, side-chain liquid-crystalline polymers, for example, have been characterized under the influence of different external fields in terms of the relative alignment of their main chain, spacer, and mesogenic groups, respectively. Although the presented data are necessarily limited to selected examples, the variety of information and its detailed quantitative character demonstrate, that FTIR/FTNIR polarization spectroscopy has certainly become an extremely valuable tool to elucidate the dynamics of segmental motion in liquid-crystalline systems.

2.9 Acknowledgments

The authors would like to thank Prof. Dr. H. Finkelmann and Dr. J. Küpfer (University of Freiburg, Germany), Prof. Dr. R. Zentel and Dr. T. Öge (University of Wuppertal and University of Mainz, Germany), Prof. Dr. B. Jordanov (University of Sofia, Bulgaria), Dr. N. C. R. Holm, Dr. S. Hvilsted, Dr. M. Pedersen and Dr. P. S. Ramanujam (Riso National Laboratory, Denmark), Dr. F. Andruzzi (CNR, Pisa, Italy), Dr. M. Paci, Dr. E. L. Tassi and Prof. Dr. P.-L. Magagnini (University of Pisa, Italy), and Dr. E. Happ (University of Ulm, Germany), Dr. C. Hendann (Deutsche Post, Darmstadt), Dr. M. Czarnecki (University of Wroclaw, Poland) for helpful discussions, individual experimental and theoretical data, and the supply of polymer samples. Financial and instrumental assistance by the Deutsche Forschungsgemeinschaft, Bonn, Germany, the Ministerium für Wissenschaft und Forschung, Düsseldorf, Germany, Fonds der Chemischen Industrie, Frankfurt, Germany, and the European Community via contract BRE2.CT93.0449 is also gratefully acknowledged.

2.10 References

[1] Vertogen G., de Jeu, W. H., *Thermotropic Liquid Crystals, Fundamentals*, Berlin: Springer Verlag, 1987.
[2] Koswig, H. D., *Flüssige Kristalle*, Berlin: VEB Verlag, 1984.
[3] Donald, A. M., Windle, A. H., *Liquid Crystalline Polymers*, Cambridge: Cambridge University Press, 1992, pp. 38–41.
[4] Finkelmann, H., in: *Liquid Crystallinity in Polymers: Principles and Fundamental Properties*: Ciferri, A. (Ed.), Weinheim: VCH Verlagsgesellschaft, 1991; pp. 315.
[5] McArdle, C. B., in: *Side Chain Liquid Crystal Polymers*: McArdle, C. B. (Ed.), Glasgow: Blackie & Son Ltd., 1989; pp. 357.
[6] Wendorff, J. H., Eich, M., *Mol. Cryst. Liq. Cryst.* **1989**, *169*, 133–166.
[7] Nakamura, T., Ueno, T., Tani, C., *Mol. Cryst. Liq. Cryst.* **1989**, *169*, 167–192.
[8] Demus, D., Goodby, J., Gray, G. W., Spiess, H.-W., Vil, V. (Eds.), *Handbook of Liquid Crystals*, Weinheim: Wiley-VCH, 1998.
[9] Carfagna, C. (Ed.), *Liquid Crystalline Polymers*, Oxford: Pergamon, 1994.
[10] Chilton, J. A., Gossey, M. T. (Eds.), *Special Polymers for Electronic and Optoelectronics*, London: Chapman & Hall, 1995.
[11] Kelly, S. M., *Liq. Cryst. Today* **1996**, *6*(4), 1.
[12] Schadt, M. in: *Liquid Crystals*: Baumgartel, H., Frank, E. U., Grunsbein, W. (Eds.), Stegemeyer, H. (Guest Ed), New York: Springer, 1994, Chapter 8, p. 195.
[13] Chilton, J. A., Gossey, M. T. (Eds.), *Special Polymers for Electronics and Opto-electronics*, London: Chapman & Hall, 1995.
[14] Siesler, H. W., Holland-Moritz, K., *Infrared and Raman Spectroscopy of Polymers*, New York: Marcel Dekker, 1980.
[15] Hvilsted, S., Andruzzi, F., Kulinna, C., Siesler H. W., Ramanujam, P. S., *Macromolecules*, **1995**, *28*, 2172.
[16] Shilov, S., Okretic, S., Siesler, H. W., Czarnecki, M. A., *Appl. Spectrosc. Revs.*, **1996**, *31*(1&2), 125–165.

[17] Hoffmann, U., Pfeifer, F., Okretic, S., Völkl, N., Zahedi, M., Siesler, H. W., *Appl. Spectrosc.* **1993**, *47*, 1531–1539.
[18] Siesler, H. W., in: *Oriented Polymer Materials*: Fakirov F. (Ed.), Zug: Hüthig & Wepf, 996, pp. 138–166.
[19] Siesler, H. W., in: *Advances in Applied Fourier Transform Infrared Spectroscopy*: McKenzie, M. W. (Ed.), Chichester: Wiley & Sons Ltd., 1988; pp. 189–246.
[20] Siesler, H. W., in: *Advances in Polymer Science*, Berlin: Springer Verlag, 1984; 64, pp. 1– 77.
[21] Bräuchler, M., Boeffel, C., Spiess, H. W., *Makromol. Chem.* **1991**, *192*, 1153.
[22] Menzel, H., Hallensleben, M. L., Schmidt, A., Knoll, W., Fischer, T., Stumpe, J. *Macromolecules*, **1993**, *26*, 3644–3349.
[23] Buffeteau, T., Natansohn, A., Rochon, P., Pezolet, M., *Macromolecules*, **1996**, *29*, 8783.
[24] Graham, J. A., Hammaker, R. M., Fateley, W. G., in: *Chemical, Biological and Industrial Applications of Infrared Spectroscopy*: Durig, J. R. (Ed.), Chichester: Wiley Interscience, 1985; pp. 301–334.
[25] Souvignier, G., Gerwert, K., *Biophys. J.* **1992**, *63*, 1393.
[26] Uhmann, W., Becker, A., Taran, C., Siebert, F., *Appl. Spectrosc.* **1991**, *45*, 390.
[27] Palmer, R. A., Spectroscopy **1993**, *8*, 26.
[28] Palmer, R. A., Chao, J. L., Dittmar R. M., Gregoriou, V. G., Plunkett, S. E., *Appl. Spectrosc.* **1993**, *47*, 1297–1310.
[29] Nakano, T., Yokoyama, T., Toriumi, H., *Appl. Spectrosc.* **1993**, *47*, 1354–1366.
[30] Gregoriou, V. G., Noda, I., Dowrey, A. E., Marcott, C., Chao, J. J., Palmer, R. A., *J. Polym. Sci. B*, **1993**, *31*, 1769–1777.
[31] Lafrance, C. P., Nabet, A., Prud'homme R. E., Pézolet, M., *Can. J. Chem.* **1995**, *73*, 1497.
[32] Maier, W., Saupe, A., *Z. Naturforschung* **1959**, *14a*, 882.
[33] Maier, W., Saupe, A., *Z. Naturforschung* **1961**, *16a*, 816.
[34] Anderle, K., Birenheide, R., Eich, M., Wendorff, J. H., *Makromol. Chem. Rapid. Commun.* **1989**, *10*, 477.
[35] Natansohn, A., Rochon, P., Gosselin, J., Xie, S., *Macromolecules* **1992**, *25*, 2268.
[36] Kulinna, C., *Dissertation*, Universität Essen, 1994.
[37] Chatelain, P., *Bull. Soc. Fr. Miner.* **1943**, *66*, 105.
[38] Blinov, L. M., *Electro-Optical and Magneto-Optical Properties of Liquid Crystals*, Chichester: Wiley-Interscience, 1983.
[39] Frank, F. C., *Discuss. Faraday Soc.* **1958**, *25*, 19.
[40] Geary, J. M., Goodby, J. W., *J. Appl. Phys.* **1987**, *62*, 4100.
[41] Castellano, J. A., *Mol. Cryst. Liq. Cryst.* **1983**, *94*, 33.
[42] Jérôme, B. in: *Handbook of Liquid Crystals*: Demus, D., Goodby, J., Gray, G. W., Spiess, H.-W., Vil, V. (Eds.), Weinheim: Wiley-VCH, 1998, Vol. 1, p. 535.
[43] Geary, J. M., Goodby, J. M., Kmetz, A. P., Patel, J. S., *J. Appl. Phys.* **1987**, *62*, 4100.
[44] Toney, M. F., Russel, T. P., Logan, J. A., Kikuchi, H., Sands, J. M., Kumar, S. K. *Nature* **1995**, *374*, 709.
[45] Happ, E., University of Ulm, Sektion Polymere, personal communication, 1994.
[46] Kaneko, E., *Principles and Applications of Liquid Crystal Displays,* Tokio: KTK Scientific Publ., 1987.
[47] Finkenzellner, U., *Spektrum der Wissenschaft*, **1990**, *8*, 54.
[48] Schadt, M., Helfrich, J., *Appl. Phys. Lett.* **1971**, *18*, 127.
[49] Gelhaar, T., Pauluth, D., *Nachr. Chem. Tech. Lab.* **1997**, *45*(1), 9.
[50] Schadt, M., *Phys. Bl.* **1996**, *7/8*, 695.
[51] Kaito, A., Wang, J. K., Hsu, S. L., *Anal. Chem. Acta* **1986**, *189*, 27.
[52] Masutani, K., Yokota, Y., Furukawa, Y., Tasumi, M., Yoshizawa, A., *Appl. Spectrosc.* **1993**, *47*, 1370–1375.
[53] Okretic, S., *Dissertation*, Universität Essen, 1995.
[54] Shilov, S. V., Okretic, S., Siesler, H. W., *Vib. Spectrosc.* **1995**, *9*, 57.
[55] Urano, T., Hamaguchi, H., *Chem. Phys. Lett.* **1992**, *195*, 287.
[56] Urano, T., Hamaguchi, H., *Appl. Spectrosc.* **1993**, *47*, 2108.
[57] Gregoriou, V. G., Chao, J. L., Toriumi, H., Palmer, R. A., *Chem. Phys. Lett.* **1991**, *179*, 491.

[58] Czarnecki, M. A., Katayama, N., Ozaki, Y., Satoh, M., Yoshio, K., Watanabe, T., Yanagl, T., *Appl. Spectrosc.* **1993**, *47*, 1382–1385.
[59] Powell, J. R., Krishnan, K., Yokoyama, T., Nakano, T., in: *Proc. 9th Int. Conference on Fourier Transform Spectroscopy*: Bertie, J. E., Wieser, H. (Eds.), Proc. SPIE 2089, **1993**; 428–429.
[60] Noda, I., *Appl. Spectrosc.* **1993**, *47*, 1329–1336.
[61] Noda, I., *Appl. Spectrosc.* **1990**, *44*, 550–561.
[62] Michl, J., Thulstrup, E. W., *Spectroscopy with Polarized Light.* Solute Alignment by Photoselection in Liquid Crystals, Polymers and Membranes, Weinheim: VCH, 1986.
[63] Gotz, S., Stille, W., Strobl, G., Scheuermann, H., *Macromolecules*, **1993**, *26*, 1520.
[64] Hendann, C., *Dissertation*, Universität Essen, Germany, 1997.
[65] Heilmeier, G. H., Zanoni, L. A., *Appl. Phys. Lett.*, **1968**, *13*, 91.
[66] Bahadur, B., *Mol. Cryst. Liq. Cryst.* **1983**, *99*, 345.
[67] Sackmann, E. in *Applications of Liquid Crystals*, Meier, G., Sackmann, E., Grabmaier, G. (eds.), New York: Springer-Verlag, 1975.
[68] Schumann, C., in *Handbook of Liquid Crystals*, Kelker, H., Hatz, R. (eds.), Weinheim: Verlag Chemie, 1980.
[69] Scheffer, T. J., Nehring, J., in *Physics and Chemistry of Liquid Crystal Devices*, Sprokel, J. G., (ed.), New York: Plenum Press, 1979, pp. 173.
[70] Uchida, T., Wada, M., *Mol. Cryst. Liq. Cryst.* **1981**, *63*, 19.
[71] Bahadur, B. in *Liquid Crystals–Applications and Uses*, Bahadur B. (ed.), Singapore: World Scientific, 1992.
[72] Jordanov, B., Okretic, S. Siesler, H. W., *Appl. Spectrosc.* **1997**, *51*(3), 447–449.
[73] de Barny, P., Dubois, J. C., in: *Side Chain Liquid Crystal Polymers*: McArdle, C. B. (Ed.), Glasgow: Blackie & Son Ltd., 1989; chap. 5.
[74] Lagerwall, S. T. in *Handbook of Liquid Crystals*, Demus, D., Goodby, J., Gray, G. W., Spiess, H.-W., Vill, V. (eds.), Weinheim: Wiley-VCH, Vol. 2B, 1998, 515–664.
[75] Dijon, J., in *Liquid Crystals–Applications and Uses*, Bahadur B. (ed.), Singapore: World Scientific, 1992.
[76] Goodby, J. W., Blinc, R., Clark, N. A., Lagerwall, S. T., Osipov, M. A., Pikin, S. A., Sakurai, T., Yoshino, K., Zeks, B., *Ferroelectric Liquid Crystals: Principles, Properties and Applications*, Philadelphia: Gordon and Breach, 1991.
[77] Kapitza, H., Poths, H., Zentel, R., *Makromol. Chem. Macromol. Symp.*, **1991**, *44*, 117.
[78] Brehmer, M., Zentel, R., Wagenblast, G., Siemensmeyer, K., *Makromol. Chem.*, **1994**, *195*, 1891.
[79] Schönfeld, A., Kremer, F., Hofmann, A., Kühnpast, K., Springer, J., Scherowsky, G., *Makromol. Chem.* **1993**, *194*, 1149.
[80] Poths, H., Zentel, R., *Liq. Cryst.* **1994**, *16*, 749.
[81] Shilov, S. V., Skupin, H., Kremer, F., Gebhard, E., Zentel, R., *Liq. Cryst.* **1997**, *22*(2), 203–210.
[82] Verma, A. L., Zhao, B., Jiang, S. M., Sheng, J. C., Ozaki, Y., *Phys. Rev. E*, **1997**, *56*(3), 342.
[83] Shilov, S. V., Okretic, S., Siesler, H. W., Zentel, R., Öge, T., *Macromol. Rapid Commun.* **1995**, 16.
[84] Öge, T., Zentel, R., *Makromol. Chem. Phys.* **1996**, *197*, 1805–1813.
[85] Eich, M., Wendorff, J. H., Reck, B., Ringsdorf, H., *Makromol. Chem. Rapid. Commun.* **1987**, *8*, 59.
[86] Fischer, Th., Läsker, L., Stumpe, J., Kostromin, *J. Photochem. Photobiol. A: Chem.* **1994**, *80*, 453–459.
[87] Wiesner, U., Antonietti, M., Boeffel, C., Spiess, H. W., *Macromolecules* **1990**, *8*, 2133.
[88] Stumpe, J., Müller, L., Kreysig, D., Hauck, G., Koswig, H. D., Ruhmann, R., Rübner, J., *Makromol. Chem. Rapid. Commun.* **1991**, *12*, 81.
[89] Haitjema, H. J., von Morgen, G. L., Tan, Y. Y., Challa, G., *Macromolecules* **1994**, *27*, 6201–6206.
[90] Fischer, T., Läsker, L., Czapla, S., Rübner, J., Stumpe, J., *Mol. Cryst. Liq. Cryst.* **1997**, *298*, 213–220.
[91] Wiesner, U., Reynolds, N., Boeffel, C., Spiess, H. W., *Makromol. Chem. Rapid. Commun.* **1991**, *12*, 457.

[92] Wiesner, U., Reynolds, N., Boeffel, C., Spiess, H. W., *Liq. Cryst.* **1992**, *11*, 251.
[93] Natansohn, A., Rochon, P., Pezolet, M., Audet, P., Brown, D., To, S., *Macromolecules* **1994**, *27*, 2580.
[94] Buffeteau, T. and Pézolet, M., *Appl. Spectrosc.* **1996**, *50*, 948.
[95] Ramanujam, P. S., Holme, C., Hvilsted, S., Pedersen, M., Andruzzi, F., Paci, M., Tassi, E. L., Magagnini, P., Hoffmann, U., Zebger, I., and Siesler, H. W., *Polym. Adv. Technol.* **1996**, *7*, 768.
[96] Tassi E. L., Paci, M., Magagnini, P. L., *Mol. Cryst. Liq. Cryst.*, **1995**, *226*, 135.
[97] Tassi E. L., Paci, M., Magagnini, P. L., Yang, B., Francescangeli, O., Rustichelli, F., *Liq. Cryst.* **1998**, *24*(3), 457–465.
[98] Holme, N. C. R., Ramanujam, P. S., Hvilsted, S., *Appl. Opt.*, **1996**, *35*, 4622.
[99] Zebger, I., Kulinna, C., Siesler, H. W., Andruzzi, F., Pedersen, M., Ramanujam, P. S., Hvilsted, S.; *Makromol. Chem. Macromol. Symp.*, **1995**, *194*, 159.
[100] Kulinna, C., Hvilsted, S., Hendann, C., Siesler, H. W., Ramanujam, P. S., *Macromolecules* **1998**, *31*, 2141–2151.
[101] Rochon, P., Gosselin, J., Natansohn, A., Xie, S., *Appl. Phys. Lett.* **1992**, *60*, 4.
[102] Kneppe, H., Schneider F. in *Handbook of Liquid Crystals*, Demus, D., Goodby, J., Gray, G. W., Spiess, H.-W., Vill, V. (eds.), Weinheim: Wiley-VCH, Vol. 2A, 1998, 142–169.
[103] Brand, H. R., Finkelmann, H., in *Handbook of Liquid Crystals*, Demus, D., Goodby, J., Gray, G. W., Spiess, H.-W., Vill, V. (eds.), Weinheim: Wiley-VCH, Vol. 3, 1998, 277–302.
[104] Onogi, S., Asada, T., in *VIII International Congress of Rheology*, Astarita, G., Marrucci, G., Nicolais, L. (Eds.), New York: Plenum Press, 1980, Vol. I, 127.
[105] Schmidt, H. W., *Angew. Chem. Int. Ed. Adv. Mater.*, **1989**, *28*, 940.
[106] Dumon, M., Zentel, R., Kulinna, Ch., Siesler H. W.; *Liquid Crystals*, **1995**, *18*(6), 903.
[107] Hessel, F., Finkelmann, H., *Polymer Bulletin*, **1985**, *14*, 375.
[108] Keller, P., Hardouin, F., Mauzac, M., Achard, M., *Mol. Cryst. Liq. Cryst.*, *155*, 171.
[109] Finkelmann, H., Kock, H. J., Gleim, W., Rehage, H., *Makromol. Chem. Rapid Commun.* **1984**, *5*, 287.
[110] Finkelmann, H., Kock, H. J., Rehage, H., *Makromol. Chem. Rapid Commun.* **1981**, *2*, 317.
[111] Küpfer, J., Finkelmann, H., *Makromol. Chem. Rapid Commun.* **1991**, *12*, 717–726.
[112] Siesler, H. W., *Makromol. Chem. Macromol. Symp.* **1992**, *53*, 89–103.
[113] Siesler, H. W., *Makromol. Chem. Macromol. Symp.* **1991**, *52*, 113.
[114] Ameri, A., Siesler, H. W., *J. Appl. Polym. Sci.*, 1998 (in press).
[115] Wu, P., Siesler, H. W., Dal Maso, F., Zanier, N., *Analusis Magazine*, **1998**, *26*, 45–48.
[116] Kundler, I., Finkelmann, H., *Macromol. Chem. Phys.*, **1998**, *199*, 677–686.
[117] Finkelmann, H., Küpfer, J., *Macromol. Chem. Phys.* **1994**, *195*, 1353–1367.
[118] Kundler, I., Nishikawa, E., Finkelmann, H., *Macromol. Symp.*, **1997**, *117*, 11.
[119] Bladon, P., Warner, M., Terentjev, E. M., *Macromolecules*, **1994**, *27*, 7067.
[120] Weilepp, J., Brand, H. R., *Europhys. Lett.*, **1996**, *34*(7), 495.
[121] Küpfer, J., Finkelmann, H., *Polym. Adv. Technol.*, **1994**, *5*, 110–115.

3 Vibrational Spectra as a Probe of Structural Order/Disorder in Chain Molecules and Polymers

G. Zerbi and M. Del Zoppo

3.1 Introduction

At present the theoretical calculation of vibrational properties of polyatomic molecules can be carried out with reasonable success with the help of large, fast computers using fully automatic computing programs. Normal mode calculations have developed along three main lines of thought. On the one hand, a major effort has been made by the pioneers of vibro-rotational spectroscopy with the aim of determining accurate ro-vibrational constants to determine harmonic and/or anharmonic vibrational force fields and the derived physical properties [1]. Once normal modes are known accurately (both fundamental frequencies and relative vibrational displacements [2, 3]), they become the essential ingredients for the treatment of vibrational intensities in infrared and scattering cross-sections in Raman [4–6] or neutron-scattering experiments [7]. A huge effort along these lines has been made by many authors in the past 50 years, with the pioneering works from the group of B.L.C. Crawford at Minnesota and of T. Shimanouchi at Tokyo having been followed by many authors. The basic mathematics and the computing programs first derived by Curtiss [8] and then by Schachtschneider and Snyder [9] and freely distributed worldwide [10] have been reproduced with different names and are used ubiquitously and on most occasions without due mention and thanks to the original authors for their great efforts.

A second school of thought has actively pursued the idea of calculating vibrational properties from '*ab initio*' quantum chemical calculations [11]. The underlying concept is that if optimal atomic wavefunctions (and if large, fast computers) are available all physical observables can be calculated *ab initio*, thus making normal-mode calculations from empirical force constants obsolete. A major effort is currently being made to assess critically the reliability of *ab initio* calculations in the prediction of spectroscopic observables [11–14].

A third group of workers has introduced the idea that vibrational force constants derived from critically determined semiempirical potentials provide an easy way to calculate vibrational properties, as well as many other structural and dynamical molecular properties. These semiempirical potentials are part of fully automated computer packages offered commercially and used widely. For expressing judgment on the reliability of these semiempirical potentials related to spectroscopic properties, one matter is to provide a qualitative description of molecular dynamics (molecules can always be made to 'wiggle' on a screen), another matter is to account quantitatively for frequencies, amplitudes, infrared (dipole derivatives), and Raman or hyper-Raman intensities (polarizability derivatives at various orders).

The latter attempt to obtain vibrational force constants is a combination of the two last methods, i.e., *ab initio* methods are used to obtain the parameters of two-body or many-body interactions generally adopted in semiempirical molecular mechanics calculations [15].

The size of the molecules that can be treated using the above-mentioned methods is quite large (20–30 atoms) and is not a problem if periodicity [16–18] or symmetry [3] are intrinsic properties of the molecule. However, the problem becomes more complex when very large and irregular molecular objects need to be studied.

In this chapter we do not wish to dwell on the problem of the usefulness and limitations of the various methods for calculating vibrational force constants. Rather, we will deal with the problems of carrying out normal-mode calculations on very large molecular systems which have no symmetry are of irregular shape, and chemically are highly disordered [19]. It appears obvious that none of the above-mentioned methods of calculation can be applied to such huge and structurally complex systems.

In nature, chemical systems with chemical and/or structural irregularities or disorder are very common and their relevance in natural phenomena is of fundamental importance. Biological as well as synthetic oligomers and polymers form a large class of compounds whose structure and properties need to be characterized qualitatively and quantitatively. Vibrational infrared and Raman spectroscopy is one of the few physical tools available to study the microstructure of these systems. Many empirical assignment and correlative and analytical studies have been carried out, but the possibility of a theoretical understanding of the vibrational properties (and the derived infrared and Raman spectra) have not been yet thoroughly explored.

In this chapter we present and discuss the dynamics and vibrational spectra of long-chain molecules with disordered structure, and discuss the current possibilities of calculating the vibrational infrared, Raman and neutron-scattering spectra of these systems.

3.2 The Dynamical Case of Small and Symmetric Molecules

Let us briefly define the elementary parameters and review the basic equations for the treatment of molecular and lattice dynamics. The concepts presented here form the background for a further discussion later in the chapter for the case of large molecular systems.

In classical molecular spectroscopy, a vibrating molecule is always represented by a set of balls connected by weightless springs. The springs represent the chemical bonds which hold atoms together through directional forces generated by the interactions of valence electrons. Since the vibrational molecular potential is unknown it is approximated by a Taylor expansion about the equilibrium geometry in terms of a set of suitably chosen coordinates. Let x_α, y_α and z_α be the instantaneous cartesian coordinates of the α-th atom of the vibrating molecule and let x_α^0, y_α^0 and z_α^0 be the corresponding equilibrium coordinates when the molecule is at rest. For simplicity let x_i ($i = 1$–$3N$) label the vibrational displacements of the type $x_\alpha - x_\alpha^0$, $y_\alpha - y_\alpha^0$ and $z_\alpha - z_\alpha^0$ for the N atoms. The generally unknown vibrational potential can be approximated by the following Taylor expansion about the equilibrium structure

$$2V = 2V_0 + 2\sum_i (\partial V/\partial x_i)_0 x_i + \sum_{ij} (\partial^2 V/\partial x_i \partial x_j)_0 x_i x_j + \cdots \qquad (3\text{-}1)$$

The zero-th order term is removed by a suitable shift of the origin, the first-order term is zero because the forces $(\partial V/\partial x_i)_0$ vanish at equilibrium, the second-order term defines a quadratic harmonic potential. In most cases of chemical interest the expansion is truncated at the second order. The analysis of the higher-order terms is outside the scope of our discussion. The truncation at the second order implies that the restoring forces are assumed to be linear with the infinitessimal displacements from the equilibrium position.

The quadratic potential can be written in matrix notation as

$$2V = \mathbf{x}'\mathbf{F}_x\mathbf{x} \qquad (3\text{-}2)$$

where $\mathbf{x}$ is the vector (and $\mathbf{x}'$ its transpose) of the 3N cartesian displacement coordinates and $\mathbf{F}_x$ is the matrix of the quadratic force constants:

$$f_{ij} = (\partial^2/\partial x_i \partial x_j)_0 \qquad (3\text{-}3)$$

Correspondingly, the kinetic energy can be written as:

$$2T = \mathbf{x}'\mathbf{M}\mathbf{x} \qquad (3\text{-}4)$$

where $\mathbf{M}$ is the diagonal matrix of the atomic masses. Only $3N - 6$ of the x_i are independent because the cartesian coordinates are related by the Eckart–Sayvetz conditions [3].

With such a model, atoms are allowed to perform very small harmonic oscillations about their equilibrium positions and the dynamical treatment describes the normal modes Q_j of a set of coupled harmonic oscillators.

When the kinetic and potential energies are introduced into the Lagrange equation the solutions are of the form:

$$x_i^j = L_{ij} \cos \lambda_j^{1/2} t \qquad (3\text{-}5)$$

where λ_j are called frequency parameters and define the frequency of oscillation of the atoms during the j-th normal mode Q_j. During each Q_j all atoms move in phase (i.e., they go through their equilibrium positions at the same time) with frequency ν_j (in cm^{-1}) $= (\lambda_j/4\pi^2c^2)^{1/2}$ and with amplitudes $\mathbf{L}_j$.

Frequency parameters and vibrational amplitudes can be calculated by solving the secular equation

$$\mathbf{M}^{-1}\mathbf{F}_X\mathbf{L}_X = \mathbf{L}_X\Lambda \tag{3-6}$$

Six of the λ_j vanish because of the Eckart–Sayvetz conditions and describe three rigid translations and three rigid rotations of the molecule; each of the remaining $3N - 6$ nonvanishing λ_j is associated to the normal mode Q_j. Depending on the shape (symmetry) of the molecule, degenerate symmetry species may occur and some of the nonvanishing λ_j may turn out to be equal (doublets, triplets, and even multiplets with very high symmetrical molecules, e.g., fullerene). Accidental degeneracy may occur for large and asymmetric molecules when intramolecular coupling is small or zero. When all 3N solutions are considered together, the relation between cartesian displacements and normal modes is as follows:

$$\mathbf{x} = \mathbf{L}_X\mathbf{Q} \tag{3-7}$$

The matrix $\mathbf{L}_X$ is nonorthogonal because $\mathbf{M}^{-1}\mathbf{F}_X$ is not symmetric. The normalization conditions are

$$\mathbf{L}_X\mathbf{L}'_X = \mathbf{M}^{-1} \tag{3-8}$$

which ensures that

$$2T = \dot{\mathbf{Q}}'\dot{\mathbf{Q}} \tag{3-9a}$$

$$2V = \mathbf{Q}'\Lambda\mathbf{Q} \tag{3-9b}$$

i.e., normal modes are mutually independent and can be described as isolated harmonic oscillators. The shape of the vibrational motions sought in any work of vibrational assignment is described precisely by the solution of Eq. (3-6), once the intramolecular potential and the geometry of the molecule is suitably chosen.

A few observations on the treatment of molecular and lattice dynamics in cartesian coordinates are necessary.

1. When molecular dynamics is applied to the understanding of chemical intramolecular phenomena the use of chemical internal displacement coordinates as initially defined by Wilson et al. [3] are much more useful and may have a direct chemical meaning (see below).
2. When molecular vibrations are studied by quantum chemical or molecular mechanics methods, the problem is first treated in cartesian coordinates and possibly later transformed into internal coordinates. These programs generally provide the numerical values of each element of the $\mathbf{F}_X$ and $\mathbf{L}_X$ matrices; the

elements of $\mathbf{L}_x$ are simply the numbers needed to directly project on a screen the atomic vibrations for each $\mathbf{Q}_j$.
3. The treatment of vibrational infrared and Raman intensities provides quantities such as $\partial\mu/\partial x_i$ and $\partial((\alpha))/\partial x_i$ (where μ is the total molecular dipole electric dipole moment and $((\alpha))$ is the total molecular polarizability tensor). These quantities have a direct physical meaning and are directly calculated *ab initio* or sometimes by extended programs of molecular mechanics calculations [4–6].
4. When lattice dynamic calculations for one- or three-dimensional systems need to be carried out, the cartesian space is more suitable in order to describe both inter- and intra-molecular vibrations [20].

When the chemical reality of an isolated molecule needs to be considered through its vibrational spectrum, new sets of 'chemical' coordinates can be defined [3]. Let

$$\mathbf{R} = \mathbf{B}\mathbf{x} \tag{3-10}$$

define a set of internal displacement coordinates where $\mathbf{B}$ is the matrix of geometrical coefficients and

$$\mathbf{x}^{\text{vib}} = \mathbf{A}\mathbf{R} \tag{3-11}$$

where

$$\mathbf{A} = \mathbf{M}^{-1}\mathbf{B}'\mathbf{G}^{-1} \tag{3-12}$$

For a discussion of the $\mathbf{A}$ matrix, see [21]. For simplicity in this discussion we assume that a nonredundant set of 3N – 6 Rs has been defined [22]. The Rs describe bond stretching, valence angle bendings, and torsions. The vibrational potential can be rewritten in terms of Rs as

$$2V = \mathbf{R}'\mathbf{F}_R\mathbf{R} \tag{3-13}$$

and the kinetic energy matrix as

$$2T = \dot{\mathbf{R}}'(\mathbf{G}_R)^{-1}\dot{\mathbf{R}} \tag{3-14}$$

where

$$\mathbf{G}_R = \mathbf{B}\mathbf{M}^{-1}\mathbf{B}' \tag{3-15}$$

$\mathbf{G}$ contains all information on equilibrium geometry and masses. The potential energy matrix is written as

$$\mathbf{F}_R = \mathbf{A}'\mathbf{F}_x\mathbf{A} \tag{3-16}$$

Equation (3-16) is the one to be used when $\mathbf{F}_x$ is calculated by *ab initio* methods and the problem requires a treatment in terms of internal displacement coordinates R. Care must be taken in the construction of **A** [21].

The eigenvalue equation in internal coordinates becomes

$$\mathbf{G}_R \mathbf{F}_R \mathbf{L}_R = \mathbf{L}_R \Lambda \tag{3-17}$$

and the relation to normal coordinates is given by

$$\mathbf{R} = \mathbf{L}_R \mathbf{Q} \tag{3-18}$$

with the normalization condition

$$\mathbf{L}\mathbf{L}' = \mathbf{G}_R \tag{3-19}$$

As discussed in Section 3.4, many normal modes are characterized by the fact that only a few atoms belonging to a specific chemical functional group are moving, while the other do not feel the intramolecular geometrical and electronic changes which occur at a specific site of the molecule. Such *group vibrations* are then characteristic of a given functional group and allow its identification in a chemically unknown material. The existence of 'localized' group vibrations showing specific and characteristic group frequencies have fueled the development of spectroscopic correlations which have made infrared (and recently also Raman) spectra a valuable physical tool for chemical and structural diagnosis [23–26]. These correlations form the basis of many automated computing programs for the chemical identification of unknown compounds.

For the above reason, researchers felt the need to describe normal modes in terms of 'group coordinates' (or local symmetry coordinates) which can be defined by an orthogonal transformation

$$\text{Я} = \mathbf{C}\mathbf{R} \tag{3-20}$$

The expressions for $\mathbf{G}_{\text{Я}}$ and $\mathbf{F}_{\text{Я}}$ and for the corresponding secular equations are the results of a straightforward similarity transformation such as in Eqs. (3-13) and (3-14). The Japanese School of Shimanouchi has provided many force fields and vibrational assignment in terms of group coordinates [27].

When all redundancies between internal coordinates are removed, the size of the eigenvalue equation to be solved is equal to the number of normal modes $(3N - 6)$ expected for the molecule under study. If the molecule has some symmetry it belongs to a given symmetry point group **g**; group theory provides the structure of the irreducible representation Γ of **g**, i.e., the number of normal modes in each symmetry species Γ_i. By a suitable linear and orthogonal transformation

$$\mathbf{S} = \mathbf{U}\mathbf{R} \quad \text{or} \quad (\mathbf{S} = \mathbf{U}\mathbf{C}') \tag{3-21}$$

(with $\mathbf{U}' = \mathbf{U}^{-1}$) it is possible to define a new set of symmetry coordinates S which form the basis of an irreducible representation of the point group **g** [3].

The scalar quantities V and T in Eqs. (3-13) and (3-14) can be re-expressed in terms of symmetry coordinates

$$2V = \mathbf{S}'\mathbf{U}\mathbf{F}_R\mathbf{U}'\mathbf{S} = \mathbf{S}'\mathbf{F}_S\mathbf{S} \tag{3-22}$$

$$2T = \dot{\mathbf{S}}'\mathbf{U}(\mathbf{G}_R)^{-1}\mathbf{U}'\dot{\mathbf{S}} = \dot{\mathbf{S}}'(\mathbf{G}_S)^{-1}\dot{\mathbf{S}} \tag{3-23}$$

The new matrices $\mathbf{F}_S$ and $\mathbf{G}_S$ are factored into blocks with dimensions identical to the structure of the irreducible representation [3]. The corresponding secular equation becomes separated into blocks of smaller size. Numerical calculations carried out separately for each symmetry block enable to identify the normal modes which belong to a given symmetry species. This is essential when the vibrational assignment needs to be carried out [2]. Band shape analysis (in IR and Raman) of gaseous samples, Raman depolarization ratios (for gaseous or liquid/solution samples) and dichroic ratios in IR or Raman experiments in polarized light on single crystals or stretch-oriented solid samples [28] provide the experimental data supporting group theoretical predictions.

The numerical solution of the eigenvalue Eq. (3-17) is obviously greatly simplified when symmetry factoring can be carried out for molecules which possess some symmetry elements. Symmetry factoring was essential in the past (even for small molecules) when numerical processes were necessarily cumbersome and time consuming because of primitive computing technologies. In spite of the explosive development of computational tools this problem cannot yet be fully overlooked. The molecular systems to be treated have become large and the size of the secular determinant can still pose some problems.

A typical example for a modern textbook is the case of the molecule of fullerene (C_{60}) which shows a very high symmetry (point group I_h): its 174 normal vibrations are separated into smaller symmetry blocks. Since icosahedral symmetry gives rise to a large number of degenerate modes, only 46 distinct mode frequencies are expected:

$$\Gamma = 2\,A_g + 3\,F_{1g} + 4\,F_{2g} + 6\,G_g + 8\,H_g + 1\,A_u + 4\,F_{1u} + 5\,F_{2u} + 6\,G_u + 7\,H_u$$

The application of group theory for the determination of the structure of the irreducible representation (for the prediction of optical selection rules for infrared and Raman spectra) is generally easy and straightforward [2, 3]. On the contrary the construction of symmetry coordinates (i.e., the construction of the **U** matrix, Eq. (3-21)) is sometimes difficult and cumbersome for large systems especially when degenerate species occur (e.g., adamantane and fullerene). An automatic method for the construction of symmetry coordinates using computers based on the diagonalization of the $\mathbf{G}_R$ matrix has been proposed [29]. The method is based on the fact that $\mathbf{G}_R$ contains in itself all the information on the symmetry of the problem and on the redundancies, if some or all have not been previously removed. The diagonalization of $\mathbf{G}_R$ provides directly the elements of the **U** matrix (Eq. (3-21)) as follows. Let $\mathbf{G_R}$ be diagonalized by the unitary transformation $\mathbf{D}$

$$\mathbf{G_R D} = \mathbf{D\Gamma} \tag{3-24}$$

where Γ is the diagonal matrix of the eigenvalues. The new set of coordinates

$$\Sigma = \mathbf{D'R} \tag{3-25}$$

forms an irreducible representation of the symmetry point group of the molecule, thus the Σ_i form a set of symmetry coordinates and $\mathbf{D'}$ can replace $\mathbf{U}$ in Eq. (3-21).

The method is particularly useful for the automatic removal of all redundancies in structurally complex systems. Indeed if $m > 3N - 6$ is the size of the eigenvalue equation

$$r = m - (3N - 6) \tag{3-26}$$

is the number of zero eigenvalues in Eq. (3-24) which provide the required r redundancy relations

$$D'_i R = 0 \tag{3-27}$$

between the coordinates used in setting up the starting $\mathbf{B}$ matrix (Eq. (3-10)).

A fully automated computer method for the handling of group theory has been proposed and is recommended for extremely large and highly symmetrical systems with many local and cyclic redundancies [30]. The numerical methods in treating group theory and symmetry coordinates have been generally neglected, as few large and highly symmetrical molecules have required the attention of spectroscopists. At present, the need to understand the vibrations of fullerene, fulleroid, tubular molecules, dendrites, etc., may revive the use of such numerical methods.

3.3 How to Describe the Vibrations of a Molecule

The main purpose of a dynamical analysis of a molecule is to assign the vibrational transitions observed in infrared and Raman in terms of molecular motions. As mentioned above, chemical group frequency correlations have been the justification of most of the vibrational assignment. The description of the normal modes has posed some problems throughout the years.

It would seem obvious that the only way to describe the atomic motion would be that of plotting the cartesian displacements of each atom during normal mode Q_j with frequency ν_j. At present, with the availability of computer techniques and fully automated computing programs, the cartesian atomic displacements are a normal part of the output, while some programs even provide animated descriptions of the molecular vibrations which can be viewed in all directions by appropriate use of the software. Using cartesian displacement coordinates, the identification of group frequencies is not always straightforward and unambiguous. Moreover, the publication of many drawings or of many tables of numbers becomes often graphical and editorial problems.

In the past, in order to describe the normal modes much use has been made of the so-called potential energy distribution (PED). Torkington [31] has first proposed and Morino and Kuchitsu [32] have later generalized the concept of PED defined as:

$$\mathrm{PED} = \Lambda^{-1}\mathbf{JF} \tag{3-28}$$

where $J_{i,mm} = L_{mi}^2$ and $J_{i,ml} = 2L_{mi}L_{li}$.

The advantage of PED is that it can be expressed either in internal or group coordinates, thus allowing a more direct description of the main characteristic of the nomal modes.

In simple words, PED provides quantitatively (generally PED is expressed as % contributions) the extent of involvement of one or several diagonal and off-diagonal force constants in a given normal mode. The classical textbook case is that of the well-known group frequency modes of a CH_2 group. Using group coordinates (Eq. 3-20)), the modes commonly described as CH_2 antisymmetric stretch (d^-), CH_2 symmetric stretch (d^+), CH_2 scissoring (δ), wagging (w), twisting (t), and rocking (P) can be easily identified with reference to the corresponding $F_{\mathfrak{R}}$ group force constants [27]. If the same motions were expressed in terms of classical internal coordinates of CCH bending the PED would indicate unselectively that in all the four motions CCH bendings are involved without allowing any distinction of the characteristic group modes. The same situation is found in the case of the characteristic group modes of the $-CH_3$ group which are recognized by their characteristic frequencies of bending, umbrella (U), in-plane, and out-of-plane rocking.

A typical case, famous in organic vibrational spectroscopy, is represented by the work on *trans*-planar n-alkanes by Schachtschneier and Snyder [33] and by Snyder *et al.* [34]; the latter has extended his study to branched paraffins [35], to n-alkanes in the liquid phase [36] and to oligo and poly-ethers [37]. The use of PED has allowed to distinguish clearly between band progressions and the evolution of the normal modes by changing the phase-coupling (see later in this chapter).

3.4 Short- and Long-Range Vibrational Coupling in Molecules

As in this chapter the aim is to understand the vibrations of large and highly disordered and structurally irregular molecules, the basic problem to be faced is whether and to what extent normal vibrations are able to probe the molecular intramolecular environment, i.e., whether normal vibrations are mostly localized or extended over a large portion of the molecular system.

This issue has rarely been faced by molecular spectroscopists for several reasons. All those who searched for chemically useful characteristic frequencies aimed at modes strongly localized within the functional group [23–26]. In contrast, those who dealt with large systems generally considered oligomeric or polymeric molecules as structurally perfect systems, often with translational periodicity, and studied 'phonons' which, by definition, are collective phenomena [17–19]. The problem cannot be overlooked when the vibrations of large and structurally complex molecules must be understood.

The first theoretical approach seeded by the explosive development of chemical spectroscopic group frequency correlations has been presented by King and Crawford who gave the dynamical conditions under which a normal mode can be defined as 'localized' [38].

In correlative vibrational spectroscopy, a group of chemically similar molecules shows characteristic group frequencies [23–26] ν_i (say, the stretching of the carbonyl group >C=O, very strong in infrared) which exhibit systematic changes that chemical spectroscopists would like to ascribe to inductive and/or mesomeric effects by the various substituents placed in the molecules either at short or large distances from the functional group of interest. On the other hand, the observed wavenumber shifts may also originate from changes of mass and/or geometry.

Use is made here [38] of the dynamical quantities presented in Chapter 2. Let $\mathbf{G}_0$, $\mathbf{F}_0$ and $\mathbf{L}_0{}^{-1}$ be the kinetic, potential, and normal coordinate transformation matrices of one molecule of the series taken as reference [39]. In going from one molecule to a chemically similar one within the same class, one may expect small changes in the geometry ($\Delta\mathbf{G}$) or in the force constants ($\Delta\mathbf{F}$), or in both, which may be the cause of the observed changes of ν_i. The corresponding matrices of the molecule so modified can be written:

$$\mathbf{G} = \mathbf{G}_0 + \Delta\mathbf{G} \tag{3-29}$$

$$\mathbf{F} = \mathbf{F}_0 + \Delta\mathbf{F} \tag{3-30}$$

$$\mathbf{L}^{-1} = \mathbf{L}_0{}^{-1} + \Delta(\mathbf{L}^{-1}) \tag{3-31}$$

If the perturbations of $\mathbf{G}$ and $\mathbf{F}$ matrices are small it can be assumed that the eigenvectors do not change and the zeroth-order values in Eq. (3-31) are used. When the eigenvalue Eq. (3-17) is expanded in terms of the quantities in equations (3.29) through (3-31) in terms of the first-order changes $\Delta\mathbf{G}$ and $\Delta\mathbf{F}$ and the zeroth-order eigenvectors, the final expression for the (group) frequency parameter turns out to be:

$$\lambda_i = (\lambda_0)_i + \sum_{l,m=1}^{s} [(L_0)_{im}(L_0)_{li}\Delta F_{ml} + (\lambda_0)(L_0{}^{-1})_{im}(L_0{}^{-1})_{li}\Delta G_{ml}] \tag{3-32}$$

in which $(\lambda_0)_i$ is the frequency parameter of the i-th group-frequency mode of the reference unperturbed molecule, ΔG_{ml} and ΔF_{ml} are the changes of those elements of the kinetic and potential energy matrix whose contributions to the changes of λ_i are scaled by the related elements of the $\mathbf{L}_0$ and $\mathbf{L}_0^{-1}$ matrices.

According to King and Crawford [38], if λ_i has to be insensitive to changes of the molecular framework (i.e., if the motion is to be localized) the sufficient conditions are that the coupling terms in Eq. (3-32) be small, namely either

$$\partial\lambda_i/\partial(\Delta F_{ml}) = (L_0)_{im}(L_0)_{li} \cong 0 \tag{3-33}$$

$$\partial\lambda_i/\partial(\Delta G_{ml}) = (\lambda_0)_i(L_0{}^{-1})_{im}(L_0{}^{-1})_{li} \cong 0 \tag{3-34}$$

or

$$\Delta G_{ml} \cong \mathbf{0} \tag{3-35}$$

$$\Delta F_{ml} \cong \mathbf{0} \tag{3-36}$$

It is interesting to note that the problem of dynamical localization versus collectivity has never been a key issue in molecular spectroscopy and dynamics in the past 40 years as the molecules considered were of limited size, thus obviously implying the existence of collective motions throughout the whole molecule. Moreover, most of the molecules considered contained either only covalent σ bonds or, if π bonds existed, they were either isolated or with limited conjugation (e.g., aromatic molecules). The recent explosive interest in highly conjugated polyene and polyaromatic systems [40] has asked molecular dynamics to reconsider the problem of the extent of conjugation (delocalization) of π bonds along the whole molecular chain [41, 42].

At the present stage of the spectroscopic studies for organic systems, a few observations on Eqs. (3-32) to (3-36) must be made.

1. From the definition of the $\mathbf{G}_R$ matrix (Eq. (3-15)) or, more directly, from the analytical expressions of the $\mathbf{G}_R$ matrix elements provided by Wilson et al. [3], it transpires that the element $(\mathbf{G}_R)_{ij}$ connecting internal coordinates i and j may be different from zero when they possess at least one atom in common. It follows that atoms away from the functional group under study certainly will not affect the group frequency; Eq. (3-35) holds, i.e., the $\mathbf{G}$ matrix favours localization.
2. For molecules made up exclusively by σ bonds or containing also isolated and nonconjugated π bonds inductive effects are generally considered not to extend at large distances, thus limiting the distances of the possible electronic effects described by ΔF_{ml} to a few bonds around the functional group (hence to a few internal coordinates). Equation (3-36) holds and again localization seems to be favored when strong electronic delocalization over large distances is not likely to occur within the molecule.
3. When intramolecular mechanical coupling is very large (i.e., Eqs. (3-33) and (3-34) do not hold) mechanical delocalization over a certain molecular domain takes place and the concept of group frequency is lost. This generally happens in the energy range between $\approx$1500–400 cm^{-1} where skeletal stretching and bending motions are strongly coupled. In this case small mass, geometry, or electronic perturbations may induce large changes in the spectrum, as indeed observed experimentally.

4. The situation represented by Eq. (3-36) is at present of particular interest when polyconjugated molecules (polyenes and polyaromatic systems) are considered. The role played by vibrational infrared and Raman spectroscopy in this new field of material science has been essential for the understanding of the new physical and chemical phenomena associated with the existence of extended π electron delocalization. The reader is referred to specialized references for a thorough discussion on the spectroscopy of these systems [41–43]. For these systems recent studies have pointed out the occurrence of characteristic normal modes which show 'frequency dispersion' with conjugation length, i.e., the characteristic frequencies lower by adding repeating units which take part in the conjugation [41–43]. Dealing with a chain molecule, the elements $(L_0)_{im}$ and $(L_0^{-1})_{im}$ entering Eqs. (3-33) and (3-34) are necessarily large. The physically relevant problem is to discover whether, to what extent, and at which distance the elements of ΔF change by adding conjugated units. The solution of this problem would shine some light on the distance and the extent of electronic interactions in polyconjugated systems [44]. This is a challenge to spectroscopists, theoretical quantum chemists, and computational chemists.
5. It is obvious that the addition of strong electron donors or acceptors at a site away from the functional group of interest will modify the electronic environment and possibly the geometry at the site of the substitution. Certainly for particular internal coordinates, large ΔF and/or ΔG must occur, but their contribution to the frequency shift of the group frequency of interest is reduced to zero since the corresponding elements of $\mathbf{L}$ or $\mathbf{L}^{-1}$ which relate the motion of the functional group and the motions at the site of the substituent are all zero (Eq. (3-32)).

3.5 Towards Larger Molecules: From Oligomers to Polymers

Molecular spectroscopy and lattice dynamics of oligomeric and polymeric materials have been treated thoroughly in several textbooks or articles [16–18]. We wish to point out here a few fundamental concepts which form the basis of the understanding of the spectra of disordered materials which will follow in this discussion.

Let us first consider the vibrations of polymers considered as one-dimensional perfect and infinite lattices. The usual basic assumptions are the following:

1. The polymer is obtained by forming a one-dimensional chain of chemical units linked with a pre-assigned *chemical sequence*. For sake of example, we consider propylene (CH_3–CH=CH_2) as the prototype of a monomer unit capable of producing a polypropylene chain. The first step is to make a chain of polypropylene in which all monomers are chemically linked head-to-tail (*chemical*

regularity), e.g.,

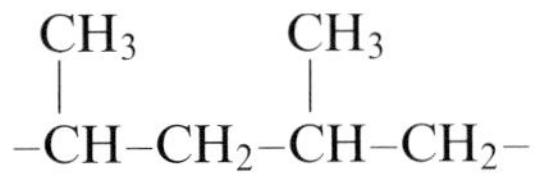

$$
\begin{array}{cccc}
CH_3 & & CH_3 & \\
| & & | & \\
-CH- & CH_2- & CH- & CH_2-
\end{array}
$$

2. During the polymerization process the catalysts is chosen in such a way that *stereo-regularity* is assured [45], e.g.,

isotactic:

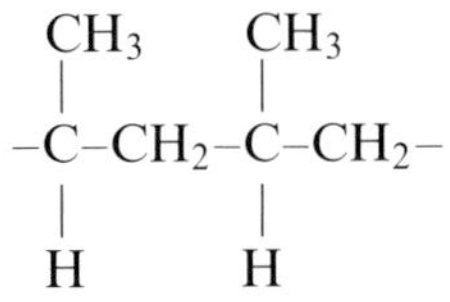

$$
\begin{array}{cccc}
CH_3 & & CH_3 & \\
| & & | & \\
-C- & CH_2- & C- & CH_2- \\
| & & | & \\
H & & H &
\end{array}
$$

3. Upon solidification from solution (or from the melt), macromolecules coil in an energetically preferred minimum energy conformation and form long chains along which *conformational periodicity* is assured [46–48]. Chains are considered long enough to be considered 'infinite'. For isotactic polypropylene, the conformational sequence is TGTGTG. The overall shape of the chain can be described as a regular helix with a threefold screw axis [45].
4. The process of solification proceeds with the formation (when possible) of crystalline material in which chains pack in a lattice with *tri-dimensional periodicity*. The conditions for perfect packing are that chains possess chemical regularity, configurational periodicity, and conformational regularity.

Since intermolecular Van der Waals-type forces which hold molecules in a crystalline lattice are very weak, when compared with intramolecular covalent forces, lattice dynamics of polymers is made easy by considering firstly the vibrations of infinite isolated polymer chains [16–18]. Three-dimensional order is considered later as a small perturbation of the normal modes of the one-dimensional chain by the weak intermolecular forces [49]. In other words, we are considering the dynamics of one chain '*in vacuo*'.

The structural characteristics of the polymer chain constructed as described above are that translational or roto-translational symmetry can be found along the chain in addition to the traditional point group symmetry operations (rotation axes, symmetry planes, inversion center, and identity) [18].

In a dynamical treatment the labeling of the internal coordinates starts from the reference repeat unit and proceeds along the chain by applying the one-dimensional translational operator which identifies the translationally equivalent internal coordinates. The structural situation is clarified if one considers that a polymer chain can be generated by application to a chemical repeat unit of a certain rototranslational operator $\psi(\theta, 1)$ which corresponds to a rotation of an angle θ about the chain axis and a translation of a fraction $1 = d/n$ of the crystallographic unit length (repeat distance) d. In general, the crystallographic repeat unit can contain z chemical units coiled in a helix with t turns [47].

Let p be the number of atoms in the chemical repeat units and let the indices n and n′ label the sites of the chemical repeat units along the polymer chain and i and l label the 3p internal co-ordinates within the repeat unit. The potential energy can be formally written in a way similar to Eq. (3-13) [50].

$$2V = \sum_{\substack{n,n' \\ i,l}} (F_R)^{nn'}{}_{il} R^n{}_i R^{n'}{}_l \tag{3-37}$$

where

$$(F_R)^{nn'}{}_{il} = (\partial^2 V / \partial R^n{}_i \partial R^{n'}{}_l)_0 \tag{3-38}$$

$$(F_R)^{nn'}{}_{il} = (F_R)^{nn'}{}_{li} \tag{3-39}$$

let $s = |n - n'|$ define the distance of interaction, which (see Section 3.6) is very important in determining the vibrations of the chain; it follows that because of translational symmetry

$$(F_R)^{nn'}{}_{il} = (F_R)^{s}{}_{il} \tag{3-40}$$

Substituting Eq. (3-40) into Eq. (3-37), one obtains

$$2V = \sum_{n,il} (F_R)^0{}_{il} R^n{}_i R^n{}_l + \sum_{\substack{n,s \\ il}} [(F_R)^s{}_{il} R^n{}_i R^s{}_l + (F'_R)^s{}_{li} R^n{}_i R^{-s}{}_l] \tag{3-41}$$

In Eq. (3-41), the first term gives the contribution of the intramolecular potential by the reference unit at the n-th site, and the second and third terms collect the contribution of intramolecular coupling of the n-th unit with the neighboring units at distances s and −s respectively.

A similar equation can be written for the kinetic energy in a way similar to what has been done for the finite molecule (Eq. (3-14)). Since the number of oscillators is taken to be infinite (the chain is considered infinite) the dynamical problem results in the writing of an infinite number of second-order differential equations, the solutions of which are of the type

$$R^{n+s}{}_j = A_j \exp[-i(\lambda t + s\varphi)] \tag{3-42}$$

It should be noted that A_j is independent of n, φ is the phase shift between two adjacent roto-translationally equivalent internal co-ordinates, and λ refers to the vibrational frequency as in the case of small molecules (see Eq. (3-5)). Physically, Eq. (3-42) tells us that the j-th coordinate at the $n + s$ site oscillates with frequency λ with an amplitude equal to the amplitude of the j-th coordinate in the reference cell at site n multiplied by a phase factor $\exp(-is\varphi)$. Attention should be paid to the fact that φ is the phase shift of two equivalent oscillators belonging to two adjacent chemical units within the crystallographic repeat unit (see above).

Following the same algebraic procedures as for the finite molecule the final secular equation to be solved is [50]

$$|\mathbf{G}_{\mathrm{R}}(\varphi)\mathbf{F}_{\mathrm{R}}(\varphi) - \mathbf{E}\Lambda(\varphi)| = 0 \qquad (3\text{-}43)$$

where

$$\mathbf{G}_{\mathrm{R}}(\varphi) = (\mathbf{G}_{\mathrm{R}})^0 + \sum_{\mathrm{s}} [(\mathbf{G}_{\mathrm{R}})^{\mathrm{s}}\, \mathrm{e}^{\mathrm{is}\varphi} + (\mathbf{G}_{\mathrm{R}}')^{\mathrm{s}}\, \mathrm{e}^{-\mathrm{is}\varphi}] \qquad (3\text{-}44)$$

$$\mathbf{F}_{\mathrm{R}}(\varphi) = (\mathbf{F}_{\mathrm{R}})^0 + \sum_{\mathrm{s}} [(\mathbf{F}_{\mathrm{R}})^{\mathrm{s}}\, \mathrm{e}^{\mathrm{is}\varphi} + (\mathbf{F}_{\mathrm{R}}')^{\mathrm{s}}\, \mathrm{e}^{-\mathrm{is}\varphi}] \qquad (3\text{-}45)$$

Equation (3-43) is of 3pth degree in λ. There are 3p characteristic roots ($\mathrm{i} = 1 \ldots 3\mathrm{p}$)

$$\lambda_{\mathrm{i}}(\varphi) = 4\pi^2 \mathrm{c}^2 \nu_{\mathrm{i}}{}^2(\varphi) \quad (\nu_{\mathrm{i}}/\mathrm{cm}^{-1}) \qquad (3\text{-}46)$$

for each value of the phase φ. The $\mathrm{r} = 3\mathrm{p}$ functions $\nu_{\mathrm{i}}(\varphi)$ can be interpreted as branches of a multiple valued function which is the dispersion relation for a one-dimensional polymer chain. The solution reached in Eq. (3-43) is analogous to that previously attained by many authors in solid-state physics for simple tridimensional lattices [51]. The important difference is that Eq. (3-43) treats the problem in internal chemical coordinates, while, generally, simple lattices are treated in cartesian coordinates (Eq. (3-6)). The function is periodic with period 2π and calculations are limited to values of φ within the first Brillouin zone [51, 52] (BZ) ($-\pi \le \varphi \le \pi$). In most cases, the results for only half of the BZ are reported, the second half being symmetrical through $\varphi = 0$.

In the spectroscopy of polymers as one-dimensional 1D lattices it is generally easier to deal with the phase shift φ at odds with solid-state physics, dealing with three-dimensional 3D crystal, which treats the whole dynamics in terms of the vector **k**. For 1D lattices we can describe the wave-motion (phonon) propagating along the 1D chain either by the phase shift φ or by the vector $|\mathbf{k}| = \varphi/\mathrm{d}$ where **k** has only one component along the chain axis $\mathrm{k_z}$. In contrast, in the spectroscopy and dynamics of 3D lattices (where intermolecular forces are active in all directions) phonons are labeled with **k** vectors with three components ($\mathrm{k_x}$, $\mathrm{k_y}$, $\mathrm{k_z}$) [51]. When **k** is used in polymer dynamics Eq. (3-43) can be easily rewritten as

$$|\mathbf{G}_{\mathrm{R}}(\mathbf{k})\mathbf{F}_{\mathrm{R}}(\mathbf{k}) - \mathbf{E}\Lambda(\mathbf{k})| = 0 \qquad (3\text{-}47)$$

It should be remembered that in a single and isolated polymer, chain atoms are allowed to move in the tridimensional space even if phonons are considered to propagate in one direction along the chain axis. This means that no neighboring intermolecular interactions are taken into account, i.e., the dynamics is that of a chain '*in vacuo*'.

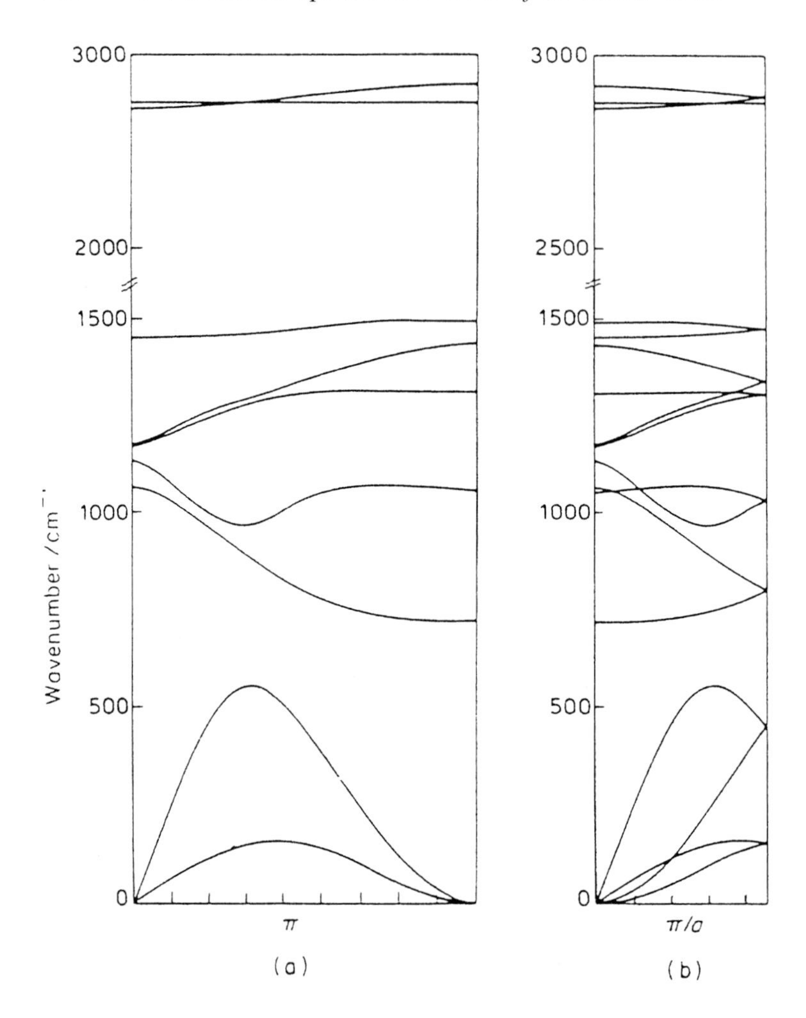

Figure 3-1. Dispersion curves of single-chain *trans* planar polyethylene. (a) ν/φ; (b) ν/k.

As an example, in Figure 3-1 we report the dispersion curves of a single chain of polyethylene in φ and **k** space. The chain of *trans*-planar polyethylene $-(CH_2)_n-$ is generated by application to a single CH_2 group of the operator $\psi(\pi, d/2)$. It follows that the crystallographic repeat unit contains two CH_2 groups. The physics and the numbers are not changed. The dispersion curves in Figure 3-1b are obtained by just halving Figure 3-1a at $\pi/2$ [53].

As an example of a more complex polymer chain we quote the case of polytetrafluoroethylene $-(CF_2)_n-$ for which one can identify the rototranslational operator $\psi(2\pi/15, d/15)$ which describes a polymer chain made up by $z = 15$ CF_2 units coiled in a helix with $t = 7$ turns. In cases such as polytetrafluoroethylene the

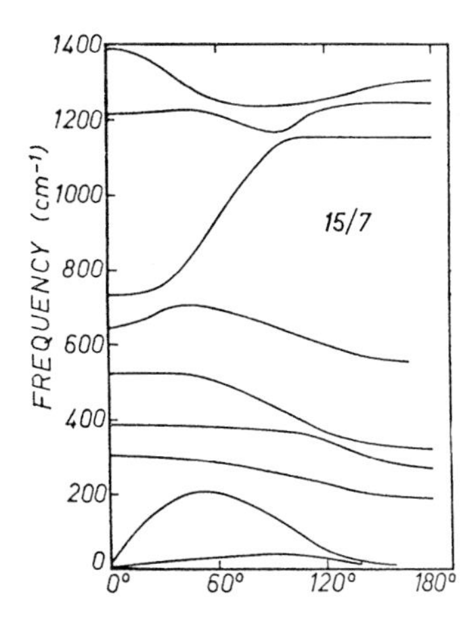

Figure 3-2. Dispersion curves ν/φ for single chain helical (z = 15, t = 7) polytetrafluoroethylene.

plot of $\nu(\varphi)$ becomes usable while $\nu(\mathbf{k})$ becomes unreadable (117 frequency phonon branches in **k** space) (Figure 3-2) [54].

Dealing with 1D chains *in vacuo* carries an important consequence on the shape of the dispersion curves if compared with the same systems considered as 3D lattices. From lattice dynamics in 3D it is known that if q is the number of atoms in the crystallographic repeat unit the dispersion relation provides 3q phonon branches, three of which are labeled as 'acoustical' and 3q – 3 as 'optical' branches. The former three have zero frequencies at the so-called Γ symmetry point in **k** space (i.e., with $k_x = k_y = k_z = 0$; see Figure 3-3) and correspond to the three nongenuine vibrations with zero frequencies which describe the three rigid translations of the whole lattice along the three cartesian coordinates [49, 55]. The 3q – 3 optical branches may interact with the electromagnetic waves according to specific selection rules and provide the system with particular optical properties (absorption and/or Raman scattering), as discussed below.

In 1D systems *in vacuo* the rigid rotation of the whole chain about its axis is not hindered. It follows that an additional zero eigenvalue is expected from Eq. (3.43) when applied to 1D systems at $\mathbf{k} = 0$. The lack of intermolecular forces for the polymer chain *in vacuo* carries an important consequence on the shape of the two acoustic branches near $\mathbf{k} = 0$. While the three acoustic branches for 3D crystals have a discontinuity at $\mathbf{k} = 0$, for 1D crystals two of the acoustical branches may approach $\nu(0)$ almost asymptotically [56]. It follows that the dynamics of 1D systems based on the model of an isolated chain at small **k** values and at very low frequencies is totally wrong and cannot be used for the interpretation of related physical properties of real systems (neutron-scattering experiments, thermodynamic properties, Debye Waller factors, mean square amplitudes, and other physical properties where thermal population is involved which is strongly affected by low-energy vibrations).

For the benefit of the discussion which follows we give here a calculated schematic example of the procedures followed in setting up the matrices $\mathbf{F}_R(\varphi)$ and

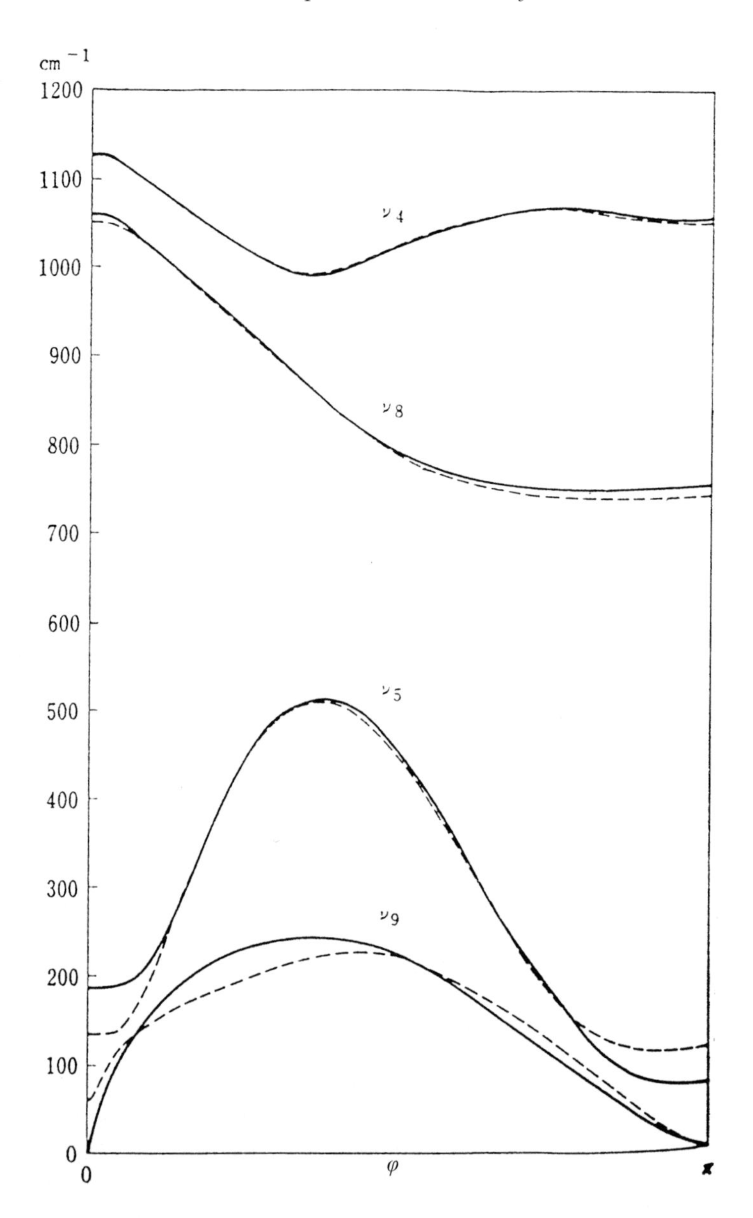

Figure 3-3. Dispersion curves of crystalline orthorhombic polyethylene with two molecules per unit cell (from [49]). Comparison with Figure 3-1 shows the splitting of the frequency branches and the shape of the acoustical branches at $\varphi \rightarrow 0$ (see text). Attention should be paid to the fact that the frequencies and the shape of the dispersion curves shown in Figure 3-1 and in this figure may differ because they have been calculated with two different force fields.

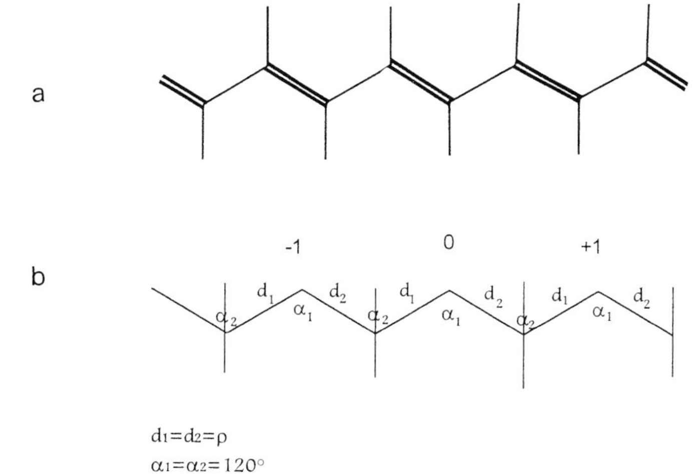

Figure 3-4. (a) Chemical structure of *trans*-polyacetylene and (b) model assumed in the calculation reported in the text. The model gives the numbering of the one-dimensional repeat unit cell and the labeling of the internal coordinates.

$\mathbf{G}_R(\varphi)$ for a simplified chain of planar *trans*-polyacetylene, $(-CH{=}CH-)_n$ where each CH group is taken as a point mass. Only in-plane motions are considered.

From the viewpoint of chemistry and materials science, polyacetylene is a prototypical relevant material in molecular electronics and shows a low (HOMO-LUMO) band gap (≈ 1.4 eV) because p_z orbitals of the carbon atoms in the sp^2 hybridization are strongly delocalized along the molecular chain (large conjugation). The distance of interaction between π bonds is yet unknown and is the subject of active research [40–42].

Figure 3-4 shows the drawing of the actual molecule as well as of the simplified model with the labeling (l) of the sites of the translational repeating unit and the labeling of the internal coordinates of the $l = 0$ unit and the equivalent ones at sites $+/-$ s. The unit cell is made up by two masses ($p = 2$) and the in-plane internal coordinates are $2 \times 2 = 4$ which is also the size of the secular equation. We expect to calculate four branches in the dispersion relation, two of which are optical and two acoustical.

We calculate explicitly the algebraic expressions of the elements of the $\mathbf{G}_R(\varphi)$ matrix which enters Eq. (3-43). The following simplifications have been accepted in such a simple calculation: (i) the groups C–H are taken as point masses of $m = 13$ daltons; (ii) C=C and C–C bond lengths are equal; and (iii) bond angles are taken equal to 120°.

	d_1^{-2}	d_2^{-2}	α_1^{-2}	α_2^{-2}	d_1^{-1}	d_2^{-1}	α_1^{-1}	α_2^{-1}	d_1^{0}	d_2^{0}	α_1^{0}	α_2^{0}	d_1^{1}	d_2^{1}	α_1^{1}	α_2^{1}	d_1^{2}	d_2^{2}	α_1^{2}	α_2^{2}
d_1^0					0	b	c	0	d	b	c	c	0	0	0	c				
d_2^0					0	0	0	0	b	a	c	c	b	0	c	c				
α_1^0					0	c	e	0	c	c	e	d	c	0	e	d				
α_2^0					c	c	d	e	c	c	d	e	0	0	0	e				

$$
\begin{aligned}
a &= 2\mu_{CH}; \\
b &= -\mu_{CH}/2; \\
c &= -\sqrt{3}/2\rho\mu_{CH}; \\
d &= 3\rho^2\mu; \\
e &= 5\rho^2\mu
\end{aligned}
\qquad (3\text{-}48)
$$

where ρ is the reciprocal of the C–C distance and μ is the reciprocal of the mass of the CH unit taken as point mass.

It is immediately seen that nonzero elements occur in the diagonal block and only in the two neighboring blocks which describe the interactions of the central unit with the nearest neighbors in position +1 and −1, i.e., the distance of interaction is s = 1, very short indeed. The construction of the $\mathbf{G}_R(\varphi)$ matrix through Eq. (3-44) yields a 4 × 4 matrix where the generic element has the following form:

$$
g_{ij}(\varphi) = g_{ij}{}^{0} + g_{ij}{}^{(1)}(e^{i\varphi}) + g_{ij}{}^{(-1)}(e^{-i\varphi}) \qquad (3\text{-}49)
$$

The opposite is the case of the potential energy matrix whose structure is symbolically given in Table 16, p. 294 of [41].

The most relevant fact is that while the $\mathbf{G}_R$ matrix is certainly truncated at the +1 and −1 units in the $\mathbf{F}_R$ matrix the distance of interaction of the zero-th unit with its s-th neighbour is yet unknown since the so-called ‘delocalization length’ needs to be determined [41–43]. The truncation at the site s of the long ribbon of interaction submatrices is then arbitrary. Many *ab initio* calculations are at present dealing with this essential problem which is common to all poly-conjugated chains presently available.

The truncation of the sum in Eq. (3-45) is then arbitrary, thus affecting the validity of the physical conclusions which may be derived. It follows that any addition or variation of terms or extension of the interactions (increase of delocalization length) will affect the numerical values of the elements of the $\mathbf{F}_R(\varphi)$ matrix, thus changing the shape of the dispersion curves [41, 42]. It is apparent that the shape of the dispersion curves is a direct consequence of the electronic properties of the poly-conjugated chains [58] (Figure 3-5).

An opposite case is found for *trans*-planar polyethylene taken as a prototype of all exclusively σ bonded chains for which conjugation is not predicted (the existence of σ-conjugated chains like polysilanes needs further studies). In the case of polyethylene, experiments or calculations indicate that the electronic interactions between σ C–C bonds fall off very quickly along the chain with $s \approx 2$ [59]. Indeed, as shown by Schachtschneider and Snyder, the normal modes of finite n-alkanes lie almost perfectly on the dispersion curves calculated for polyethylene [33–35]. No chain length effect in the $\mathbf{F}_R(\varphi)$ matrix is revealed for these classes of systems. Analogously for σ bonded tridimensional lattices like cubic diamond and silicon, a valence force field limited to a few first-neighbor interactions nicely accounts for the optical and neutron-scattering data available to date [60] (Figure 3-6).

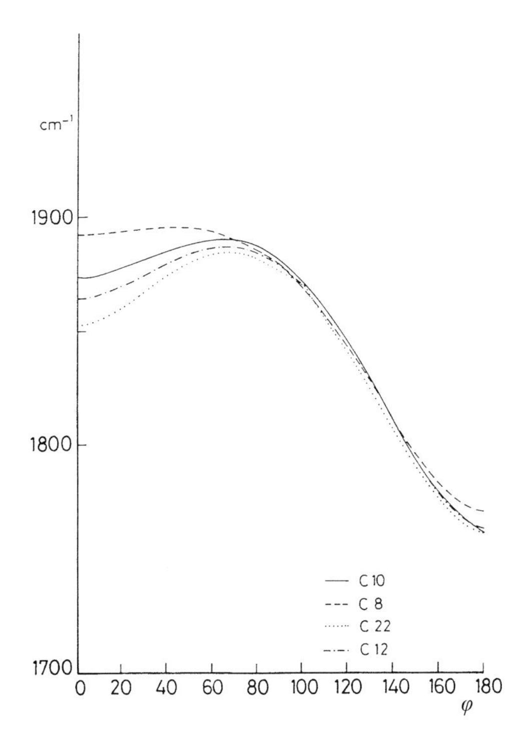

Figure 3-5. Effect of the long-range interactions (s in equation 3.45) on the dispersion curves of the frequency branch ν_3 (associated with the in-phase skeletal stretching) of *trans* polyacetylene calculated for the unsimplified and realistic equilibrium geometry. The force field has been derived by '*ab initio*' calculations on polyene oligomers with $n = 2, 4, 5, 6$, and 11 double bonds [58]. Attention should be paid to the fact that the 'softening' of the $\nu(\varphi = 0)$ mode modulated by the distance of interaction generates a 'dispersion' with chain length of a very strong and very characteristic Raman line ([41, 42]).

These observations are essential when the dynamics of disordered chains will be discussed later in this chapter.

3.6 From Dynamics to Vibrational Spectra of One-Dimensional Lattices

The experiments which provide direct information on the vibrations of one- and tridimensional lattices are infrared absorption (reflection) spectra, Raman scattering and neutron scattering. For a general discussion on the theoretical principles and experimental methods the reader is referred to the abundant literature on these

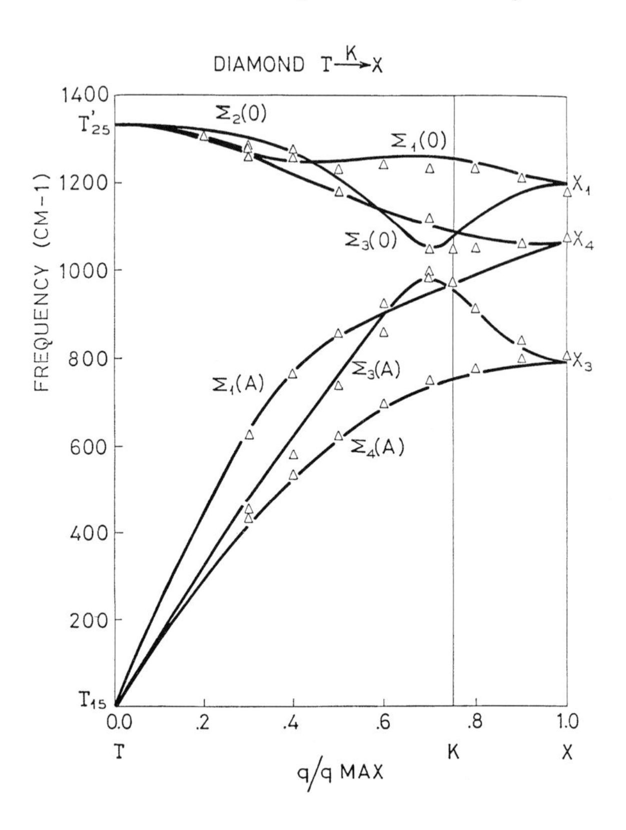

Figure 3-6. Example of lattice dynamical calculations on σ-bonded tridimenional crystals with short range interactions. Dispersion curves for cubic diamond along the $\Gamma \to (K) \to X$ symmetry direction. Experimental points from neutron-scattering experiments; dispersion curves from least squares frequency fitting of a six parameters short range valence force field (from [60]). The Raman active phonon is the triply degenerate state indicated with Γ'_{25} near 1300 cm^{-1}. Notice that at $\mathbf{k} \neq 0$ the degeneracy at Γ is removed because of the lowering of the symmetry throughout the whole BZ. Notice also the three acoustic branches for which $\nu \to 0$ at $\mathbf{k} \to \Gamma$.

techniques (e.g., refs. [16–19]). Our aim is to present here methods for extracting from the experimental data the maximum amount of information on structure and properties of these relevant materials.

The next step in our analysis is to define the experimental observables predicted by lattice dynamics. The spectroscopic selection rules are very restrictive and, of the very many phonon frequencies calculated, very few can be directly observed experimentally [16–19] in the optical vibrational spectra.

First, only phonons with $\mathbf{k} = 0$ can interact with the electromagnetic wave; thus, only the fundamental frequencies at the centre of the Brillouin zone in $\mathbf{k}$ space are potentially infrared or Raman active. Second, further restrictions on the spectroscopic activity are introduced by the symmetry of the polymer chain. Once the translational (or rototranslational) symmetry element is properly taken into account (in combination with point-group symmetry operations) a factor group isomorphous to a symmetry point group can be defined and $\mathbf{k} = 0$ phonons can be separated into symmetry species, just as in the case of finite molecules. Traditional infrared and/or Raman activities can be predicted as well as the optical transitions associated to specific dipole allowed transitions with their directional properties for oriented samples.

The physical meaning of the selection rule which states that only $\mathbf{k} = 0$ phonons can be observed spectroscopically is that the atomic displacements in translationally equivalent crystallographic units must be all in-phase. When the geometry of the polymer chain (i.e. $\psi(\theta, 1)$ and t = no. of turns of the helix within the rototranslational unit cell) is taken into account, phonons are infrared active for $\varphi = 0$ and $\varphi = \theta$ (doubly degenerate); Raman activity is achieved for phonons with $\varphi = 0$ (nondegenerate), θ and 2θ (both doubly degenerate) [15, 61, 62]. Degeneracy is clearly understood when folding of the dispersion curves is carried out at the zone boundary in going from φ to $\mathbf{k}$ space. If p is the number of atoms in the chemical repeat unit for a helical polymer, one expects in the infrared 3p − 2 totally symmetric $\|$ modes with $\varphi = 0$ and 3p − 1 doubly degenerated $\perp$ modes with $\varphi = \theta$. Coincidences with the infrared spectrum are expected in the Raman with the modes at $\varphi = 0$ and $\varphi = \theta$, while extra modes are expected at $\varphi = 2\theta$ (practically never observed because their Raman intensity is very weak).

Let us further consider the direction of the transition dipole moments in infrared for a stretch-oriented polymer sample in which it is assumed that polymer chains are aligned along the stretching direction. Phonons with $\varphi = 0$ have their transition dipole moment parallel to the chain axis which coincides with the direction of stretching ($\|$ modes); modes with $\varphi = \theta$ have transitions moments perpendicular to the chain axis ($\perp$ modes). They can easily be identified in an infrared experiment on stretch-oriented samples carried out in polarized light. Analogously, the elements of the polarizability tensor can be selectively excited by Raman scattering experiments on oriented polymer samples examined in polarized light in different scattering geometries. For a study of the Raman spectrum of a stretch-oriented polyethylene rod in various geometries see [28] and Figure 3-7 [63].

A few comments of general relevance need to be made.

The shape of the dispersion curves and the corresponding eigenvectors which describe the vibrational displacements (phonons) depend necessarily on the vibrational force field adopted in the calculations. The fitting of the experimental frequencies with the calculated ones, although being a good indication that the force field is reasonably acceptable, can by no means be taken as proof of its general reliability throughout the whole BZ. Indeed, fitting has been obtained only at two values of the phase coupling φ. The intramolecular interactions may be different at $\varphi \neq 0$ and $\varphi \neq \theta$ or 2θ, thus in principle changing the shape of the theoretically predicted dispersion curves (see, for example, [64]). Other independent experiments are needed before claiming the success and generality of the intramolecular force field.

So far, we have discussed $\mathbf{k} = 0$ phonons of the one-dimensional lattice and have

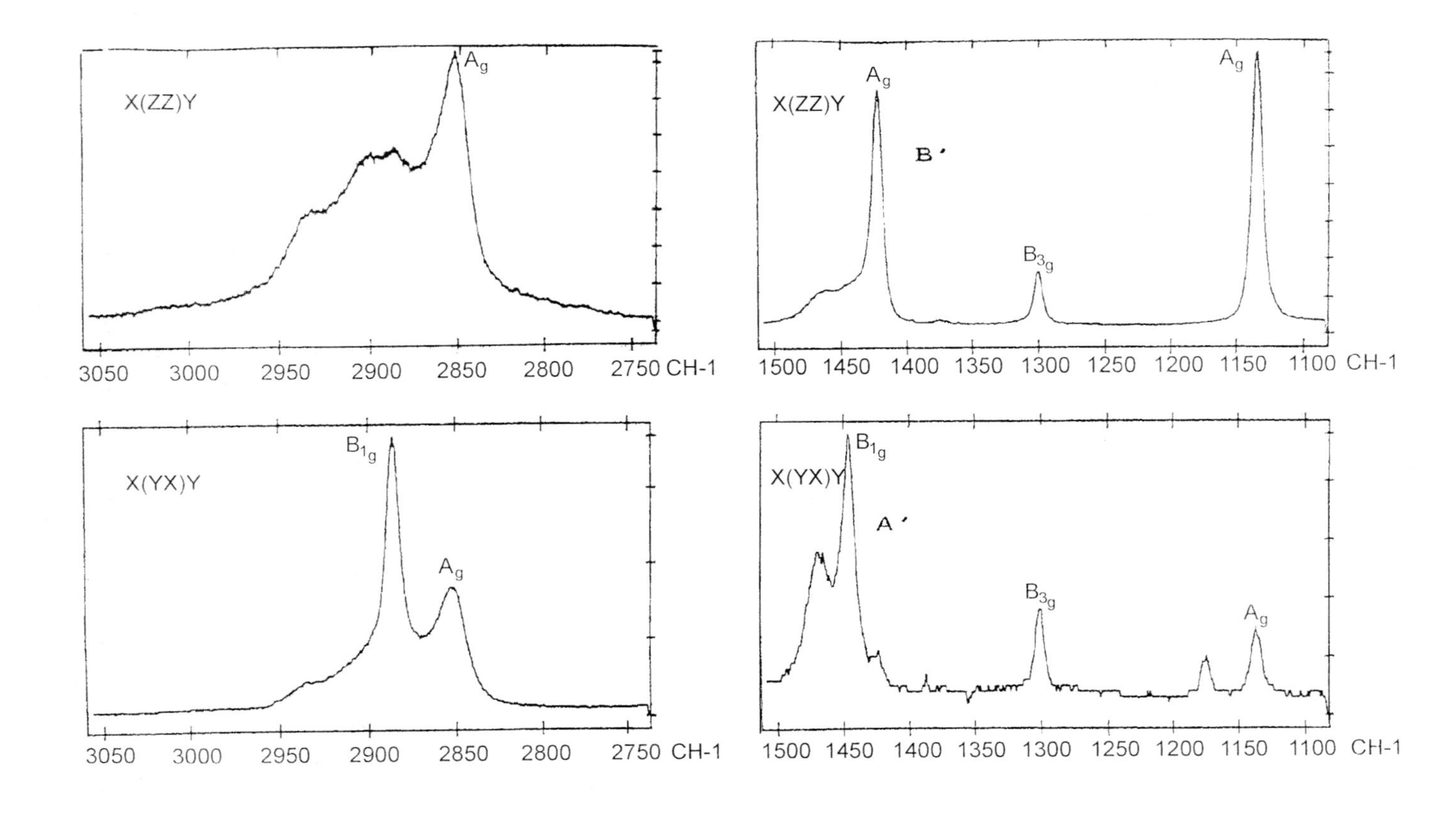

Figure 3-7. Raman spectra in polarized light of stretch-oriented high density polyethylene. Experiments were taken in different scattering geometries. Porto's notation is used in the definition of the scattering geometries. Space group symmetry species are indicated (from [63]).

neglected the possible existence of a tridimensional arrangement of the polymer chains in the solid. Experimentally, it transpires that for most of the polymers the entire observed spectrum can be accounted for in terms of the $\mathbf{k} = 0$ modes of the single infinite chain [65]. When the material is melted, or is dissolved in some suitable solvents, $\mathbf{k} = 0$ modes disappear and the spectrum becomes typical of a conformationally disordered 'liquid-like' molecule with the classical 'group frequency' modes which can be interpreted using the classical spectroscopic correlations (see Section 3.2).

The fact that, in going from the solid to the liquid state, many bands disappear has been taken by authors as an indication that these bands must be associated with the material in the crystalline state. In spite of the existence of clear and easy dynamical theories on polymer vibrations, a whole body of literature has accepted the direct correlation between $\mathbf{k} = 0$ modes of the single chain and content of material in the crystalline (3D) state.

This view is wrong both in principle and in practice. In spite of strong warning and extensive discussions (theoretical and experimental [65, 66]) the general definition of 'crystallinity bands' has been widely and uncritically accepted by the chemical polymer community and even analytical determinations on the concentration of crystalline material have been carried out on polymers which show no direct spectroscopic indications of crystallinity [65, 66]. The disappearance of such bands upon melting is simply due to the fact that the conformational regularity of the chain collapses and no 1D-translational periodicity can be found anymore over a reasonable chain length. Chemical and stereo regularities are not modified in the melt or in solution, but all optical selection rules are removed because of the lack of phase coupling between adjacent units. From the above discussion it becomes apparent that the $\mathbf{k} = 0$ bands previously discussed (generally and reasonably called 'regularity bands' [65, 66]) arise from a polymer molecule organized as a one-dimensional crystal *in vacuo*, as if there were no intermolecular lattice forces. It becomes apparent that their labeling as crystallinity bands is a conceptual error.

Certainly, the dynamical treatment can be and has been carried out for a few cases by taking into account the tridimensional arrangement of polymer chains [49]. A model of suitable intermolecular nonbonded atom–atom potential has to be chosen critically [49, 67] and calculations can be carried out using the same principles discussed previously in this chapter. The number (q) of atoms per tridimensional unit cell increases, the complexity of the BZ increases, many more phonon branches are calculated for different directions of the wave-vectors, and special symmetry directions and symmetry point in the BZ can be found depending on the symmetry of the whole lattice [60, 64]. Typical examples are given in Figures 3-3 and 3-6.

Since intermolecular forces in polymers are weak, their effects on the phonons of the whole lattice are relatively small, thus originating small splitting of few of the regularity bands [68]. Rigorously speaking, the limited splitting of the regularity bands observed for a few polymers originate from phonons at the Γ point ($k_x = k_y = k_z = 0$) of the tridimensional BZ. Such splitting can indeed be considered as crystallinity band and certainly originate from material organized in a tridimensional lattice. This occurs when intermolecular forces are strong enough, the num-

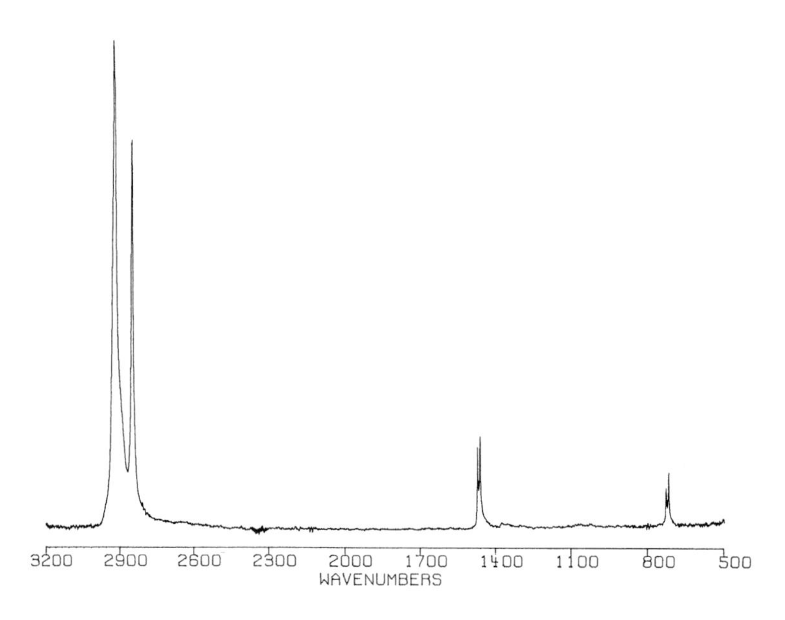

Figure 3-8. Infrared absorption spectrum of highly crystalline polyethylene where crystal field splitting is shown in the 1450 and 720 cm^{-1} range due to crystal field splitting (crystallinity bands).

ber of chains in the unit cell is appropriate, and space-group selection rules are favorable [67–69]. Among the true crystallinity bands we have also to consider the low-frequency lattice modes associated with the translation and librations of the molecules which behave almost as rigid objects moving within the 3D lattice.

The classical case of crystalline polyethylene (Figure 3-8) may be mentioned in reference also to the vibrational spectra of its model compounds consisting of a series of n-alkanes. Relatively short alkanes with an odd number of C atoms crystallize in the orthorhombic lattice with two molecules in the unit cell. Relatively short n-alkanes with an even number of C atoms crystallize in monoclinic or triclinic unit cells with one molecule per unit cell. Very few of the normal modes of orthorhombic n-alkanes show in infrared and/or Raman doublets due to phonons at Γ (i.e., true doublets due to weak intermolecular forces; such splitting is commonly known and studied as correlation field splitting [70]); the spectra of monoclinic (or triclinic) n-alkane do not show any splitting as if the molecules, in a first approximation, were *trans*-planar, but *in vacuo*. Indeed, there exists only one molecule per unit cell and no correlation field splitting can occur. Analogously, when orthorhombic n-alkanes molecules are heated just before melting they transform in a probably monoclinic unit cell (there may be different interpretation of the so-called monoclinic modification). The classical doublets (phonons at Γ) observed in infrared and Raman disappear and singlets (phonons at $\mathbf{k} = 0$) are observed as if the molecules were *in vacuo*. This problem will be matter of extensive discussion later in this chapter.

The fact which needs to be justified is that in some analytical determinations of the concentration of crystalline material in a polymer bulk, correlations are found between the intensity of the regularity bands and the data from X-ray diffraction experiments (which measures scattering from 3D periodicity). The measure of the intensity in infrared or Raman provides the amount of material organized as 1D straight chains. If the '*spaghetti*' model is accepted, the measure of the intensity of the *regularity bands* [65, 66] gives an estimate of the concentration of straight *spaghetti* in an otherwise disordered environment of 'boiled' *spaghetti* (conformationally disordered chains). Whether the chains are packed in a crystal (3D order) or exist in a sort of 'liquid crystalline' arrangement (1D order) cannot be revealed from $\mathbf{k} = 0$ regularity bands.

True crystallinity bands have been observed for only a few polymers and their origin experimentally verified by isotopic dilution studies [70]. We may quote the classical prototype case of orthorhombic polyethylene [68, 69], orthorhombic polyoxymethylene [71], and–with caution–possibly a few others [72, 73].

3.7 The Case of Isotactic Polypropylene – A Texbook Case

It is beneficial here to discuss the case of isotactic polypropylene (IPP) as a worked example of the analysis of the vibrational spectrum of a polymer molecule. Let us assume that polymerization of propene with Ziegler–Natta catalysts has produced a fully head-to-tail isotactic polymer.

Let us first consider the vibrational spectra (infrared) of IPP in the melt (Figure 3-9a) [74]. In this physical state, polymer chains possess a conformationally irregular structure like any liquid branched n-alkane. We expect the infrared spectrum to consist of relatively few bands easily identified as the group frequencies of CH_3 and CH_2 and CH groups. Other modes may be identified with caution, but their location is irrelevant to the present discussion. The experimental infrared spectrum of Figure 3-9a is in full agreement with the expectations.

It must be pointed out that the infrared spectrum of liquid IPP shows a few bands (especially the band at 973 cm^{-1}) which are neither observed in the spectrum of atactic PP nor in the spectrum of liquid methyl-branched hydrocarbons. Moreover, in the Raman spectra of molten IPP, three characteristic tacticity bands have been located at 1002, 973 and 398 cm^{-1} which allow IPP to be distinguished from the syndiotactic stereosiomer (993, 966 and 310 cm^{-1}). It follows that the 'isotactic' chiral units generate vibrations mostly localized on the asymmetric carbon atom, but with some coupling with the neighboring units. This allows the chirality of the neighbors to be probed. These bands become characteristic of the tacticity and allow us to distinguish between isotactic and syndiotactic sequences [75, 76].

Next, let the sample of molten IPP solidify. Polymer chains are allowed through their conformational flexibility to reach the minimum energy conformational struc-

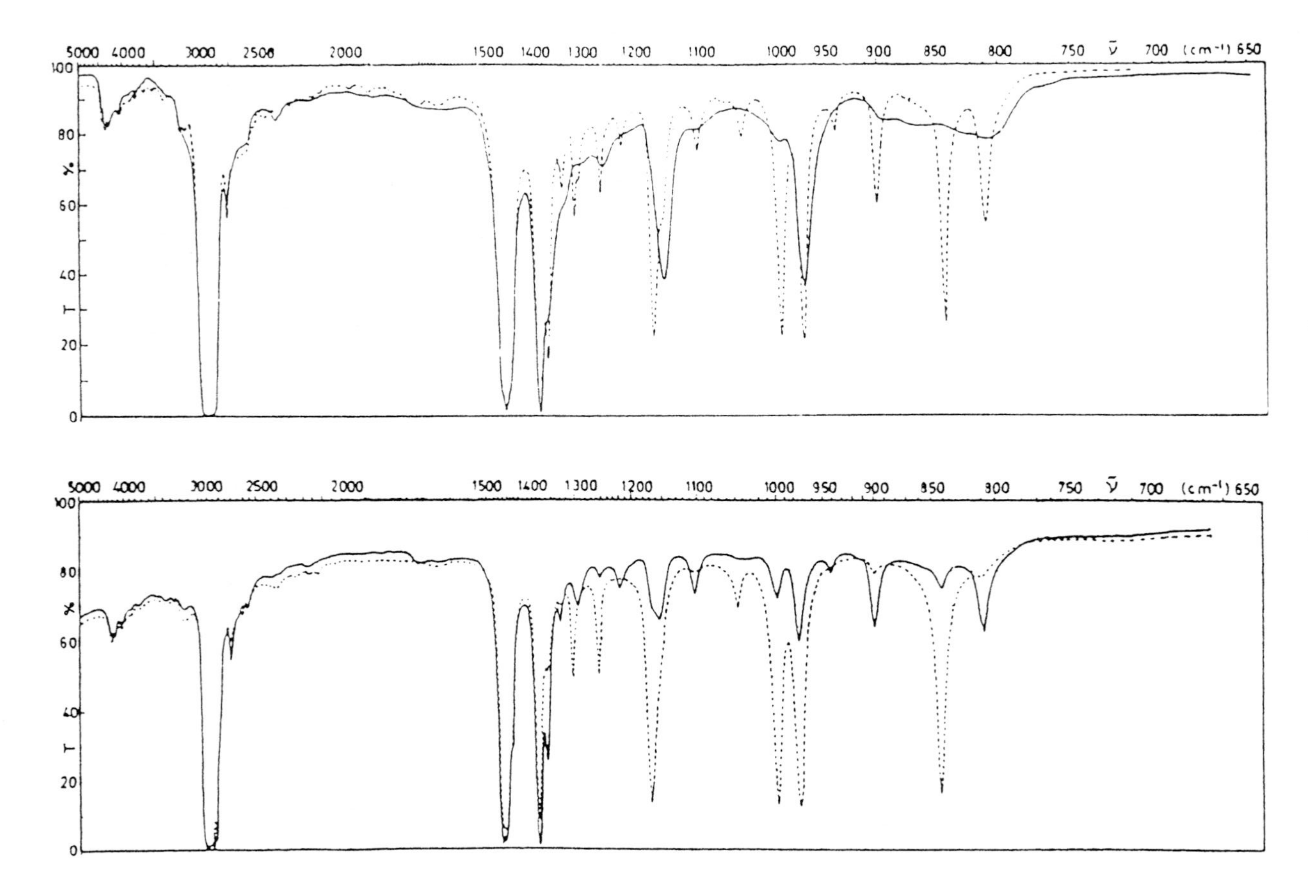

Figure 3-9. Infrared absorption spectrum of isotactic polypropylene in the various physical states. (a) Melt (—) and normal solid at room temperature (- - -) New bands appear associated to 'regularity bands' originating from **k** = 0 phonons of the chain as 1D crystal; (b) stretch-oriented film in polarized light showing which of the bands are of A(||) (- - -) or E(⊥) (—) species (from [74]).

ture which is known from calculations and experiments [77] to form a 1D periodical lattice with the conformational sequence –GTGTGT– which defines a threefold helix with $\psi = (2\pi/3, d/3)$ and one turn (t = 1). The isolated chain belongs to the point group C_3. Each chemical unit contains nine atoms and three chemical units (z = 3) are contained in the crystallographic repeat unit [77]. Thus, we expect $3 \times 9 = 27$ phonon branches in φ space or $3 \times 9 \times 3 = 81$ in **k** space.

Following the previous discussion we expect:

- 27-2 totally symmetric $\varphi = 0$ nonzero frequency modes of A species, infrared active with || dichroism and Raman active in zz polarization.
- 54-2 doubly degenerate modes (thus 27–1 frequencies) at $\varphi = \theta = 2\pi/3$ of E species infrared and Raman active ($\perp$). The phonons for $\varphi = 2\theta$ coincide with the phonons at $\varphi = \theta$ because of the periodicity of the functions $\nu(\varphi)$ (see Figure 3-10a).

The observed infrared spectrum of an isotropic sample of IPP is shown in Figure 3-9a, and clearly identifies the regularity bands associated with 1D periodicity. The assignment of the observed infrared bands to vibrations of species A, $\theta = 0°$ (||) and E, $\theta = 120°$ ($\perp$) is carried out on a stretch-oriented sample (Figure 3-9b).

Dispersion curves have been calculated (Figure 3-10a) [56] from which spectroscopically active phonons have been derived for $\varphi = 0$ and $2\pi/3$. One-phonon density of states have been calculated and compared with the few data experimentally available from neuron-scattering spectroscopy (Figure 3-10b) (see Section 3.8). In Figure 3-10a, the dispersion curves of isotactic polypropylene are reported in terms of the phase coupling φ; in the same figure the six high-energy branches (associated with the C–H-stretching modes in the 3000 cm^{-1} range are omitted. The C–H branches are flat (i.e., C–H-stretching phonons practically do not show dispersion with φ).

Next, let IPP chains crystallize in a cell with space group isomorphous with the point group C_s with four molecules per unit cell in three dimensions. In principle, one should expect $(9 \times 3 \times 4 \times 3) - 3 = 321$ vibrations in the spectrum for the spectroscopically active phonons at the symmetry point Γ (i.e. $k_x = k_y = k_z = 0$). This would mean that, in principle, each of the $25A + (2 \times 26)E$ $\mathbf{k} = 0$ bands of the single chain should split into four lines if intermolecular Van der Waals-type forces between atoms were strong enough to generate an observable splitting. Generally, such forces are very weak, especially since interchain distances are relatively large. If such splitting could be observed, true infrared crystallinity bands of crystalline IPP would be identified. The search for such splitting has not been easy and we made careful experiments at very low temperature aiming at shrinking the crystal, thus increasing the interchain interactions which should increase the correlation field splitting (i.e. $k_z = 0$ line-group modes in 1D should show splitting into multiplets due to phonons with $k_x = k_y = k_z = 0$ in 3D (phonons at the symmetry point Γ in the BZ). Moreover, bands should become sharper since 'multiphonon states' are depopulated. After much work the experiments were finally successful and the Raman spectrum of IPP at liquid nitrogen temperature shows indeed multiplet splittings due to crystallinity (Figure 3-11) [78].

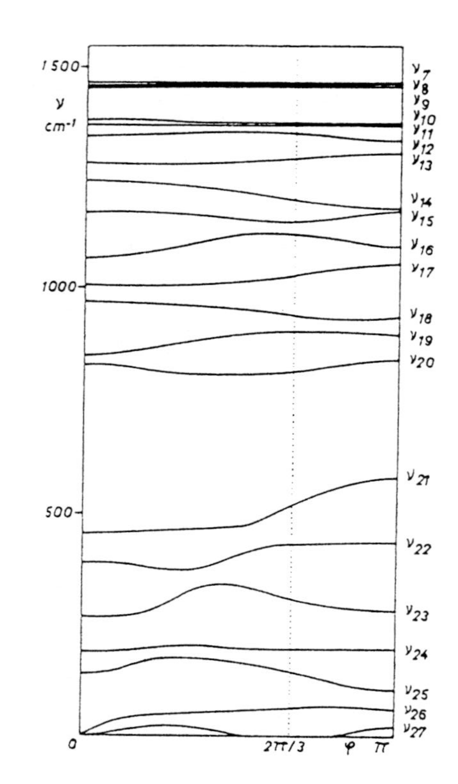

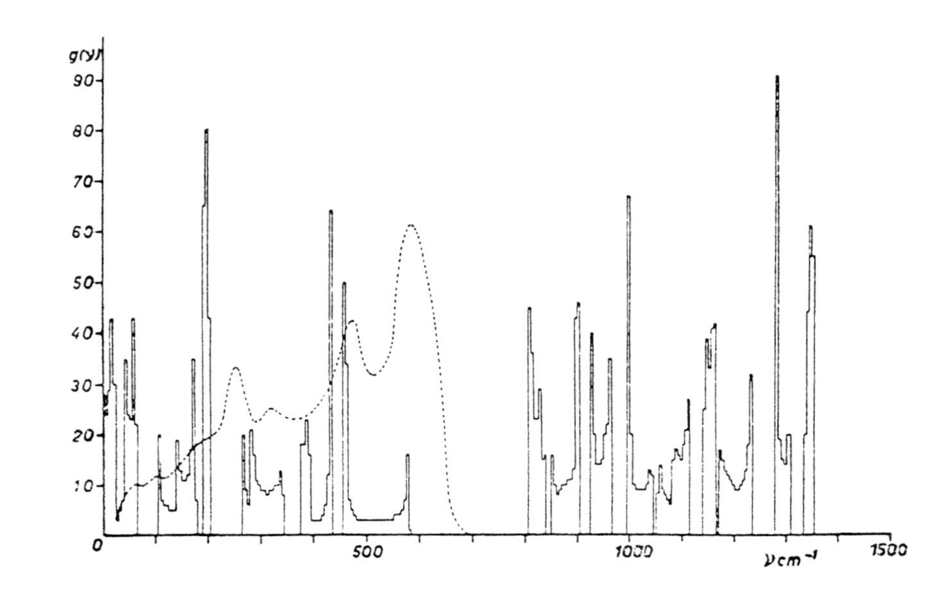

Figure 3-10. (a) Calculated dispersion curves and (b) experimental (- - -) and calculated (—) density of vibrational states of single-chain isotactic polypropylene (from [56]).

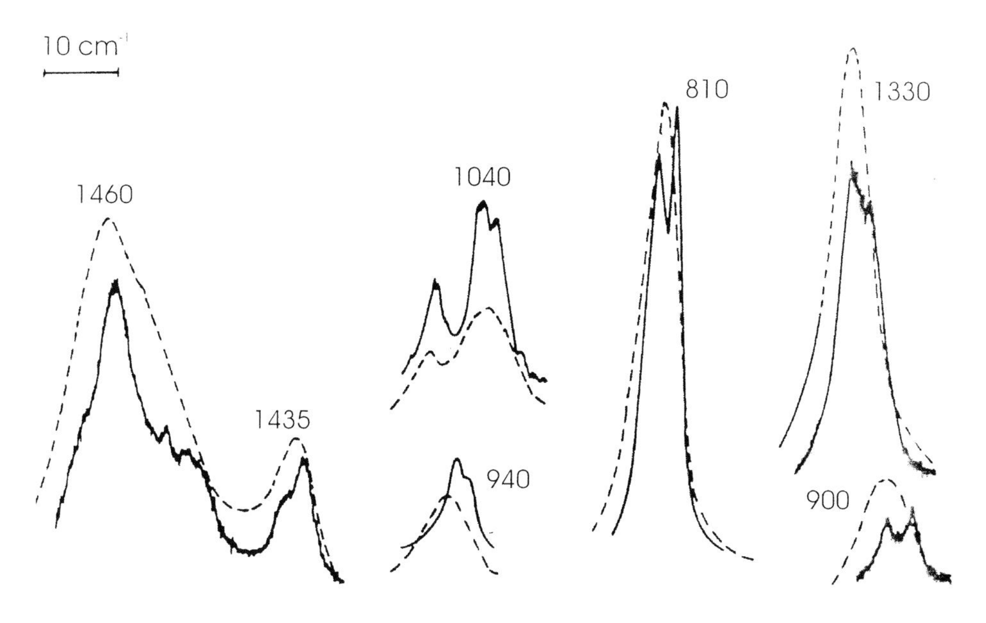

Figure 3-11. Raman spectrum of isotactic polypropylene from 'smectic' (- - -) to crystalline (—) (tridimensional order) obtained after long annealing processes at low temperatures. Raman regularity bands split into doublets (crystallinity bands) due to phonons at Γ ($k_x = k_y = k_z = 0$) (from [78]).

The case of isotactic polypropylene becomes an interesting texbook case which shows clearly and separately the many steps in the formation of a polymeric material.

3.8 Density of Vibrational States and Neutron Scattering

The next step in the development of vibrational spectroscopy of large organic molecules is to realize that other physical measurements, such as neutron-scattering spectroscopy, can provide essential information on the normal vibrations of organic molecules. This information is complementary to that provided by infrared and Raman spectroscopy which is intrinsically limited by the optical selection rules. Indeed, the limitations to the spectroscopic activity imposed by symmetry selection rules are very strict and select only a few phonons. Moreover, when some sort of molecular disordering is introduced into the system (and this is always the case), symmetry is removed, selection rules are relaxed, and the spectra show sharp absorptions (or Raman scattering) originating only from characteristic localized vibrations floating on top of a broad background absorption whose origin needs to be explored and understood.

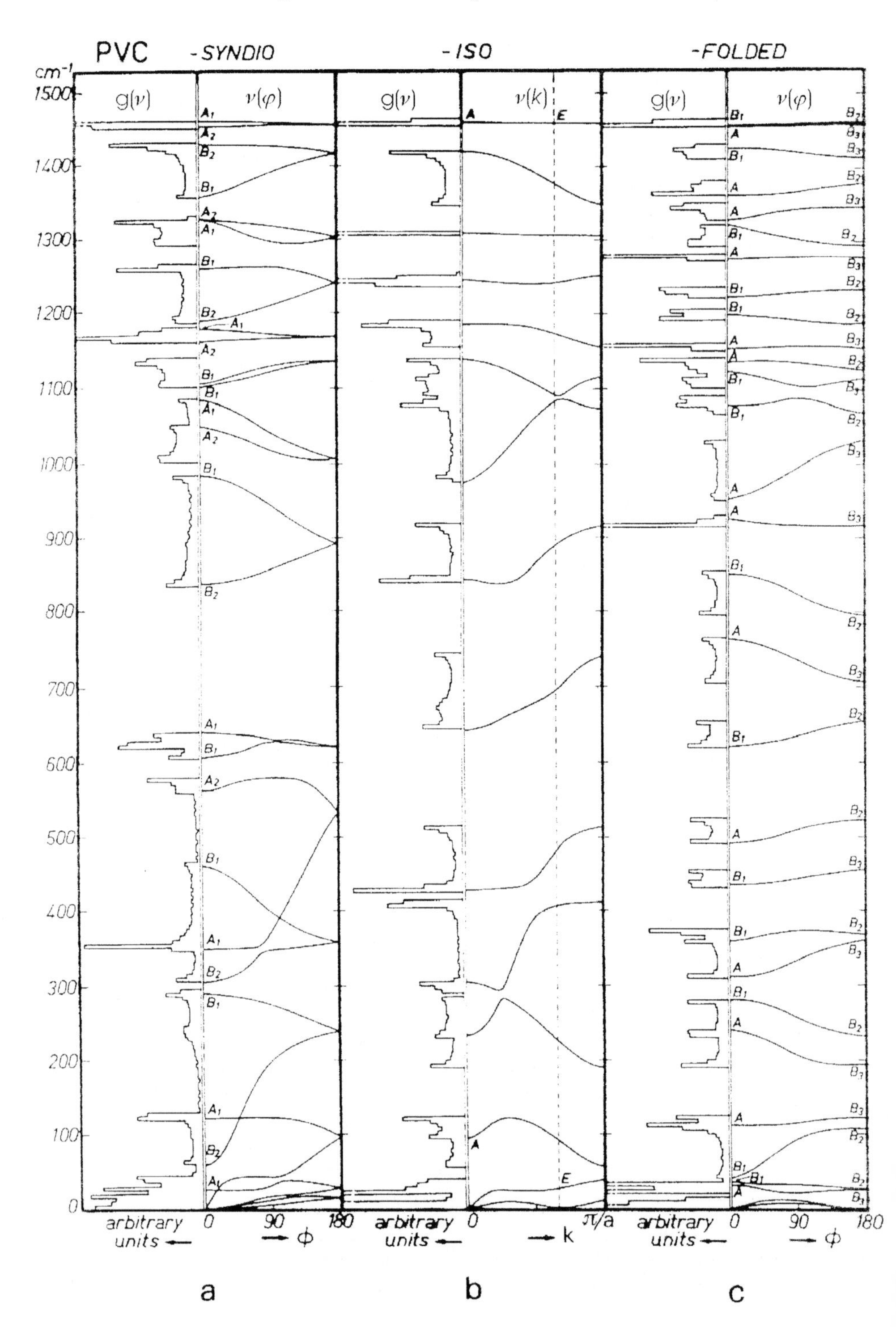
PVC
-SYNDIO
-ISO
-FOLDED
cm⁻¹
g(ν)
ν(φ)
ν(k)
arbitrary units
φ
k
π/a
a
b
c

We will introduce the basic principle and use of neutron-scattering spectroscopy by first introducing an important dynamical quantity, namely the density of vibrational states $g(\nu)$.

Let $g(\nu)$ be the fraction of normal modes with frequencies in the interval ν and $\nu + d\nu$ with $d\nu \to 0$. $g(\nu)$ can be calculated analytically only for very simple atomic chains which are never encountered when realistic polymeric materials must be studied.

Let us take a very large molecule and let $3N = C$ be the number of vibrations of the whole system; if ε is the number of frequencies in the interval $\Delta\nu = \nu + d\nu$ then

$$g(\nu) = \varepsilon(\Delta\nu)/C \tag{3-50}$$

The total density of states is evaluated as the sum over all the r branches of the dispersion relation (Eq. (3-43)). Numerical methods such as the 'root sampling method' have been proposed for simple one- or tridimensional lattices [51]. When the systems become more complex, as in the cases discussed in this chapter, the easiest numerical method is that proposed by Dean based on the application of the Negative Eigenvalues Theorem (NET) [79]. The method is extremely useful and has been extensively used in the laboratory of the writer for many polymeric cases. The details of the method are discussed in the next section.

NET is a way to calculate $g(\nu)$, i.e., it provides the number (or the fraction) of vibrational energy states or normal modes in a chosen φ (or $\mathbf{k}$) interval. $g(\nu)$ can be plotted as an histogram whose accuracy depends on the frequency interval $\Delta\nu$ chosen in the numerical calculation (numerical resolution of the computer experiment). This will be used later in this chapter in the study of disordered systems.

Let us consider $g(\nu)$ of an ideal infinite polymer and examine the case of single-chain polyvinylchloride in its three possible structures (threefold helix isotactic, *trans*-planar syndiotactic, folded syndiotactic) [80]. For easier understanding, Figure 3-12 shows both dispersion curves and $g(\nu)$ for this polymer in its three possible structures. It is apparent that the shape of $g(\nu)$ is determined by the shape of the dispersion curves which in turn depend on the vibrational force field used in the calculation of $\nu(\varphi)$. Flat sections of $\nu(\varphi)$ give rise to strong 'singularities' in $g(\nu)$. The shape of the dispersion curves and of electronic and/or vibrational $g(\nu)$ for both electronic bands and vibrational bands have been the subjects of extensive theoretical treatments [51]. For infinite and perfect systems, the shape of $g(\nu)$ is characterized by different types of singularities which are related to the dimensionality of the systems and to many other physical factors. We restrict our analysis to the case of polymeric materials which can be considered as one-dimensional lattices. Our discussion can only be sketchy and qualitative, but it is aimed at pointing out the features most relevant to the problems of disorder treated in this chapter.

Figure 3-12. Phonon dispersion curves $\nu(\varphi)$ (a, c), $\nu(\mathbf{k})$ (b) and density of vibrational states $g(\nu)$ from 0 to 1500 cm^{-1} of three possible models of a chain of polyvinylchloride showing the IR–Raman active and the inactive 'singularities' corresponding to flat sections of the disperion curves (from [80]).

The following observations should be kept in mind.

1. Because of the shape of the calculated $\nu(\varphi)$, which may show many maxima and minima (due to the vibrational coupling intrinsic to the system and to the force field used), $g(\nu)$ will also show corresponding strong or weak singularities. Some of these singularities necessarily coincide with $\mathbf{k} = 0$ spectroscopically active phonons, while other will not be seen in the optical spectrum of a perfect system, but will (in principle) be observed in the neutron-scattering spectra and possibly in the optical spectra of a partially disordered polymer. In a disordered material, no symmetry selection rules are active and all modes may gain some activity. The vibrational spectra of a disordered material can thus be considered the mapping of $g(\nu)$ dipole (or polarizability) weighted. The $g(\nu)$ calculated with NET or any other numerical or analytical method is, instead, dipole (polarizability) unweighted. Any comparison with the experimental $g(\nu)$ must be made with great caution.
2. As discussed in Section 3.5 for tridimensional lattices, $\nu \to 0$ for $\mathbf{k} \to 0$ for the three acoustical branches with a positive slope and with a discontinuity at $\mathbf{k} = 0$ (Γ). The shape of $g(\nu) \to 0$ has been matter of strong interest in physics for the calculation of specific heat and related thermodynamic quantities. As already mentioned for one-dimensional lattices *in vacuo*, some of the acoustical branches tend asymptotically to zero. The derived calculated $g(\nu)$ shows a very strong singularity at $\nu = 0$. Such a singularity is meaningless and only due to the limitation of the molecular modes adopted in the calculations (1D lattice) which is unable to account for intermolecular forces. Also, for real polymer samples the experimental $g(\nu) \to 0$ for $\mathbf{k} \to 0$ as expected for classical crystals since they do interact with the neighboring chains with very weak intermolecular forces.
3. Additional singularities both in calculated and experimental $g(\nu)$ are expected at very low energies ($\approx$0–30 cm^{-1}) originating from very low-frequency 3D-lattice phonons. These singularities may coincide with the optical phonons in infrared and/or Raman spectra at very low frequencies.

As already anticipated, a complementary experimental technique for deriving information on the dynamics (frequencies and vibrational amplitudes) of polymers or of materials in general is the use of inelastic neutron-scattering techniques (INS). After a long development time, during which experiments were difficult and provided limited information, the instruments in a few specialized centers recently began to provide detailed data covering the whole spectrum. Thus, we predict a 'renaissance' of INS techniques for the studies of molecular and lattice dynamics.

For a thorough discussion of INS spectroscopy we refer the reader to specialized publications [81] (see also other references quoted below); here, we restrict ourselves to the basic principles of the technique connected especially with the dynamic quantities which are of interest in optical spectroscopy.

Let a beam of cold neutrons made mono-energetic with wave-vector $\mathbf{k}_0$ be shone onto a sample, and let the beam be scattered inelastically and incoherently. The outgoing neutrons have wave-vector $\mathbf{k}_f$ and are collected and detected by suitable devices. Since the energy of the thermal neutrons is of the same order as that of lattice and molecular vibrations, neutrons impinging on the sample may gain or lose energy in a way proportional to $g(\nu)$.

It has been shown [82] that the scattering cross-section of a polycrystalline material is approximately the same as that of cubic crystals. We are, at present, concerned with one-phonon processes and neglect the possibility that multiphonon processes may occur (also in vibrational spectra we have neglected anharmonic effect with the consequent appearance of overtones, combinations, and hot bands).

The one-phonon scattering cross-section for a cubic lattice can be written as [83, 84]:

$$d^2\sigma_{incoh}/d\Omega\, d\omega = A|\mathbf{k}_f|/|\mathbf{k}_0|e^{-2w}(\hbar\mathbf{K}^2/2M)[g(\omega)/\omega] \times [\delta(\omega_f - \omega_0 + \omega)/(e^{\beta\omega} - 1)] \tag{3-51}$$

In Eq. (3-51), and in a few equations which will follow the vibrational frequencies are not expressed as ν in wavenumbers, but as ω in Hz. In Eq. (3-51), A is the scattering cross-section of the scattering nucleus (since the H atom has a very large cross section, namely $A_H = 82.5$ barn, $A_C = 5.5$ barn, molecules containing hydrogen are very suitable for neutron experiments), $|\mathbf{k}_f|$ and $|\mathbf{k}_0|$ are the moduli of the wave-vectors of the scattered and incident neutron respectively, e^{-2W} the Debye–Waller factor or temperature factor, $\mathbf{K} = \mathbf{k}_f - \mathbf{k}_0$, M the mass of the unit cell, $g(\omega)$ the desired density of vibrational states, and $\beta = \hbar/2k_BT$ (where k_B and T are the Boltzman factor and the temperature respectively).

It is clear that thermal population is strongly contributing to the scattering cross-section with the Debye–Waller factor and the term β. Measurements must then be carried out at very low temperature.

The experimental $g(\omega)$ (or $g(\nu)$), while showing qualitatively all the features of the calculated $g(\omega)$, cannot yet be quantitatively compared because the vibrational amplitudes (generally called polarization vectors in neutron-scattering circles) of the various normal modes (phonons) are not yet taken into account. The situation is similar to the features of the vibrational spectra of an isotropic sample compared with the features of the spectra in polarized light of an oriented anisotropic sample.

Let us take a stretch-oriented sample of a polymer subject to a neutron beam with incident wave-vector k_0. A more complete expression of the scattering cross-section is the following [82, 84]:

$$d^2\sigma_{incoh}/d\Omega d\omega = \sum_{n,r} A_n(|\mathbf{k}_f|/|\mathbf{k}_0|)e^{-2W_n}(\hbar K^2/2M_n)(1/N) \times \sum_j [\coth(\beta\omega_j^r) + 1]/2\omega_j \times |\mathbf{e}_k \cdot \mathbf{C}_j^{n,r}|^2 \delta(\omega_j - \omega_0 - \omega_j^r) \tag{3-52}$$

Some of the terms appearing in Eq. (3-52) are already defined for Eq. (3-51). The new terms are the following: A_n is the scattering cross-section of the n-th nucleus of the unit cell, e^{-2W_n} the Debye–Waller factor for the n-th nucleus of mass M_n, ω_j^r the frequency of the j-th normal mode of the r-th branch in the dispersion relation. The most important quantity is $\mathbf{e}_k \cdot \mathbf{C}_j^{nr}$ which gives the projection along $\mathbf{k}_0$ of the atomic displacement **C** of atom n for the j-th normal mode of the r branch with frequency ω_j^r. The sum is extended to all the atoms n in the unit cell and to all branches r in the dispersion relation. The Debye–Waller factor W_n of each atom n has an explicit expression [82, 84] which again contains quantities defined in Eqs. (3-51) and (3-52).

In spite of the seemingly complex expressions given in Eqs. (3-51) and (3-52) for the scattering cross-sections, all terms appearing in these equations are known from dynamics and from experiments, and have been defined in the previous sections of this chapter. Automated computer programs have been added to the classical programs for normal coordinate calculations of polymers for the calculations of the INS spectra.

The reason of requiring the experiments to be carried out on stretch-oriented samples mounted with known geometry with respect to the incoming neutron beam is the existence of the $\mathbf{e}_k \cdot \mathbf{C}_j^{n,r}$ term which indicates that the scattering cross-section is determined by the directional interaction of the neutrons with wave vector $\mathbf{k}_0$ and the direction of the atomic displacements. This situation is very similar in the experiments of infrared spectroscopy in polarized light with a stretch-oriented material. In dichroism experiments the absorption intensity is proportional to $|\boldsymbol{\mu}_i \cdot E|^2$, i.e., to the square of the projection of the molecular dipole moment change $\boldsymbol{\mu}$ during the i-th normal mode onto the electric field of the light beam.

The $g(\omega)$ measured with INS experiments on stretch-oriented materials is then 'amplitude-weighted' and is 'directional'. The directionality of the displacements associated with a given normal mode can be determined by placing $k_0 \parallel$ and $\perp$ to the chain direction.

The history of polymer neutron spectroscopy can be divided in two different time domains. The enthusiasm for INS experiments in polymers reached its highest peak in the 1970s, but the experiments have been restricted to a few polymers with a limited number of data [84, 85] which could hardly solve particular problems in polymer dynamics. The availability of modern instruments with higher resolution and with the possibility of removing most of multiphonon scattering has presently revived the field and vibrational spectroscopists await with great interest for the revival of this important field of research.

In Figure 3-13 we show the most recent INS results reported by Parker [86] on oriented polyethylene at 30 K obtained with the high-resolution, broad-band spectrometer (TFXA) at ISIS pulsed spallation neutron source at the Rutherford Appleton Laboratory, Chilton, UK. Notice that the whole spectral range 0–4000 cm^{-1} range has been explored with considerable resolution, especially in the lower-energy region.

3.9 Moving Towards Reality: From Order to Disorder

3.9.1 Basic Concepts

In reality, polymeric materials never possess the perfect chemical stereochemical and conformational regularity which were assumed at the beginning of Section 3.5.

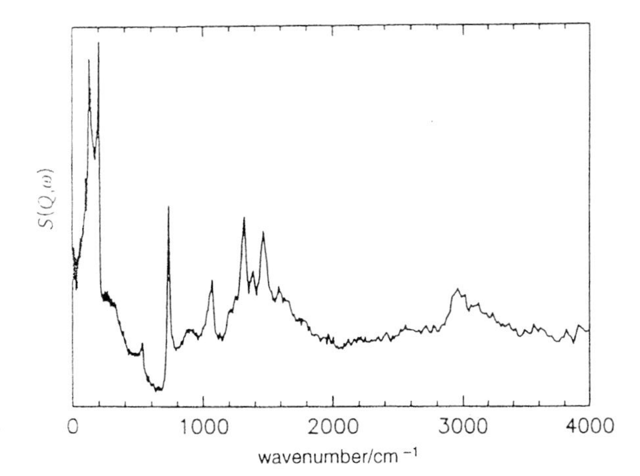

Figure 3-13. Inelastic neutron scattering (INS) spectrum of polyethylene at 30 K (from [86]).

The ideal infinite polymer chain *in vacuo* has allowed to construct dynamical theories which describe ideal spectroscopic properties of the polymer determined through strict optical selection rules and obtained by ideally clean experiments. In reality, polymerization reactions never yield perfect chemical linking and perfect stereo-regularity; these chemical and stereochemical defects necessarily introduce conformational distortions as to avoid steric encumbrance. Moreover, the complex processes of crystallization from melt or solution of a very long and flexible object implies the generation of various kinds of morphologies which reflect particular packing of 'conformationally' ordered chains. New concepts such as chain folding, looping, tie molecules, fringed micelles, etc., are necessarily introduced for the description of the building up of the supermolecular structure of a polymer chain. It follows that any polymer material which has been allowed to crystallize consists of a fraction of matter crystallized in a 3D-lattice and a fraction of 'conformationally irregular' material [46, 87].

In polymer material science it is essential to know the detailed independent structural characteristics (and the relative concentrations) of the crystalline and 'amorphous' (or irregular) fractions of a polymer sample. All mechanical, visco-elastic or, in general, the physical properties of real polymeric materials depend on the relative fraction of crystalline versus non crystalline material.

Since the spectroscopic manifestations of molecular disorder are many and very characteristic, the vibrational spectrum has become a useful probe for molecular order–disorder which can be described at the level of a few Ångstroms.

Spectroscopists are asked to find ways to account for the new spectral features associated to disorder, thus making vibrational spectra useful for the practical structural characterization of a real polymer material. Obviously, the theories presented in the previous sections of this chapter must be adapted or re-worked in order to understand the spectra of disordered organic polymers.

The prototypical case of an experimental observation to be accounted for is the spectrum of a sample of orthorhombic polyethylene [88] (Figure 3-14). Let us con-

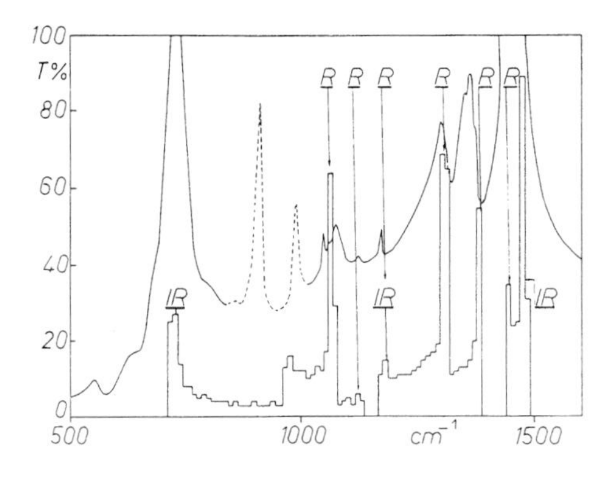

Figure 3-14. Comparison of the calculated g(ν) of single-chain polyethylene with the infrared spectrum of an extended chain of polyethylene.

sider the experimental infrared spectrum of high-density crystalline polyethylene (Figure 3-14). The infrared active phonons at Γ for an orthorhombic lattice with two chains per unit cell show the predicted crystallinity bands as doublets originating from the $\mathbf{k} = 0$ modes of the former regularity bands for a chain with a 1D symmetry operator $\psi(\theta = \pi, d/2)$. In Figure 3-14, the infrared spectrum is superposed to the singularities of the calculated g(ν) while the Raman active peaks are indicated in the figure. It is quite clear that not all the peaks in the observed vibrational infrared or Raman spectrum find a one-to-one correspondence with the singularities in the calculated g(ν). It follows that, in addition to the 'perfect' part of the all-*trans* polyethylene chain, extra structural features must exist whose existence is indicated by the additional peaks in the infrared or Raman spectra. It will be shown that the extra absorption or scattering are associated with various kinds of conformational defects.

In this section we proceed as follows.

1. First, we take the perfect and infinite polymer chain and truncate the chain in increasingly shorter segments and predict theoretically and watch experimentally the effects of such truncation. The structural perfection is retained.
2. Any kind of chemical and/or structural disorder can be introduced into the chain. We chose the various plausible kinds of disorder, their population and distribution, and try to predict the infrared, Raman and neutron-scattering spectra by suitable theories and computational techniques.
3. Calculated theoretical and experimental cases will be presented.

3.9.2. Finite Chains

The theoretical modeling of the vibrations of molecular chains with finite length has been nicely treated by Zbinden [18] and by Snyder and Schachschneider [9, 34]. While the approach by Zbinden is mathematically complete, but it applies to simplified monoatomic chains, quite away from chemical reality, the model by Snyder and Schachschneider is more directly applicable to real molecules and will be quickly mentioned here. We restrict our discussion to the most common case of σ-bonded chains, which, as discussed in Section 3.5, do not permit long-range electronic coupling.

Let η be the number of identical chemical units in a chain segment (different chemical groups at either ends are neglected) and let each chemical unit consist of n atoms; we treat here the case of a model molecule with free ends [18]. Let us take one chemical unit and its $3 \times n$ oscillators (which, for sake of clarity, we describe here as internal (R) or group coordinates Я). Because of elastic coupling with its neighbor at either side, each oscillator couples with its identical neighbors and generates η normal modes which can be described as waves characterized by a phase coupling $\varphi = j\pi/\eta$ (where $j = 0, 1, 2, \ldots \eta - 1$). Their η frequencies $\nu_j(\varphi)$ lie at finite φ values along the dispersion curve of the corresponding infinite polymer with identical chemical, stereochemical, and conformational structure. Each vibration corresponds to a given phase coupling defined by the number of chemical repeat unit which make up the finite chain. The introduction of phase coupling implies the fact that within such short chains, nomal modes describe quasi regular waves (quasi phonons) similar to those which propagate in an infinite chain. The remaining modes which do not lie on the branches of the dispersion curves must be associated with the vibrations localized on the end groups and can be identified by the constant values of the frequencies when chain length is changed. It follows that:

(i) if η is the number of repeat units making up the finite chain (neglecting the two end groups) one expects to find a sequence of bands (or band progression) which lie on each of the dispersion branches of the corresponding polymer. Let us take for example the n-alkane n-nonadecane ($CH_3(CH_2)_{17}CH_3$). Let us focus on the vibrations of the segment consisting of $\eta = 17$ CH_2 groups all in *trans* conformation. Each CH_2 generates $3 \times 3 = 9$ vibrations (CH_2 antisymmetric and symmetric stretch, CH_2 bending, wagging, twisting, rocking, antisymmetric and symmetric C–C stretch, CCC bending, and C–C torsion). Each of these 'oscillators' generates 17 waves, each of which is characterized by a phase coupling $\pi j/17$. The corresponding frequencies lie on the dispersion curves of a single chain of infinite all-*trans* polyethylene. The band progressions observed for n-nonadecane will be discussed in Section 3.19.

(ii) The frequency range spanned by each band progression depends on the extent of intramolecular coupling. If the dispersion of the frequency branch is small, the progression is squeezed, many bands overlap (e.g., the stretching vibrations of C–H groups), and no progression can be observed experimentally. When intramolecular coupling is large, band progressions are clearly observed ($\sim$1100–720 cm^{-1} is the frequency range covered by CH_2 rockings in all-*trans* n-alkanes).

(iii) The intensity in infrared and/or Raman of each of the bands within the progressions depends on the dipole (polarizability) changes associated with the corresponding vibration (quasi phonon with $\varphi_j \approx j\pi/\eta$). We have previously seen that for an infinite chain only $\mathbf{k} = 0$ phonons are active and all other are inactive, as dipoles or polarizability changes cancel out in the case of perfect phonon waves. In going from short to longer chains, optical selection rules tend to the limit of the infinite case and we expect to find in the spectra of the finite, but long, chains clear (generally, but not always strong) quasi $\mathbf{k} = 0$ phonons while the other members of the band progression will show quickly decreasing intensity. This is the typical case observed in the series of n-alkanes by Snyder and Schachschneider [9, 34].

(iv) It is obvious that if normal-mode calculations have permitted calculation of the dispersion curves for the infinite polymer, the identification of the band progressions in finite chains with identical structure is made easier. In contrast, if a systematic study is carried out on a series of molecules of the same chemical class, but with increasing chain length, the experimental identification of the band progressions allows experimental determination of the phonon dispersion branches from which a reliable force field can be derived. Vibrational spectroscopy then becomes a complementary tool (sometimes unique) to the extremely more expensive and elaborate neutron-scattering techniques. The whole work has been very clearly and precisely explained and successfully applied by Snyder and Schachschneider [9, 34].

3.9.3 Polymer Chains with Structural Defects

In order to model the reality of a polymeric material we need first to introduce in the calculations the various energetically possible structural defects. The *types of defects* to be considered are the following:

- *Chemical defects*, e.g., head-to-head linking in an otherwise head-to-tail chain. A typical case is the real structure of the chain of polyvinylfluoride $(CH_2{-}CF_2)_n$ which contains a sizeable fraction of undesired head-to-head defects [89]. Once the chemistry of a given polymer is approximately known, other kinds of defect structures can be envisaged. A typical case is often found for isotopically substituted chains when substitution is not ideally complete.
- *Stereochemical defects*, e.g., syndiotactic configurations in an otherwise isotactic chain.
- *Conformational defects*, e.g., *gauche* conformers in an otherwise all-*trans* chain structure.

It is obvious that, because of intramolecular interactions, the introduction of some kind of chemical and/or stereochemical defects forces also the introduction of conformational defects.

Next, the model requires the definition of the *concentration and distribution* (e.g., Bernoullian, Markovian, etc.) of defects. If a small concentration of defects with a random distribution is considered, defects most probably are isolated in the host polymer 1D lattice. When the concentration increases, even a random distribution generates both isolated defects and a distribution of 'islands' of various lengths (see, for instance, [90, 91]).

When such variables are well defined, calculations require the construction of the usual dynamical matrix and the solution of the corresponding eigenvalue equation.

It becomes apparent that as the translational symmetry of the chain is removed by the existence of defects, any periodicity is lost and Eq. (3-43) cannot be used. Ratter, Eq. (3-17) must be used which corresponds to the dynamical case of a finite molecule. Since in this case our molecular model is huge and has no symmetry, the size of the secular equation to be solved becomes extremely large. Moreover, such types of study require the freedom to try many models with different kinds, concentrations, and distributions of defects. Last, but not least, in order to play a game as close as possible to reality in these studies, molecular models must be composed of as many monomer units as possible.

In solid-state physics, the vibrations of very simple lattices containing a small concentration of simple defects have been treated with sophisticated analytical treatments using Green's functions [92]. Even if some authors have bravely tackled an analytical solution of the dynamical problem of disordered polymers [93], they were forced to introduce into the molecule such drastic structural simplifications that the flavor of chemistry has been lost and the theoretical molecular models have become, again, too unrealistic.

For our complex molecular systems the problem must be solved by numerical methods. The problem of the solution of a huge secular equation producing thousands of vibrational frequencies can be solved by the application of the so-called 'Negative Eigenvalue Theorem' (NET) originally proposed by Dean [79] who aimed at the calculation of the vibrational spectra of disordered ice. We think that the original work by Dean did not receive due acknowledgement; indeed, his methods has been widely applied (with little due reference to Dean) in solid-state physics and in molecular theories whenever the eigenvalues corresponding to vibrational or electronic states of huge and disordered systems had to be calculated. (As an example of the calculation on the electronic states of disordered systems, see [94]).

Dean's method calculates the density of states (vibrational or electronic) in a given energy range. The whole energy range can be spanned by the calculation and the complete $g(\nu)$ is obtained and can be plotted as an histogram to be compared with the experimental spectrum. The accuracy of the histogram can be improved by narrowing the steps ($\Delta\nu$) in the energy interval in each cycle of the calculation (i.e., by improving the resolution of the numerical experiment).

Dean's method works well for band-matrices which are always found in the case of polymer chains or in the case of tridimensional lattices with short-range interactions. The reader is referred to the original papers for a discussion of the method, its advantages and limitations [91, 95].

Once the histogram of $g(\nu)$ is plotted, one needs to find the eigenvalues for such huge matrices corresponding to the approximate eigenvalues comprised in a given energy range. The problem has been solved with the application of the Wilkinson's inverse iteration method [96] (hereafter referred as to IIM; see for instance [97]).

The comparison of the calculated $g(\nu)$ with the actual experimental infrared, Raman and neutron-scattering spectra requires some care and possibly more refined calculations with improved resolution. $g(\nu)$ gives one information, namely how the very many normal modes are clustered in a given frequency range. From IIM we may also learn the shape of the normal modes at a chosen frequency. Nothing is known about the dipole transition moments (infrared intensities) or Raman or neutron-scattering cross-sections unless specific calculations are carried out using other models for the electronic distribution (dipole derivatives, polarizability derivatives, and vibrational amplitudes respectively [97]). Care must then be taken in carrying out the comparison between calculated $g(\nu)$ and observed spectra.

3.10 What Do We Learn from Calculations?

We have just outlined the way theoreticians in polymer dynamics face the problem of the understanding of the vibrational spectra of polymers in their realistic state. The reader interested in the theoretical aspects will find in the references quoted all indications which will enable him or her to grasp the theory and to carry out the calculations.

However, a larger number of reader do not wish to play with calculations, but simply to know what we have learned about polymer structure and spectra using these calculations and to apply quickly and directly the knowledge acquired to their own problems. We report here the concepts of general validity in polymer spectroscopy which were derived from calculations and which can be applied to any polymer (for a general discussion, see [16, 19, 66]).

The reader is certainly familiar with the traditional group–frequency approach generally used in the past 50 years for the chemical application of the infrared spectra (Section 3.4). Correlation tables and books have been written which discuss group–frequency correlations and provide the way to carry out a chemical diagnosis from the vibrational spectrum (so far, the infrared spectra have enjoyed great popularity; the reader is advised to extend their interest to the very useful and easily available Raman spectra).

As already discussed in Section 3.4, from the dynamical viewpoint chemical correlations are based on the fact that there are a few vibrations (normal modes) which are *largely localized* on the functional group and give rise to medium/strong absorption bands in infrared (scattering lines in the Raman) at specific and well-isolated frequencies. The occurrence of such bands in the spectrum implies the existence of such a functional group in the molecule under study. On the other hand, when the vibrations involve a large molecular domain (*collective motions*) the observed bands are no longer so characteristic of a given group of atoms.

The vibrations of disordered polymer chains follow the same kind of conceptual path which, on the other hand, was also found earlier by physicists in the study of lattice dynamics of very simple disordered lattices [98].

Let us make the problem simpler by considering briefly the small molecular unit which contains the defect as an isolated entity consisting of n atoms which generate 3n normal vibrations that occur somewhere in the vibrational spectrum. Next, we insert just one defect unit into the otherwise perfect polymer chain and allow the whole system to display its own new dynamics through the observed (or calculated) vibrational spectra.

The dynamics (new frequencies and displacements, see Eq. (3-17)) of the defect

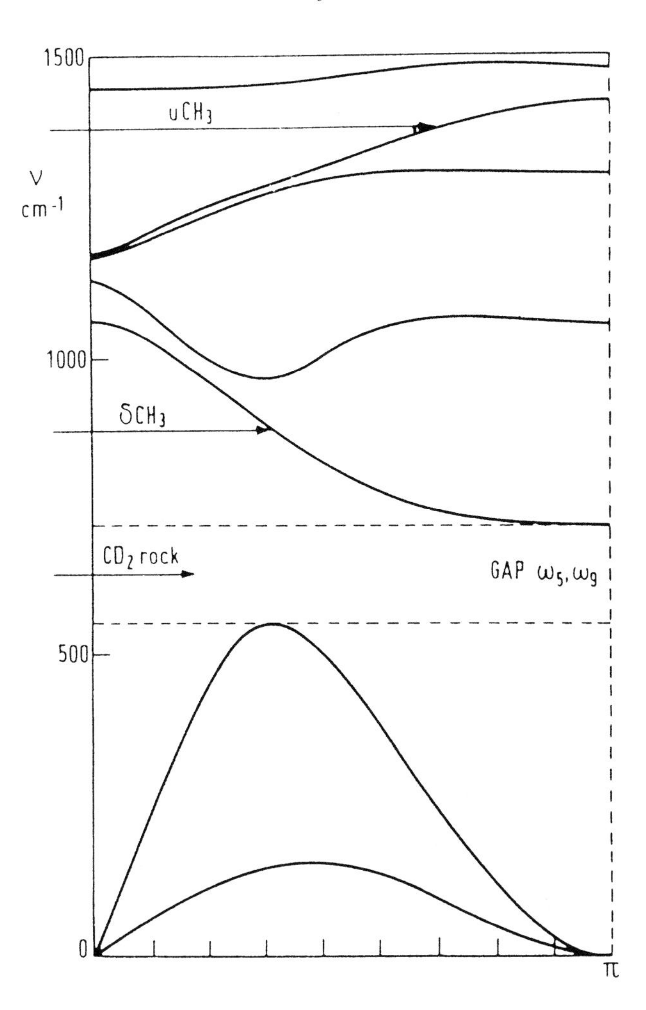

Figure 3-15. Example of a dynamical situation of a polymehylene chain $-(CH_2)_n-$ which contains defects consisting of CD_2 groups and of CH_3 units at either ends. The motion of CD_2 rocking occurs in a gap and does not couple with the host lattice, thus generating localized modes; on the contrary the frequencies of both external deformation and umbrella motions of the CH_3 group occur in the frequency range spanned by the dispersion curves of the host lattice; coupling takes place and resonance modes are generated.

embedded in the polymer chain is determined by the masses and the geometrical arrangement of its vibrating atoms (**G** matrix, Eq. (3-15)) and by the new intramolecular forces, i.e., by the new force constants (type, strength, and distance of intra-unit and inter-unit coupling (**F** matrix, Eqs. (3-2), (3-3) and (3-16)).

Calculations carried out so far on the simplest prototypical classes of polymer describe substantially three types of phenomena (Figure 3-15):

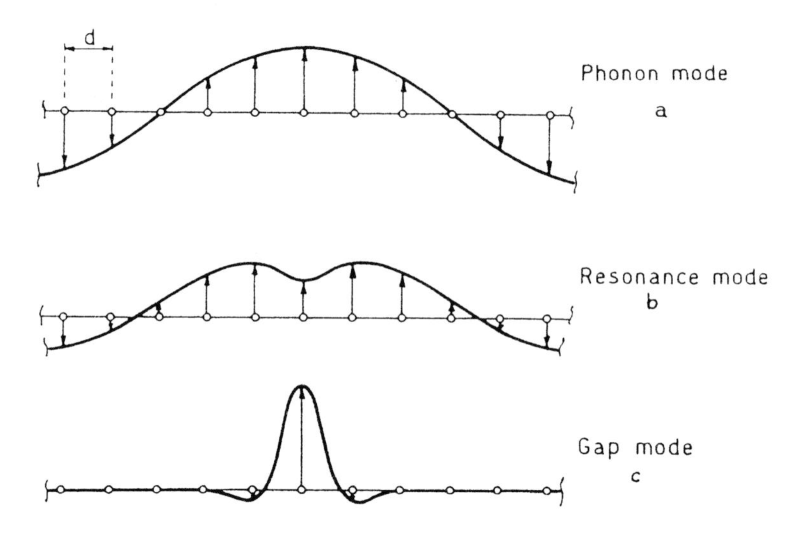

Figure 3-16. Dynamics of a one-dimensional lattice containing one defect. Vibrational displacements of each unit are rotated of 90° for a better display of the normal modes. (a) unperturbed phonon of the host lattice; (b) resonance mode; and (c) gap/local mode (from [66]).

1. Some of the normal modes of the defect strongly couple with the normal modes (phonons) of the polymer and generate collective motions which lose all the characteristic of the motion of the isolated unit. These normal modes are generally labeled as '*resonance modes*' (Figure 3-16) because their frequencies are close enough to those of the host lattice that they generate mechanical resonance with some of the normal modes of the polymer. This is generally found when the frequency of the separated unit happens to occur within the range of frequencies spanned by one of the dispersion curves of the perfect host polymer 1D lattice.
2. When the frequency of the separated unit differs strongly from the frequencies spanned by the dispersion branches of the polymer chain, the motion of the unit cannot couple with the motions of the host lattice, it cannot be propagated and generates a mode strongly localized in space (at the site of the defect within the polymer) and in energy (at a characteristic specific frequency). These modes are generally labeled as '*gap modes*' or '*out-of-band modes*', since the frequency of the isolated units occurs either in a gap between dispersion branches or above the highest frequency branch respectively. Out-of-band modes in organic polymers can only be observed in the case of perdeuterated polymers which contain H atoms as defects. In this case, C–H stretching modes occur at frequencies higher than any other normal modes of the polymer.
3. Compared with the results from the dynamics of simple impure 3D lattices, in the case of 1D lattices (polymers) localized modes (or quasi-localized modes) can sometimes also be generated if the frequency of the defect unit occurs within the frequencies spanned by a dispersion branch of the perfect host lattice. As the

polymer chain is conformationally flexible it may well occur that, for a certain geometrical arrangement of the atoms, the motion of the defect unit happens to be orthogonal (or quasi-orthogonal) to the displacements of the atoms during a certain vibration of the host lattice. It has been shown [99] that in an n-alkane chain the CH_2 wagging of the central unit in a conformational sequence such as –(G)–CH_2–(G)– generates a pseudo-localized CH_2 wagging motion almost fully decoupled from the vibrations of the host lattice, even if the frequency occurs in the frequency range spanned by the CH_2 wagging dispersion branch [99].

The observation of characteristic gap modes in disordered polymers becomes then the basis for the use of the vibrational spectra as useful probes for the structural characterization of these systems.

It should be remembered that the above discussion has been presented by considering only one single defect embedded in a host polymer chain. When a distribution of defects is considered, defects may occur close enough to couple mechanically. Splitting and changes of the gap modes (frequencies and intensities) may occur and are indicated by the calculations (see, for instance [100]).

3.11 A Very Simple Case: Lattice Dynamics of HCl–DCl Mixed Crystals

It is considered that the simplest possible realistic case on which our theories can be tested, and which could act as a guide to the reader for the discussions which follow, are the studies of mixed crystals of HCl and DCl [97]. Solid HCl and DCl undergo a first-order phase transition at 98.4 °K, and 105.0 °K, respectively. Below the transition temperature, X-ray and neutron diffraction studies show the existence of an ordered face-centered orthorhombic structure which contains planar zig-zag chains consisting of HCl (DCl) molecules linked to each other by hydrogen bonds (for a list of references on the physical properties of solid HCl, see [97]). Our model of the chain consists of a pure infinite isolated chain of the type ···H–Cl···H–Cl···H–Cl··· (Figure 3-17). Using the techniques presented in the previous sections we have

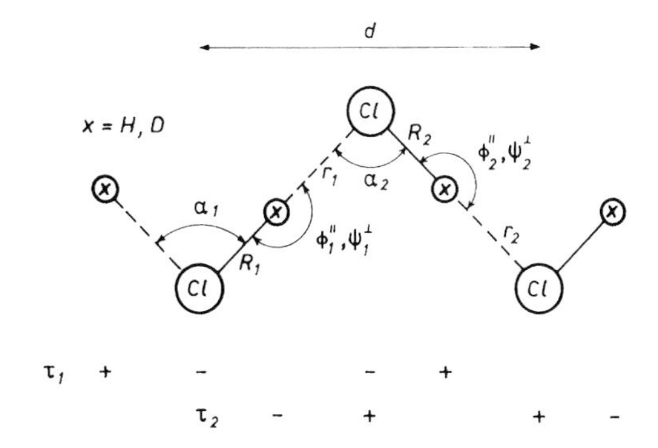

Figure 3-17. Geometry and internal coordinates (in-plane and out-of-plane) for crystalline HCl or DCl in the orthorhombic modification. Only one single chain has been considered.

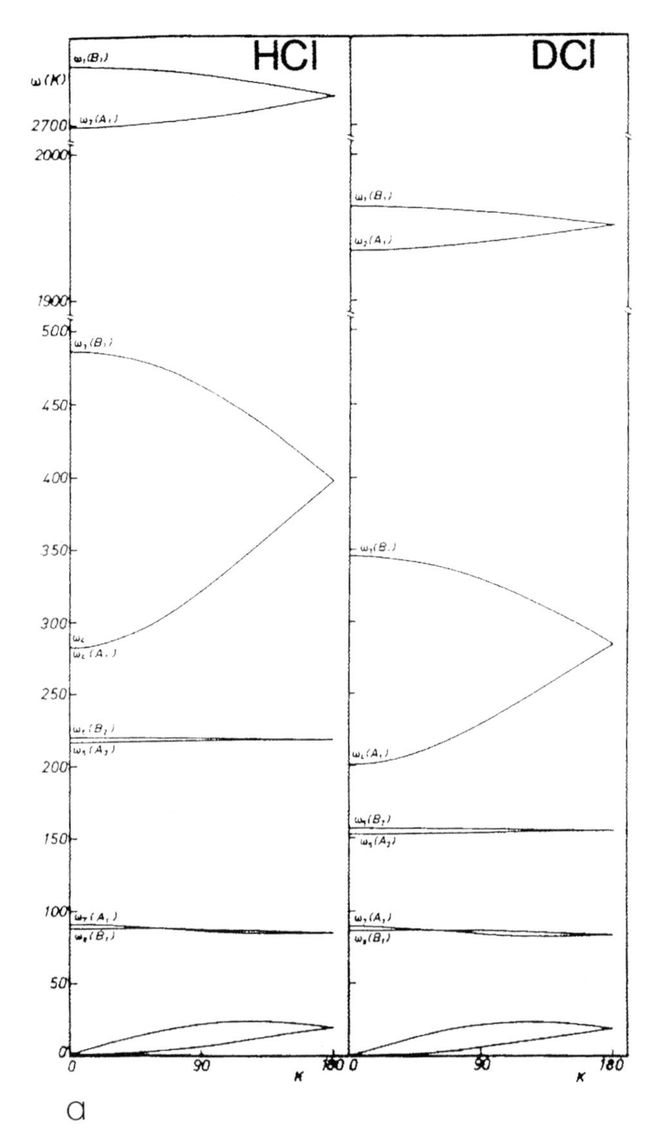

Figure 3-18. Dynamical data for the perfect chain of HCl and DCl based on a valence force field. (a) Dispersion curves; (b) $\mathbf{k} = 0$ normal modes; (c) densities of vibrational states.

calculated dispersion curves (Figure 3-18a), density of vibrational states $g(\nu)$ (Figure 3-18b) and vibrational displacements for $\mathbf{k} = 0$ phonons both for pure HCl and DCl (Figure. 3-18c). Vibrational infrared intensities have also been calculated based on a simple fixed point-charge model [97].

As a second step, we generated models of chains of HCl containing defects of DCl. Calculations are made easier because isotopic substitution does not change the force field. Several mixtures were considered and we will focus here only on three cases of mixed chains with 3%, 25% and 45% DCl in HCl. Chains were made up by 600 chemical units.

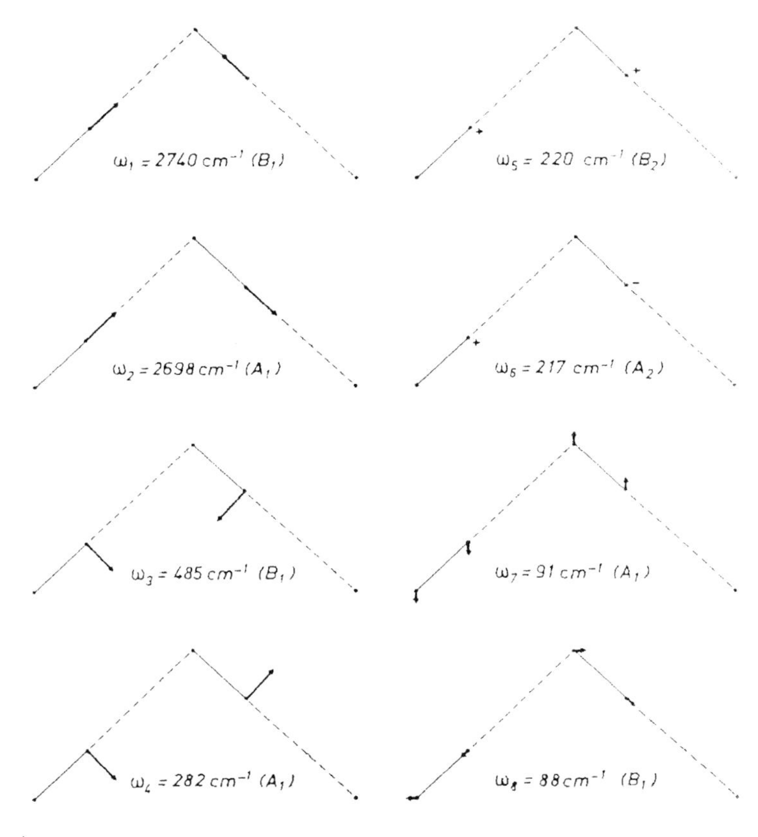

Figure 3-18 (b)

Suitable experimental procedures were envisaged in order to obtain such materials and in order to record the infrared spectra to be compared with the calculated ones.

Figure 3-18a immediately indicates that the HCl stretching of molecules isolated in a DCl lattice can generate localized out-of-band modes, while DCl stretchings in a HCl host lattice generate typical localized gap-modes. This can occur for other lower-frequency modes for both molecules. The ω_3–ω_4 branches span frequency ranges with partial overlapping between HCl and DCl lattices, thus possibly generating resonance modes. Also, the vibrations of ω_7–ω_8 may couple. No coupling can occur between the out-of plane phonons of HCl (ω_5–ω_6) and the in-plane phonons of DCl (ω_3–ω_4) since they are orthogonal.

The histograms calculated for the three isotopically impure chains are given in Figure 3-19 and show a complicated and almost inextricable pattern. Some details are enlarged in Figure 3-20.

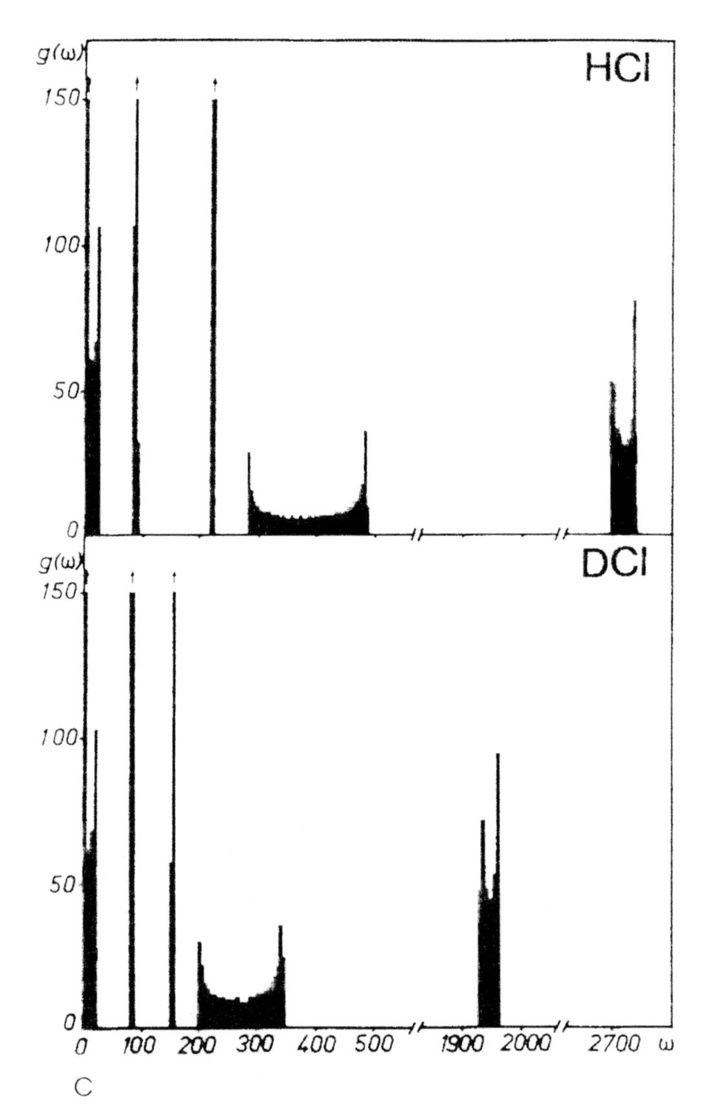

Figure 3-18 (c)

The object of the next step is to apply IIM to identify the origin of the various peaks calculated in the $g(\nu)$ of the mixtures. First, we consider the random distribution of 25% DCl defects in a host lattice of HCl (Figure 3-21); the total number of units in this calculation is 100. Let Z be the number of chemical units within the defect. Moreover, for sake of completeness, let us also associate qualitatively to each motion its corresponding dipole moment changes. Let us assume that each bar in Figure 3-21 gives not only the vibrational amplitude, but also the dipole moment

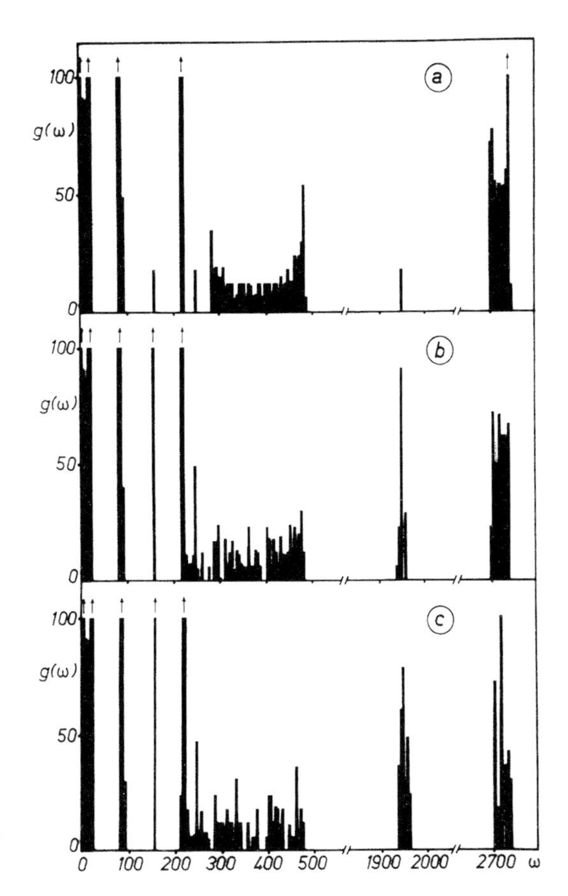

Figure 3-19. Density of vibrational states for a single mixed chain of (a) 3%, (b) 25%, and (c) 45% DCl in HCl (chains of 600 chemical units).

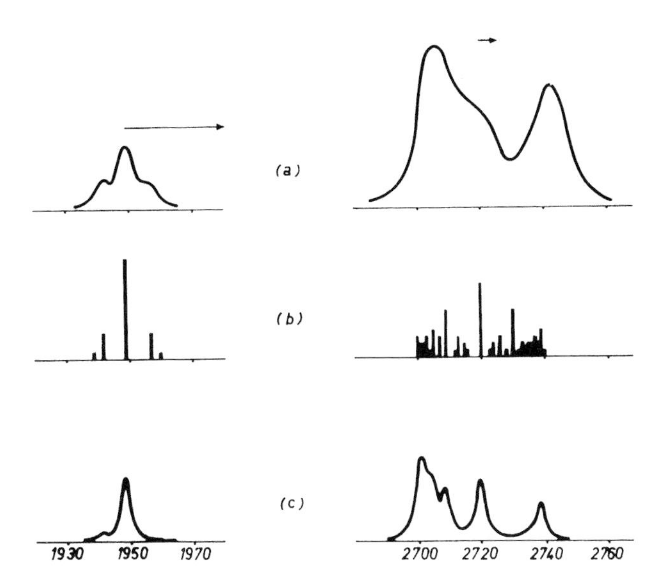

Figure 3-20. (a) Experimental infrared spectrum in the HCl and DCl stretching regions of a mixed HCl/DCl crystal containing 25% of DCl; (b) g(ν) and (c) dipole-weighted g(ν) assuming a Lorentzian band shape with $\Delta\nu_{1/2} = 2$ cm^{-1}. The observed spectra have been shifted as indicated for sake of qualitative comparison.

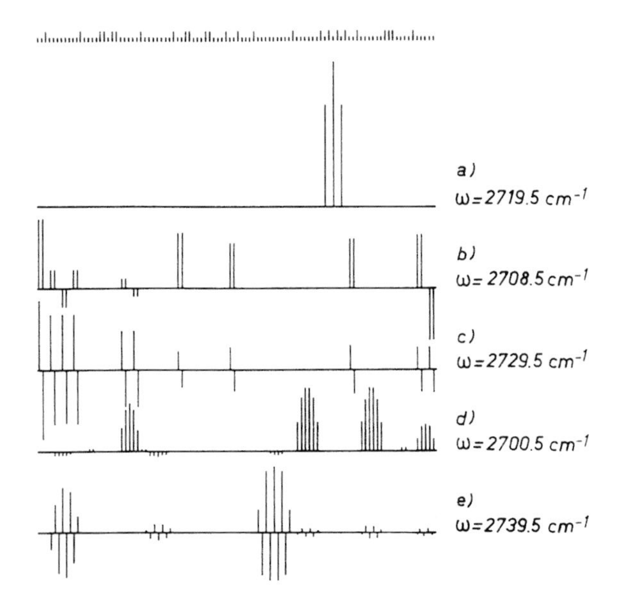

Figure 3-21. Sample eigenvectors in the HCl stretching region for a single chain of mixed 25% DCl in HCl. The top line gives the distribution of defects. Vibrational displacements are rotated of 90° for an easier display of the vibrational motions.

changes. The vectorial sum of the dipole moment changes is related to the absorption intensities. We can then qualitatively predict also the intensity of each of the calculated modes. The distribution of defects given at the top of the figure indicates the existence in the chain of 14 DCl isolated units ($Z = 1$), +4 doublets ($Z = 8$) +1 triplet ($Z = 3$). Application of NET, IIM, and intensity coinsiderations indicate that at $\nu = 1948.76\ \text{cm}^{-1}$ only the single DCl molecules as isolated entities perform fully localized vibrations at the site where they are located with strong intensity. At $\nu = 1948.5$ and $1948.68\ \text{cm}^{-1}$, the four single units separated only by one HCl spacer feel some intramolecular coupling and generate localized modes with some phase coupling, the former has strong intensity and the latter zero intensity. Next, at $\nu = 1941.5\ \text{cm}^{-1}$ only isolated doublets generate an in-phase motion (strong) of the two molecules. The out-of-phase motion occurs at $\nu = 1956.5\ \text{cm}^{-1}$ (inactive). Finally, the motions characteristic of the triplet occur at $\nu = 1938.5$ (strong) and $1959.9\ \text{cm}^{-1}$ (weak). It is also apparent that in this frequency range the atoms which belong to the host lattice do not move (i.e., the motion of the DCl defect is not propagated throughout the chain of the host HCl lattice). The concept we wish to stress here is that the cases just examined are typical examples of modes localized in space and energy characteristic of particular 'defects' or of 'islands' or 'clusters' of defects. Moving to the HCl stretching range (near $2700\ \text{cm}^{-1}$) the analysis of the composition of the chain from localized modes can be completed. For the same chain considered in Figure 3-22 the distribution of HCl sequences is the following: 4 ($Z = 1$), 4 ($Z = 2$), 3 ($Z = 3$), 2 ($Z = 4$), 2 ($Z = 5$), 2 ($Z = 6$), 1 ($Z = 7$), 1 ($Z = 8$) and 1 ($Z = 9$). From Figure 3-22, the 'island analysis' carried out using the IIM method allows the following frequencies characteristic of each island to be located

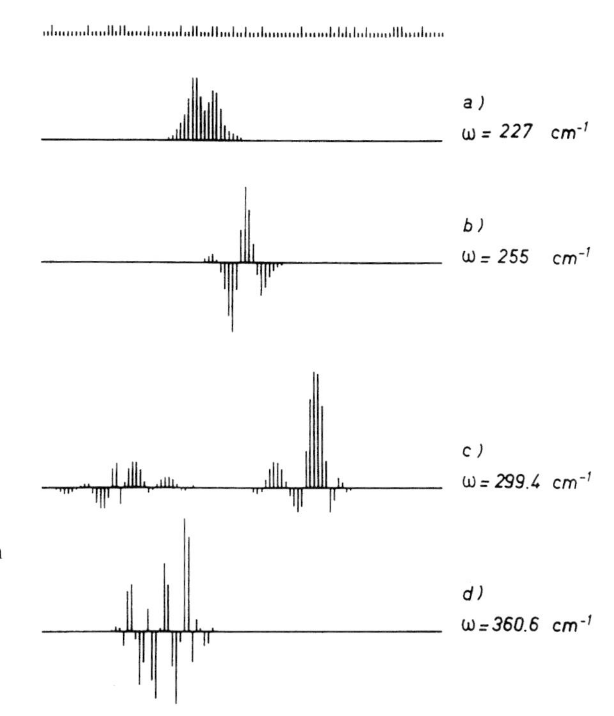

Figure 3-22. Sample eigenvectors in the in-plane bending region for a single chain of mixed 25% DCl in HCl. The distribution of defects is the same as that given in Figure 3-21.

unquestionably: Z = 1, 2719.5 cm^{-1} (strong); Z = 2, 2708.5 cm^{-1} (weak or inactive) and 2729.5 cm^{-1} (inactive); Z = 5–6, 2700.5 cm^{-1} (strong) and Z = 8–9, 2739.5 cm^{-1} (inactive).

An example of resonance modes which cannot provide any signal characteristic of a given defect is given in Figure 3-23 for the same chain as that considered in Figure 3-21 and 3-22. In this case the bending modes are considered where (see above) frequency branches overlap. Because of intramolecular coupling, collective vibrational waves extending along large domains of the chain are generated and do not provide any useful and characteristic signal which may contribute with a weak and broad background.

It is realized that the situation found for impure simple 3D lattices is found also for the single chain considered here and very likely may be found for other polymers containing an increasing concentration of defect. In the HCl (or DCl) stretching region these mixed crystals behave as 'two-mode system' or 'persistence-type' system. 'Amalgamation-type' behavior is observed in the bending range where sections of density of states overlap both for HCl and DCl [98].

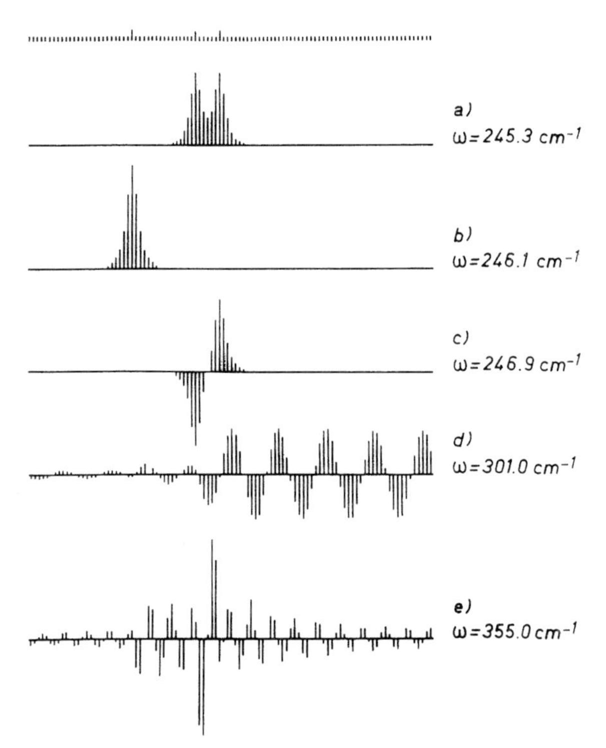

Figure 3-23. Examples of resonance modes in the bending region for a single chain of mixed 25% DCl in HCl.

3.12 *CIS–trans* Opening of the Double Bond in the Polymerysation of Ethylene

The method just outlined was used, at the earlier stages of these kinds of study [101], for the elucidation of the mechanism of polymerization which was, at that time, an unsolved puzzle. The chemical problem can be stated briefly in the following way: when ethylene is polymerized using Ziegler–Natta catalysts a linear polymer is formed. It was known that the catalysts acts on the double bond of the ethylene unit and the polymethylene chain can grow on both sides of the ethylene unit once the double bond has been opened. The opening, however, can occur in *cis* or *trans* configuration.

H H H ↑ H
C–C C–C
H ↓ ↓ H H ↓ H

cis *trans*

The question which remained to be answered for more than a decade was whether the opening of the double bond occurs in *cis* or *trans*. The understanding of this reaction mechanism was of fundamental importance for the development of other similar reactions for other classes of materials.

One way to solve the problem was to polymerize deuterated derivatives of ethylene with the same Ziegler–Natta catalysts. If *cis* di-deutero ethylene (monomer A) is polymerized and the opening of the double bond is *cis*, the resulting polymer should be rich in the so-called '*erithro*' units

```
H H        H D D H
C–C   →   C–C–C–C
D D        D H H D

 a          1  2
```

If the opening is *cis* for the case of monomer B (*trans*-dideuteropolyethylene) the reaction should yield a polymer rich in the so-called '*threo*' units.

```
D H        H H D D
C–C   →   C–C–C–C
H D        D D H H

 b          1  2
```

The opposite should occur if the catalyst forces the double bond to open in *trans* configuration.

The polymers were made by chemists, but the identification of the configuration of the deuterium and hydrogen atoms remained unsettled for a long time. The vibrational spectra were recorded, but their interpretation was impossible on the basis of the classical techniques of polymer spectroscopy.

An example of the procedure followed in the identification of the spectra is the following.

Let the configurations of monomeric units be indicated as follows

```
H D        D H        H H        D D
C–C–  →  –C–C–  →  –C–C–  →  –C–C–  →
D H        H D        D D        H H

 I          II         III        IV
```

From the viewpoint of calculations a computer program which generates a sequence of random numbers was used in order to generate a random sequence g(ν)

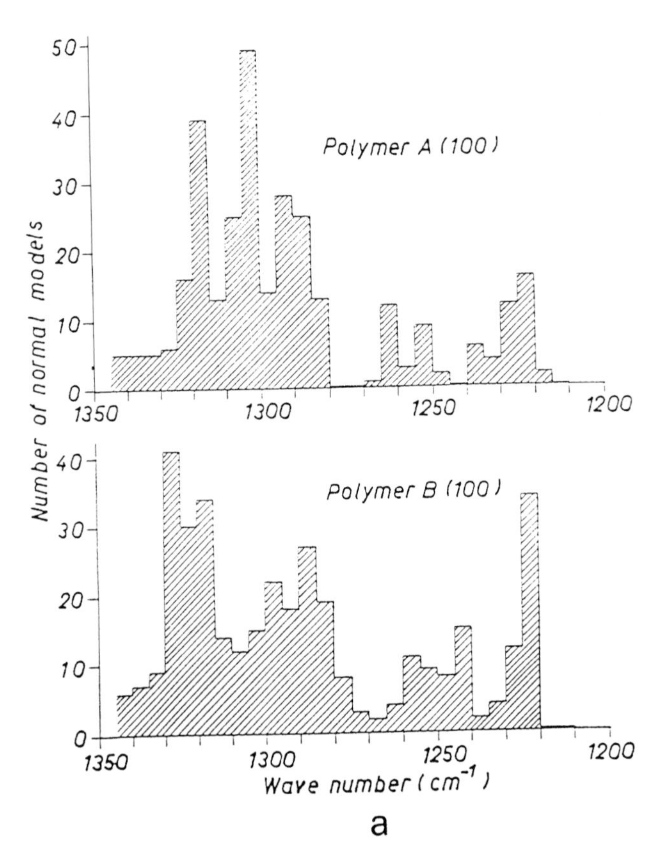

Figure 3-24. (a) Calculated density of vibrational states of polymers A and B with 100 monomeric unit. (b) Observed infrared spectra of poly(*cis*-CHD=CHD) and poly(*trans*-CHD=CHD).

for both chains with 50/50 concentration of a and b. The sequence is the following:

bbaabbaabbbbaabaaabababa
bbbbaabaaabbbabbbbaaabba
aaababababbbaababababbbaabaa
ababbabbbabbaaababaaaabbb

The two cases considered here are: (i) polymer A, generated by replacing a with unit I and b with unit II; and (ii) polymer B, obtained by replacing a by unit III and b by unit IV.

Each chain contains 600 atoms. The corresponding $\mathbf{G}_R$ and $\mathbf{F}_R$ matrices (of dimension 1800×1800) were constructed and NET applied for the calculation of $g(\nu)$. The calculated $g(\nu)$ (Figure 3-24a) compared with the experimental infrared spectra (Figure 3-24b) clearly shows that the calculated $g(\nu)(A)$ corresponds to

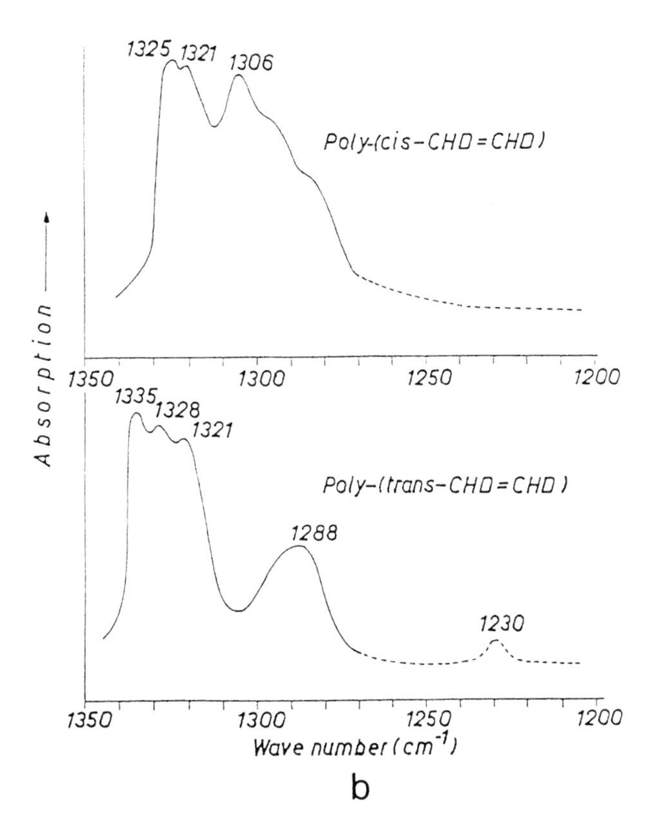

Figure 3-24 (*continued*)

the infrared spectrum of poly(*cis*-CHD=CHD) and g(ν)(B) corresponds to the experimental spectrum of poly(*trans*-CHD=CHD).

It can then be concluded that the polymerization by Ziegler–Natta catalysts occurs with the *cis* opening of the double bond.

3.13 Defect Modes as Structural Probes in Polymethylene Chains

The strongest effort by our group and by other authors was aimed at revealing the structure and molecular dynamical evolution of the very many classes of technologically and scientifically relevant molecules consisting of n-alkanes or n-alkyl long chains. We report below some of the results obtained and will show some applications.

3.13.1 Mass Defects

Let us introduce a CD_2 group as a defect in n-alkanes (CH_3–$(CH_2)_nCH_3$) or n-alkyl chain ($CH_3(CH_2)_xX$ or in any polymethylene chain. This is a typical case of 'mass defect' which leaves the force constant matrix (Eq. 3-3)) unchanged because of Born–Oppenheimer approximation. The site where the CD_2 group is located can be changed along the chain. Chemically, this can be easily achieved and selectively deuterated n-alkane or alkyl derivatives are either commercially available or can be (and have been) easily synthesized.

Calculations have shown that the frequency of the rocking motion of the unit CD_2 embedded in a CH_2 host lattice changes by changing the conformation of its surroundings [102, 103]. The calculated and observed frequencies for the CD_2 rocking motion when the adjacent torsional angles are changed are reported in Table 3-1. These motions are classical well-localized gap-modes as they occur in the frequency gap between the CH_2 rocking and the CCC bending dispersion branches of the host lattice.

Table 3-1. Calculated and experimental CD_2 'gap frequencies' for selectively deuterated nonadecane as function of the torsional angle about C–C bonds.

	ν (cm^{-1})	
	Calculated	Observed
4 3 2 1		
–CH_2–CH_2–CH_2–CD_2–CH_3		
T T	615	622
G T	615	622
T G	658	658
G G	654	
13 12 11 10 9 8 7		
CH_2–CH_2–CH_2–CD_2–CH_2–CH_2–CH_2		
T (T T) T		
G (T T) T	~615	622
G (T T) G		
G′ (T T) G		
T (T G) T		
G (T G) T	~650	650
G (T G) G		
G′ (T G) T		
T (G G) T		
T (G G) G	~677	
G (G G) G		

3.13.2 Conformational Defects

The study of the conformational defects in polymethylene chains has been tackled some time ago in a systematic way by the school of Snyder by normal mode analysis of short-chain molecules (see, for example [104]) and by our group at Milano using NET [95, 101]. The results by both techniques are very similar. In the previous discussion in Figure 3-14 we have pointed out that a few peaks in the calculated $g(\nu)$ for infinite *trans* polyethylene do not find corresponding infrared and/or Raman lines to be assigned to $\mathbf{k} = 0$ motions. These additional experimental peaks are just an evidence of the existence of extra structures which we can identify as conformational defects using the theory we are presently discussing.

The main characteristic defect modes calculated (and observed) for *trans* planar segments are the following:

1. *GG defect.* The wagging vibration of a CH_2 group located between two adjacent *gauche* conformations $-CH_2(G)-CH_2(G)-CH_2-$ wagging. The calculated (and observed) frequency lies near 1355 cm^{-1}.
2. *GTG′ defect.* CH_2 wagging. $-CH_2(G)-CH_2(T)-CH_2(G')-$. This defect introduces two wagging modes because the two CH_2 groups are joined by a bond in *trans* conformation and are partly decoupled from the host lattice by the *gauche* bonds. The intra-defect coupling generates out-of-phase and in-phase modes calculated (and observed) near 1375 and 1306 cm^{-1}.
3. *GGTGG defect.* This defect is the one which possibly exists on the surface of the single crystal lamellae of polyeythylene if the chain re-entry forms a tight fold along the [200] crystallographic plane [105]. This defect generates practically the three lines expected for GG and GTG′ defect in the CH_2 wagging region and an additional *quasi* out-of band mode (near 714 cm^{-1}) just outside the $\mathbf{k} = 0$ limiting rocking mode of the CH_2 group [99].
4. *Head-to-head defects of the type: $-RCH-CH_2-CH_2-RCH-$.* The CH_2 rocking mode of a sequence of two CH_2 groups between two heavy boundaries is calculated and observed near 850 cm^{-1} in agreement with empirical correlations derived from the study of ethylene–propylene copolymers [106].
5. *Chain end modes in n-alkyl chains.* When the group $-CH_2-CH_2-CH_2-CH_3$ in such chains has TTT conformation, the methyl 'umbrella deformation' (U) is calculated and observed near 1375 cm^{-1} and the external deformation mode δ CH_3 occurs near 890 cm^{-1}. According to the previous discussion in this chapter both motions should be considered end-group modes whose frequency should remain constant when the length of the chain increases. This is indeed the case if the chains keep the all-*trans* structure. However, since both motions occur in a frequency range spanned by the dispersion curves of the 1D lattice they couple with the vibrations of the bulk and their coupling extends for a few CH_2 units along the chain. If the conformation of the chains near the ends changes, such intramolecular coupling changes and the frequencies of the 'end groups' may change. Table 3-2 reports the values of the calculated and observed frequencies when chain ends are conformationally distorted in the TG and GT conformations.

Table 3-2. Characteristic calculated and experimental 'in band' frequencies for the conformers $-CH_2^{(T)}CH_2^{(G)}CH_2-CH_3$ and $-CH_2^{(G)}CH_2^{(T)}CH_2-CH_3$ at the end of the n-nonadecane molecule.

	ν (cm^{-1})	
	Calculated	Observed
TG	1348	1342 (IR)
	879	866 (R), 866 (IR)
	769	766 (IR)
GT	1367	
	851	844 (R), 845 (IR)

3.14 Case Studies

3.14.1 Conformational Mapping of Fatty Acids Through Mass Defects

It becomes apparent that the use of selective deuteration becomes a powerful tool for the mapping, site by site, of the conformation of polymethylene chains. This has been applied for the conformational mapping of n-alkane and of fatty acids. The idea was first proposed by Snyder and Poor for n-alkanes [102] and was later applied to the case of fatty acids [107] and has settled the issue on the conformation in the solid state which has long been matter of great debate [108].

Fatty acids are typical of a class of compounds for which the existence of pre-melting phenomena has been claimed with the formation of 'liquid-like' structures for the molecules while still in the solid. The structure of these systems has been studied by many workers. The molecules in the solid are dimeric since pairs of carboxyl groups face each other and are held together by hydrogen bonds. Moreover, complex polymorphic phenomena have been revealed depending on the way crystallization is carried out (from the melt, from solution, from the kind of solvent, etc.) [107, 109]. More recent works based on the study of LAM modes in the Raman suggested that the molecules of fatty acids, independent of the crystalline form, are never all-*trans* planar and contain somewhere along the chain a *gauche* distortion [108]. These information were at odds with our studies on LAM modes of the same materials which were shown to be fully *trans*-planar [110, 111].

The use of defect gap modes using selectively deuterated materials have definitely solved the problem. We have recorded the infrared spectra of solid perhydrogenated stearic and palmitic acids and of a few selectively deuterated derivatives. For palmitic acid, single CD_2 group were placed in position 2, 3, or 4 (carbon atom 1 is that of the carboxyl group); for stearic acid, the single CD_2 group was placed in position 2, 3, 4, 5, 6, 7, or 13. As an example of the way we proceeded, Figure 3-25 illustrates the infrared spectra of: (i) normal undeuterated palmitic acid (Figure

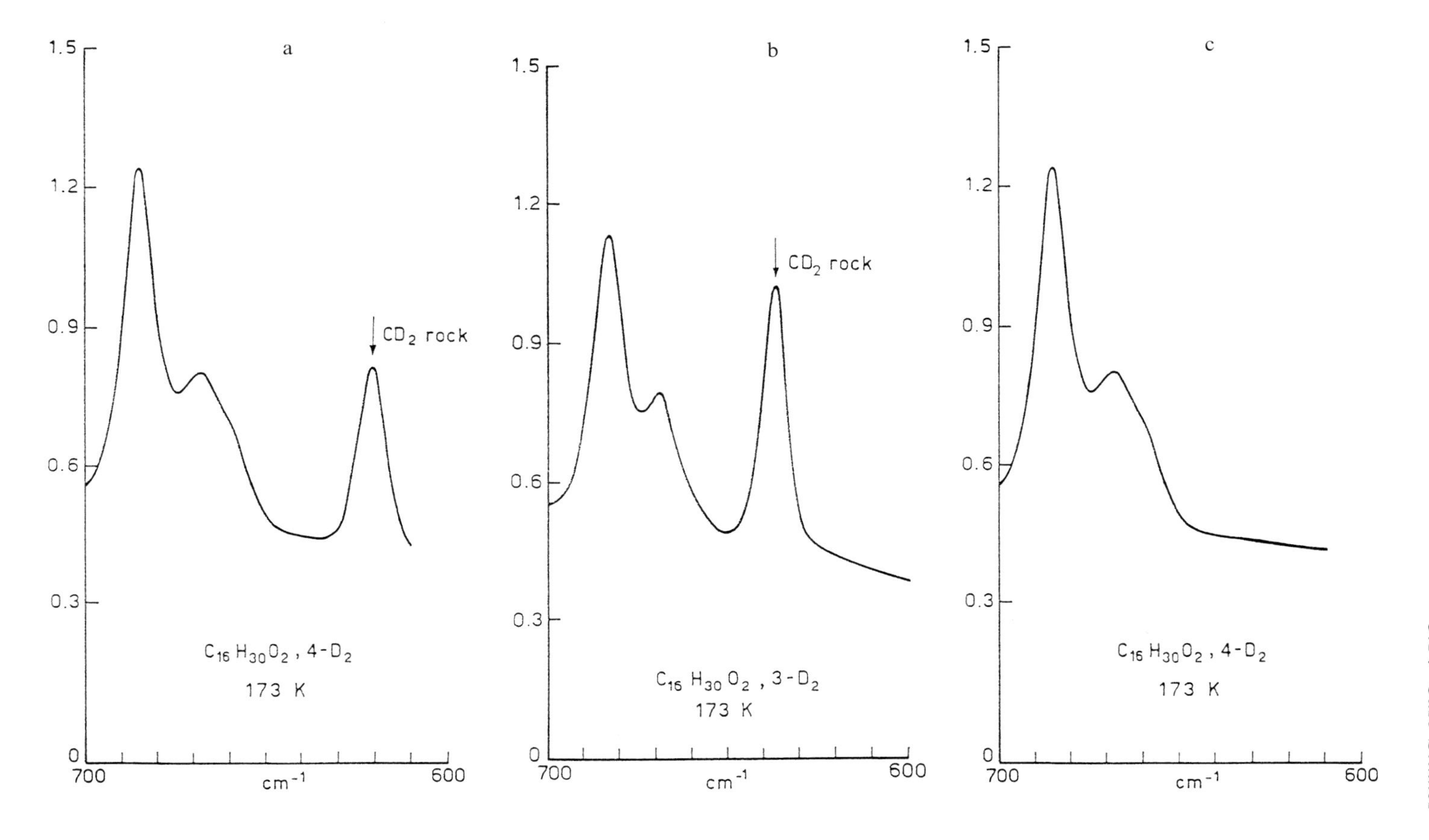

Figure 3-25. Spectroscopic evidence of *trans*-planarity of palmitic acid from the gap modes of selectively deuterated derivatives. (a) Palmitic acid, (b) palmitic acid 3-d_2 and (c) palmitic acid 4–d_2.

3-25a); (ii) palmitic acid 3-D_2 (Figure 3-25b); and (iii) palmitic acid 4-D_2 (Figure 3-25c). The clear band which appears near 620 cm^{-1} is the gap-mode indicating unquestionably the existence of the local conformation CH_2(T)–CD_2(T)–CH_2 both in positions 3 and 4. Identical spectra have been observed along the whole chain, thus confirming the whole *trans* structure of palmitic acid in the solid state [107, 110, 111].

3.14.2 Molecular Mobility and Phase Transitions in Thermotropic Liquid Crystals from the Spectroscopy of Defect Modes

The issue is the description of the molecular changes accompanying the phase transitions in systems which show liquid crystalline phases. While much is known on the morphological changes, little is known on what happens to the molecular structure through the various phase transitions.

We use defect mode spectroscopy for aiming at the structural changes of prototypical thermotropic liquid crystals such as alkylcyanobiphenyl and polyesters.

3.14.2.1 Case 1: Cyano-alkyl biphenyls

Let us first consider dodecyl-cyanobyphenyl which shows two phase transitions at 48 °C (crystalline–semectic A) and at 58.5 °C (smectic A–isotropic). The analysis of the whole spectrum is reported in the original work [112] and we focus here only on the variation with temperature of the conformational structure of the dodecyl side chain which we hope to reveal with the study of the temperature-dependent vibrational spectrum in the 1420–1280 cm^{-1} range where defect modes are expected to occur.

Figure 3-26 reports such spectra, together with the difference spectra obtained in the following way. Let $s(j+1)$ be the spectrum recorded at temperature $T(j+1)$ and $s(j)$ the spectrum taken at temperature $T(j)$. Then $\Delta T = T(j+1) - T(j)$ is the temperature gradient in our experiment. From Figure 3-26, while drastic changes are observed from the crystalline to smectic A phase, the spectra do not change much from smectic A to isotropic. The difference spectra highlight such changes and allow the defect modes GTG′, GTG, GG, and end-TG to be identified. The following conclusion can be easily reached: (i) the *trans*-planar chain exists up to the crystal–smectic A transition; (ii) at the crystal–smectic A transition the *trans*-planar chain collapses in a conformationally distorted geometry, as indicated by the appearance of all defect modes listed above; (iii) the observed upward shift of the umbrella deformation mode of the CH_3 group from 1376.5 to 1378 cm^{-1} indicates that the environment at the end of the chain is changed; and (iv) no great conformational changes occur from the smectic A to the isotropic phase; the cyano-biphenyl core must then supervise this phase transition (as also shown by the vibrational spectrum). Another case treated is that of 4-cyano, 4′octyloxy biphenyl [113].

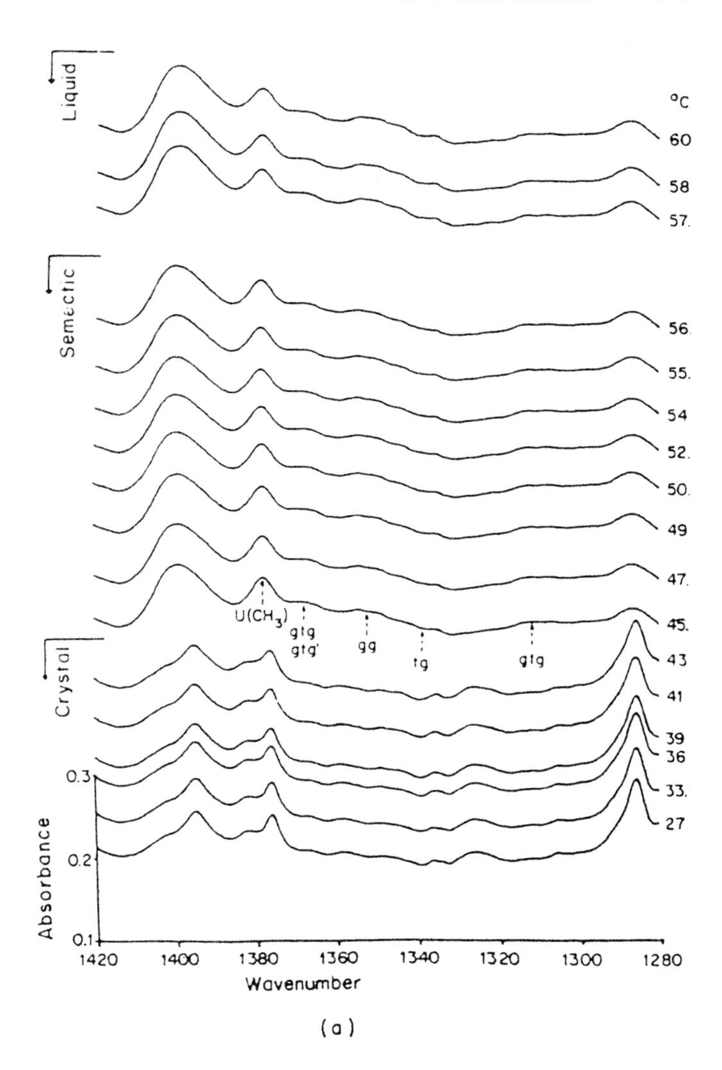

Figure 3-26. Dodecylcyanobiphenyl. (a) Temperature-dependent infrared spectra in the defect mode frequency range. (b) Experimental spectroscopic evidence from difference spectroscopy that at the phase transition crystal → smectic the conformation of the dodecyl residue changes from all-*trans* to 'liquid-like' (the most common defect modes appear clearly). From smectic to liquid phases no drastic conformational changes are observed (see text).

3.14.2.2 Case 2: Liquid Crystalline Polymers: Polyesters

We consider the case of the polymer $-[(CH_2)_{10}-OOC-C_6H_4-OOC-C_6H_4-COO-C_6H_4-COO]_n-$ (hereafter referred as HTH-10) whose phase transitions are the following: K–S 221 °C; S–I 260 °C. From the overall analysis of the temperature-

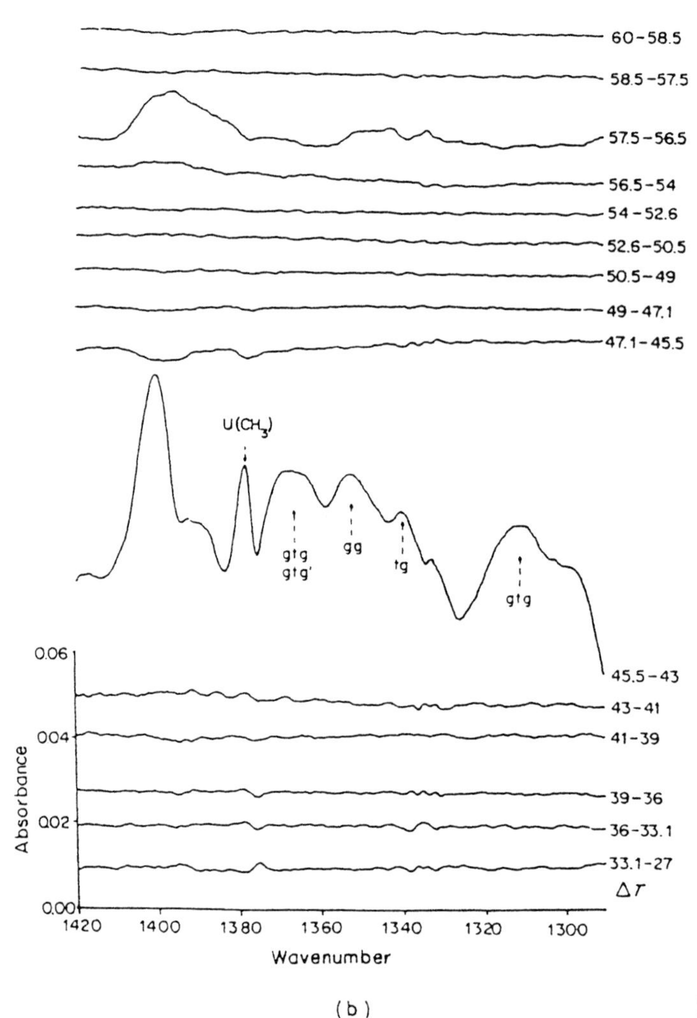

Figure 3-26. (*continued*)

dependent Raman and infrared spectra [114] it could be concluded that, in the K phase, most of the decamethylene segments are in the *trans*-planar conformation and that upon heating *gauche* rotamers are generated. The identification of what kind of conformational defects occur in the disordering process when the temperature increases can only be attempted by the study of the temperature-dependent spectrum in the traditional CH_2 wagging range. We restrict here our analysis to the temperature-dependent spectra in the usual defect modes range from ~ 1400 to ~ 1300 cm^{-1}. We aim at revealing the conformational evolution through phase transition of the segments consisting of 10 CH_2 groups. Figure 3-27 illustrates the temperature-dependent infrared spectrum in the CH_2 wagging range plotted in

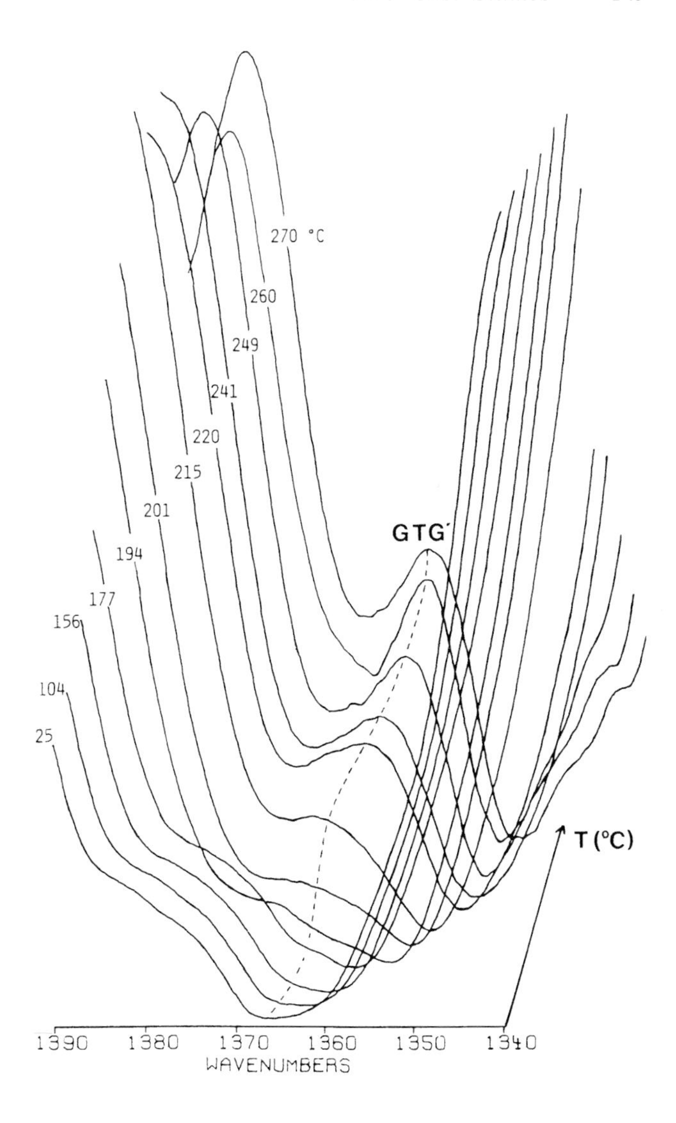

Figure 3-27. Liquid crystalline polyester with decamethylene spacers. Temperature-dependent infrared spectrum in the defect modes frequency range (see text). The spectra and shifted one from the other as in a tridimensional plot.

three dimensions for the sake of clarity. It becomes apparent that the *trans*-planar structure (no defect modes in this frequency range) evolves with increasing temperature with the generation of only the defect mode near 1365 cm^{-1} due to GTG′ defects. No absorption due to GG defects (band at 1353 cm^{-1}) is observed, i.e., sharp kinks are neither generated in the smectic nor in the isotropic phases. The intensity of the GTG′ defect modes increases sharply at the K–S transition, remains on a smooth *plateau* until the S–I transition occurs, and increases steeply in the I phase. It has to be noted that even in the I phase the GG defect mode is still practically not observable.

The mechanism on a molecular level which is derived from the vibrational spectra can be summarized as follows: at room temperature the sample contains a sizable concentration of ordered material in which the decamethylene spacers are predominantly in the *trans*-planar structure. When the K–S transition is approached the main spectroscopic signal observed is that associated to the GTG′ modes. It is known that the GTG′ defect introduces a kink in the chain but does not change the trajectory of the polymethylene sequence keeping both arms at either side of the defect parallel. The overall effect is that the length of the decamethylene segment simply shrinks. It follows that in the S phase the spacers shrink because of the introduction of conformational kinks, inducing a sort of shearing motion between the large aromatic mesogenic groups. The lack of any sizable concentration of GG structures even in the I phase leads to the interesting conclusion that the molecular structure in the I phase is not much different from that in the S phase. No indication is found of 'liquid-like structures' observed in other polymethylene systems in the liquid phase. It has been proposed [114] that one way to reconcile the observation of macroscopically isotropic phase with the observation on the molecular level is to envisage that the isotropic phase consists of a collection of microdomains whose directors are randomly oriented. However, inside the microdomains polymer molecules are organized as they are in the S phase.

3.14.2.3 Case 3: Chain Folding in Polyethylene Single Crystals

It is known that chain folding occurs in polyethylene crystals; the fold structure, however, has been a subject of interest and controversy [105]. It was obvious for polymer spectroscopists to use vibrational infrared and/or Raman spectra to try to contribute to the understanding of the shape of the fold (i.e., the molecular conformation) within the fold at the surface of polyethylene single crystals.

The large amount of spectroscopic work carried out by many authors is sumarized in [115]. Here, we wish to focus specifically on the contribution given by defect mode spectroscopy considering both calculations and experiments. The issue is whether the surface of polyethylene single crystals consists mostly of loosely looped folds or also of a fraction of chains with a tight adjacent re-entry. In the former case, 'liquid like' defects such as GG, GTG, GTG′ etc. are expected and can be recognized in the vibrational spectrum. If tight fold re-entry exists defects of the type GGTGG must occur at the surface of the lamellar single crystal.

The first question which needed to be answered in a less qualitative way was where to locate the vibrations localized on the GGTGG defect.

Calculations have been carried out on a long polymethylene all-*trans* host molecule containing evenly spaced GGTGG defects. NET and IIM and the 'island analysis' were applied [99]. The most meaningful results are presented below.

Two highly localized modes are calculated to lie near 1361 and 1358 cm^{-1} and correspond to the out-of-phase motions of the two CH_2 groups located between two C–C bonds in *gauche* conformation. These two modes are largely localized at the defect while the other portion of the chain remains practically still. It is interesting to note that the defect modes in the CH_2 wagging region near 1350 cm^{-1} show that in this spectral range the GGTGG defect behaves as if it were made up of GG and GTG defects moving almost independently.

In the lower-frequency region defect modes are calculated near 1105, 900, and 714 cm^{-1}. The mode near 714 cm^{-1} is of particular interest since it lies just below the limiting $\mathbf{k} = 0$ mode located at 720 cm^{-1} of the CH_2 rocking dispersion curve and is just at the edge of a large energy gap which extends as far as the cut-off of the branch ω_5 (see Figure 3-3). From IIM, we learn that the defect mode near 714 cm^{-1} originates from a cluster of several CH_2 groups, including the defect, which is performing mainly a rocking motion. From the shape of the normal mode this near-the-edge mode cannot yet be considered a localized gap mode since it shows a nonnegligible degree of cooperativity. Similar cases of near-the-edge resonances were found for polyethylene in the lower frequency range and for polyvinylchloride containing head-to-head defects [90].

The modes calculated near 1105 and 900 cm^{-1} are the results of a complicated mixture of several internal coordinates, but can be described mainly as C–C resonance modes. They are likely to be observed in the Raman spectra.

The results of the calculations were checked with the experimental infrared and Raman spectra of the cyclic hydrocarbon $C_{34}H_{68}$ whose structure determined by Newman and Kay [116] indicates that two segments of 12 CH_2 groups in a *trans*-planar zig-zag conformation are joined by two GGTGG defects. Coincidences between experiments and calculations are striking. However, a critical analysis needed to be made ([99, 115]) on whether these signals are characteristic only of the GGTGG defect and may allow such a defect to be disentangled from the signals of other conformational defects possibly existing in a realistic polyethylene single crystal. The conclusion reached is that calculations on several model systems and experiments on $C_{34}H_{68}$ suggest [99] that a simultaneous appearance in the infrared spectrum of two bands near 1342 cm^{-1} and one near 700 cm^{-1} (certainly below the in-phase limiting CH_2 rocking mode of the *trans*-planar chain) can be taken as evidence for the existence of GGTGG defects in polyethylene single crystals. The other calculated modes occur at frequencies where other modes of the host lattice may also occur. Calculations were also extended to predict the intensity pattern for the defect modes in the CH_2 wagging range [117, 118]. Figure 3-28 compares the experimental spectrum with that calculated using the force field proposed by Shimanouchi [119].

Experiments of difference spectroscopy combined with band deconvolution processes were presented and conclusions were not unambiguous [115]. On the other hand, the arbitrary decisions taken when applying band deconvolution techniques sometimes make issue more complex. We have tried to avoid band deconvolution and proceeded directly to careful experiments of difference spectroscopy on sectored polyethylene single crystals. The principle on which the experiments was based is the following. From the study of sectored polyethylene single crystals it is known that sectors correspond to regions where molecular chains are crystallized in the

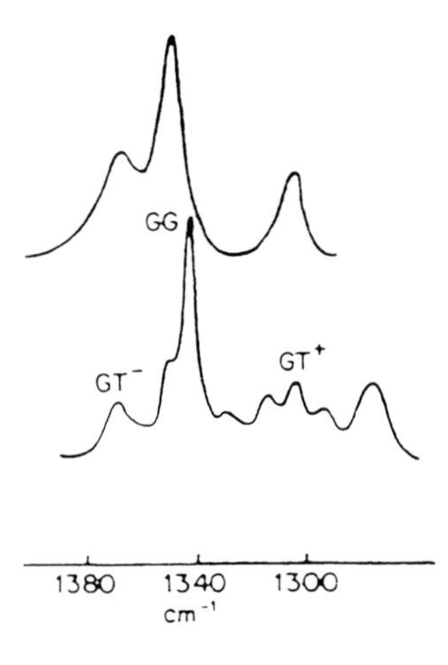

Figure 3-28. Comparison between the calculated and experimental infrared spectrum in the CH_2 wagging region of the GGTGG conformational defect. Calculations have been made the with force field by Shimanouchi [119]. The intensity parameters were developed in the laboratory of the author.

{110} and {100} crystallographic planes. By changing the crystallization conditions, sectored crystals with varying ratios between the {110} to {100} sectors can be obtained. Chain-folding studies has also indicated that tight-fold re-entry with GGTGG defects can only occur when chain crystallize in the {100} planes.

Sectored crystals were kindly provided to us by Professor A. Keller with the following % of {100} faces: 0, 31, 35, 45, and 55. The experiment of difference infrared spectroscopy we attempted is sketched Figure 3-29. We aimed at isolating the absorption to be associated with the GGTGG possibly existing on the surface of the {100} fraction of the sectored crystals. The experiments were not easy and after many attempts the final spectrum obtained (Figure 3-30) is compared with that of the cyclic hydrocarbon $C_{34}H_{60}$. The matching of the two spectra is striking and does suggest that, in agreement with calculations, on the surface of polyethylene single crystals obtained in particular crystallization conditions tight fold re-entry with GGTGG defects may indeed occur.

3.14.2.4 Case 4: The Structure of the Skin and Core in Polyethylene Films (Normal and Ultradrawn)

In addition to the defect modes presented in the previous sections it has been shown recently by both calculation and experiments that generally when *gauche* defects exist in polyethylene a broad band arises in the CH_2 bending range near 1440 cm^{-1} [120]. The band is very broad, possibly because it originates from a distribution of nonequilibrium *gauche* structures constrained by the sample morphology (Figure 3-31). Close to such absorption one observes for crystalline orthorhombic n-alkanes and polyethylene the doublet due to crystal field splitting (see Section 3.6). Let a and b label the two components observed near 1473 and 1463 cm^{-1} respectively (Figure 3-31). It is known that the setting angle θ between the planes defined by the *trans*-planar chain skeleton in the lattice is approximately 42° [121]. Such value of θ determines the intensity ratio $I_a/I_b = 1.2333$ between the two components of the crystal doublet of the CH_2 bending and CH_2 rocking in orthorhombic n-alkane and

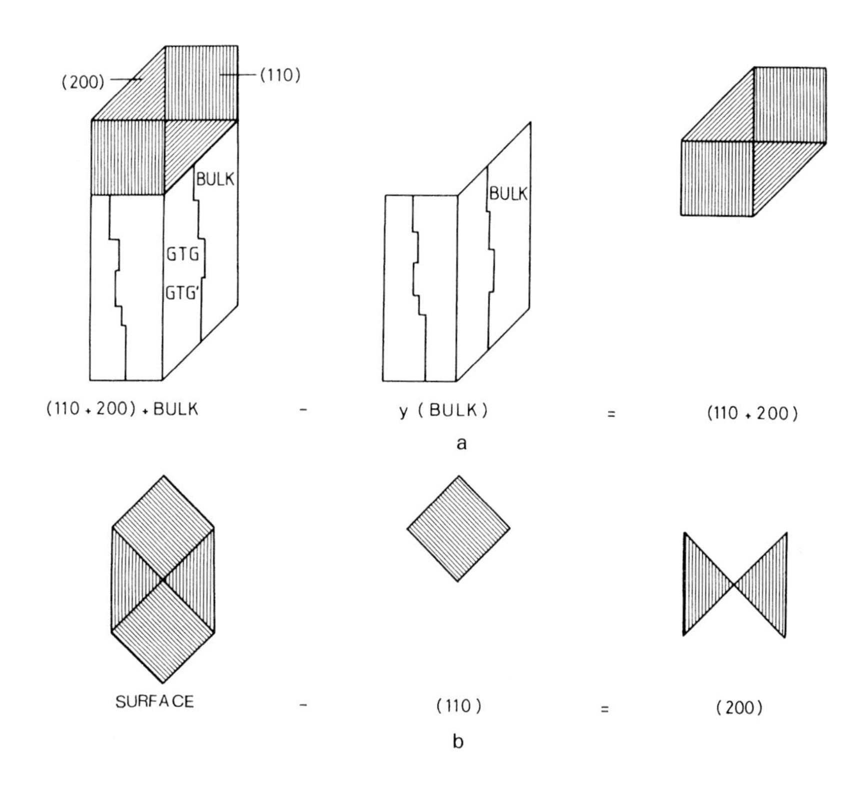

Figure 3-29. Scheme of the experiments of difference spectroscopy on sectored single crystals of polyethylene. (a) single crystal – bulk = (110) + (200) surface of the sectored crystals; (b) (110) + (200) surface – (110) surface = (200) surface.

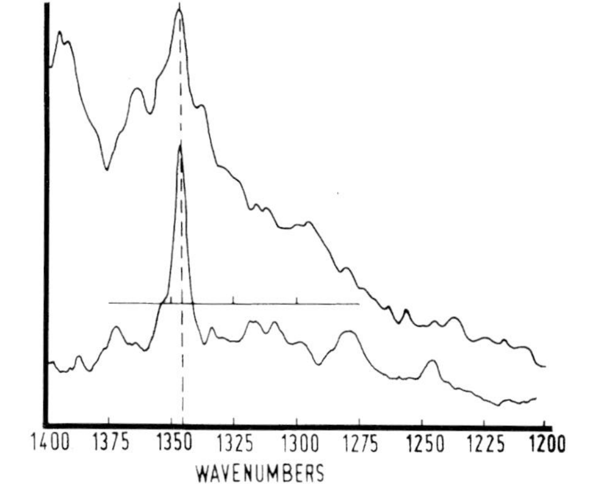

Figure 3-30. Comparison between the infrared spectrum in the GGTGG defect mode region of the crystalline cyclic molecule $C_{34}H_{68}$ and the result of the difference spectroscopy sketched in Figure 3-29.

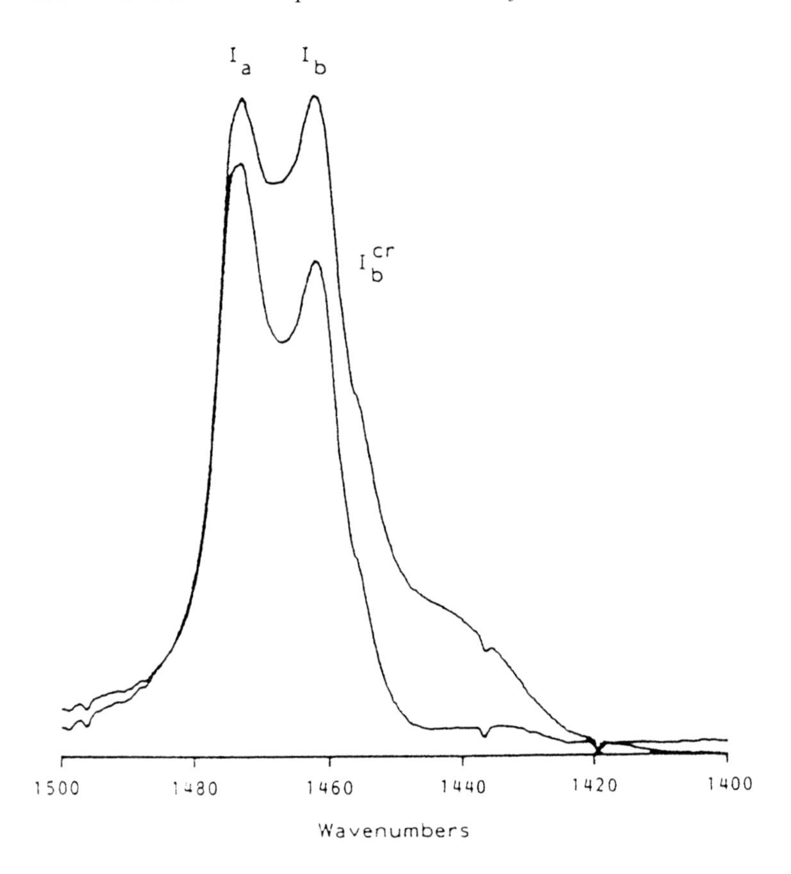

Figure 3-31. Infrared absorption spectrum of a commercial film of polyethylene before and after subtraction of the spectrum of the defect mode component near 1440 cm^{-1}. After subtraction of the defect mode the ratio between component a and b of the correlation field splitting approached the ideal value for a setting angle of 42° (see text).

polyethylene [122]. Certainly the wing at the higher frequency side of the defect mode overlaps with at least line b of the doublet, thus adding more intensity to line b and removing any physical meaning to be observed intensity ratio between the two components. However, by difference spectroscopy it is possible to remove the absorption due to the conformationally irregular fraction.

This observation is the basis for the determination of the amount of material with *gauche* structure in any sample consisting of polymethylene chains partially crystalline and partially conformationally disordered amorphous.

This method has been applied to experiments of multiple internal reflection infrared spectroscopy on thin films of commercial polyethylene [120]. It is known that 'optical microtomy' can be achieved by changing the incidence angle of the incoming beam. Structural depth profiling has then be performed thus allowing to probe the ordered–disordered structure of such films (Figure 3-32).

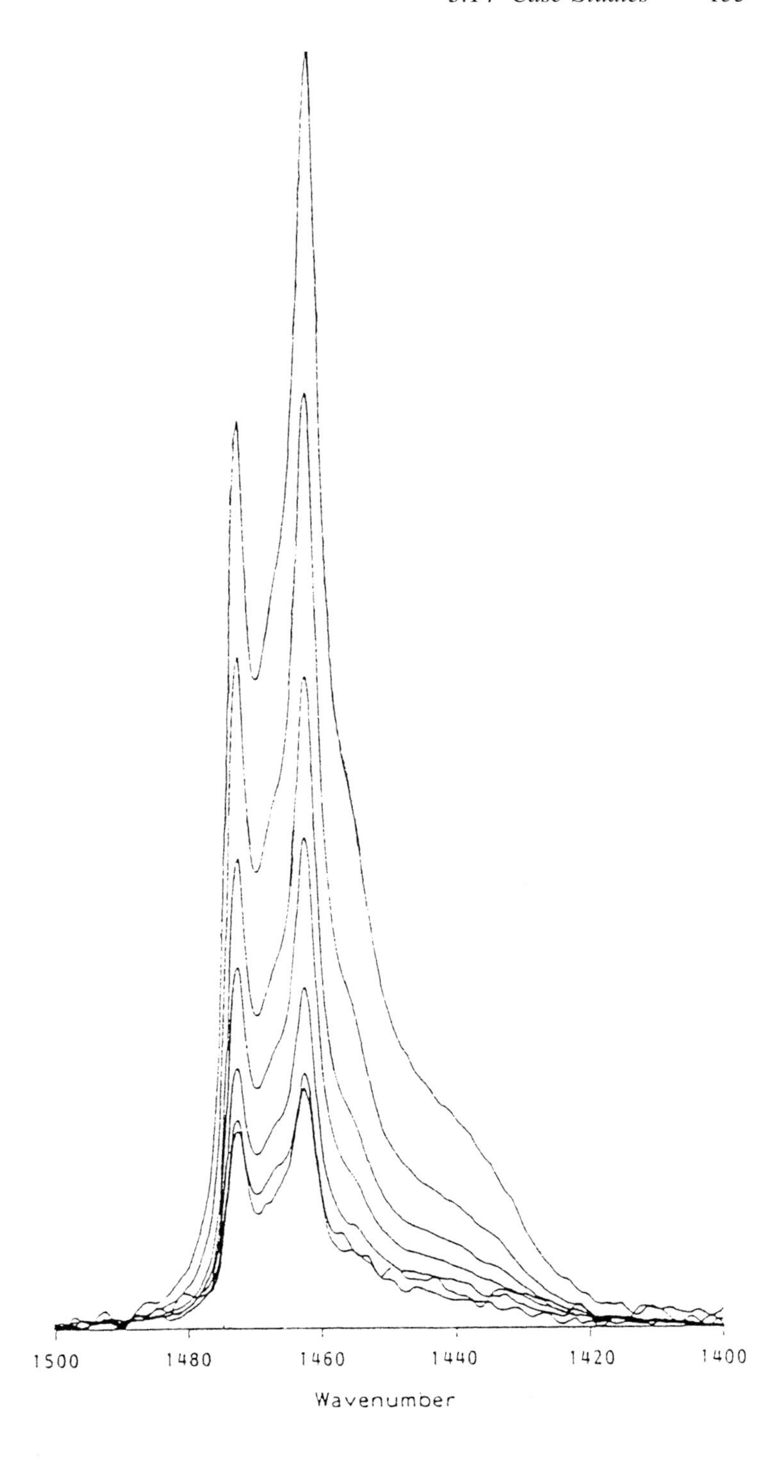

Figure 3-32. Multiple internal reflection spectrum in the CH_2-bending region at different angle of incidence of a 56 μm-thick film of commercial polyethylene. The spectrum shows the increase of the fraction of conformationally irregular material by increasing penetration depth.

It has been found that the outer skin of the film is much more crystalline (~80%) and the crystallinity decreases when probing inside the film reaching a concentration of ~60% crystallinity approximately 5–8 μm inside the film [120].

The above analysis combined with defect mode spectroscopy carried out in infrared (absorption in polarized and unpolarized light, and multiple internal reflection) and Raman has been applied to the structural characterization at the molecular level of ultra-drawn films of high-density polyethylene (UD-HDPE) [123]. The comparative study has allowed to describe the structural situation of UD-HDPE films as follows:

(i) The skin of the film within ~10 μm is highly crystalline and orthorhombic.
(ii) UD-HDPE film is a multiphase system with the topological distribution of the phases changing in going from the surface into the core.
(iii) The orthorhombic phase exists both on the surface and in the core, while the monoclinic phase starts at a penetration depth $d_p > \sim 10$ μm.
(iv) Both the orthorhombic and the monoclinic phases are highly anisotropic and are oriented along the draw direction.
(v) An additional phase is identified from $d_p > 10$ μm into the core. This phase consists of sequences of CH_2 groups in *trans* conformation. The orientation of the chains is not parallel to the draw direction. This additional phase exists at low draw ratios and decreases its concentration qualitatively at higher draw ratios with respect to the orthorhombic component.
(vi) GTG′ kinks are organized in anisotropic domains, seemingly as a result of collective phenomena of orientation during stretching.
(vii) The chain end methyl groups find themselves in a disordered environment similar to that found in liquid n-alkanes. The existence of liquid-like droplets or of disordered domains around the CH_3 groups may be envisaged.
(viii) GG defects are observed and may originate from intrinsic liquid-like structures, from the disorder around the CH_3 heads and from some kind of chain folding between crystallites or within polymer lamellae.
(ix) The concentration of conformational defects is comparatively small when normal solid polyethylene is considered. The concentration of defects decreases when the draw ratio increases, thus showing that conformationally disordered and coiled chains unwind towards a better alignment and perfection of the *trans*-planar molecules.

All these information have been compared with other physical data which help in the understanding of the molecular structure of such technologically relevant materials.

3.14.2.5 Case 5: Moving Towards More Complex Polymethylene Systems

The role of long alkyl chains in natural phenomena and in industrially relevant systems is well recognized. Due to the flexibility of the polymethylene segments these systems show many phase transitions which are not yet clearly understood at the molecular level. Many authors have focused on the vibrational infrared and Raman spectra as probes for understanding the structure and dynamics of these systems.

We have tried to analyze a few cases with our spectroscopy of defect modes and mention here some of the systems studied.

1. *Fatty acids*. Generally, fatty acid crystallize as dimers held together by hydrogen bonds between the two carboxyl groups which face each other in a head-to-head arrangement. In other sections of this chapter the contributions to the understanding of the 'static' structure given by the spectroscopy of mass defects and by LAM spectroscopy have been already presented. The case of fatty acids with long polymethylene chains has been studied [124] focusing on the conformational evolution which originates the phase transitions. FTIR spectra clearly and distinctly show that when thermal energy is provided to stearic acid in the crystalline state first interlamellar $CH_3 \cdots CH_3$ distances increase (as seen in the case of n-nonadecane; see Section 3.19) and hydrogen bonds then relax. Simultaneously, intermolecular distances increase with resulting lateral expansion of the lattice. These phenomena become relevant approximately 10° below melting. Few conformational kinks are generated below melting and only close to the melting the concentration of GTG, GTG′, and GG defects increases rapidly bringing the system to a conformational collapse in the liquid phase [124]. However, most of the conformational disorder occurs on the lamella surface, thus adding more evidence to the existence of a general phenomenon of 'surface melting' which emerges from our studies on many chain molecules [125]. These results give a detailed description of the mechanism of phase transition in fatty acids which was vaguely hinted by NMR experiments [126].
2. *Bilayer systems*. The bilayer organic perowskytes were obviously used as simple models of biological bilayered phospholipids and biomembrane systems. We have studied [127] the system $[n\text{-}CH_3(CH_2)_nNH_3]_2MnCl_4$ (hereafter referred to as C14Mn) which in the solid forms two coexisting, but parallel, layers; a similar molecule with Zn (C14Zn) in place of Mn forms two interpenetrating or 'intercalated' layers. The conformational flexibility and phase transitions of these two classes of molecules turns out to be different. DSC data indicate for C14Mn a main phase transition in the range 74–85° which is described by our conformational studies as involving the formation exclusively of GTG′ kinks which do not alter the conformational trajectory of each alkyl stem. No GG defects are observed and clusters of ordered and disordered chains coexist over a large temperature range (67–80 °C). By simulation of the spectra it is found that, on the average, at 78 °C only one GTG′ kink per chain is formed; between 65 and 78 °C chains with only one kink are formed and arrange themselves in clusters. It follows that disorder proceeds cooperatively throughout the system.

   The behavior with temperature of C14Zn is totally different [127]. The main phase transition at 100 °C is accompanied by an abrupt generation of conformational defects at 99 °C. Above melting, chains have reached almost a liquid-like structure. Correlations between chains do not exist. However, from the temperature dependence of the factor group splitting of the CH_2 bending and

CH_2 rocking modes it is observed that the transition is 'prepared' since the intermolecular forces between chains in C14Zn (weaker than in C14Mn) quickly weaken and the thermal expansion gives more freedom for torsional fluctuations of the alkyl chains, thus leading to conformational collapse.

Two classes of mechanisms which lead molecules to phase transition and melting have been found, namely: (i) nucleation of conformational disorder from clusters whose size continuously increases, thus affecting later all chains; and (ii) cooperative expansion of the whole lattice caused by an increase with temperature of the mean amplitudes of torsional fluctuations which reaches a threshold level after which conformational collapse occurs.

In nature, many other polymethylene systems–also of practical relevance–can be found which show phase transitions that need to be accounted for. More theoretical and experimental spectroscopic work needs to be done. We have studied the spectra of intercalated layered alkylammonium zirconium phosphates [128] and of the solutions of decylammonium chloride [129]. The spectroscopic data present unusual features with temperature which need further theoretical and experimental studies using the techniques presented in this chapter.

3.15 Simultaneous Configurational and Conformational Disorder. The Case of Polyvinylchloride

The cases treated in the previous sections were restricted to the study of the few cases of mass defects and of the very many cases of conformational defects. Most of the relevant polymers, however, may contain configurational disorder which carries as a consequence also some conformational disorder. This case has been considered within the theoretical approach discussed in this chapter.

The mathematical aspects for a general case of polymers with defects in tacticity were first worked out, formulae are available and can be applied to any case. We have applied these mathematical techniques to the case of polyvinylcloride (PVC) which is known to be strongly configurationally disordered. Our work was done when the problem of the structure of PVC was highly controversial.

First, the dynamics of ordered PVC was treated in terms of the three energetically most likely ordered chain structures [80], namely: (i) planar extended syndiotactic with (–TTTT–) conformation; (ii) helical isotactic, threefold helical structure (–GTGTGT–); and (iii) folded syndiotactic (–TTGGTTGG–) (Figure 3-12). Phonon dispersion curves and density of vibrational states were calculated for the three models using the most reliable force fields available at that time (Figure 3-12). The infrared and Raman activity of the $\mathbf{k} = 0$ phonons for each perfect chain are the following: model (i), $z = 2$, $t = 1$, 12 atoms per crystallographic unit cell, $\psi(\pi, d/2)$, C_{2v} point group, $9A_1$ (ir $\parallel$, R) $+7A_2$ (R) $+9B_1$ (ir $\perp$, R) $+7B_2$ (ir $\perp$, R); model (ii) $z = 3$, $t = 1$, 18 atoms per unit cell, $\psi(2\pi/3, d/3)$, C_{3v} point group, 16A (ir $\parallel$, R) +

17E (ir ⊥, R); (iii) $z = 4$, $t = 1$, 18 atoms per unit cell, D_2 point group, 17A (R) $+17B_1$ (ir ∥, R) $+17B_2$ (ir ⊥, R) $+17B_3$ (ir ⊥, R).

The knowledge of the dispersion curves, of the spectroscopic activity of the $\mathbf{k} = 0$ phonons and their precise frequencies has allowed us to locate the energy gaps, where to expect localized defect modes, and to predict where the activation of the density of states singularities (strong peaks in $g(\nu)$ with no activity in IR and/or Raman) due to lack of symmetry because of disorder could generate extra absorption or scattering.

Since we take for granted that the actual samples of PVC contain geometrical as well as configurational defects at non-negligible concentrations, calculations were extended [90] with the introduction of: (i) conformational defects as isolated units or as possibly interacting multiple units; and (ii) isolated or possibly interacting configurational defects. Of necessity, when configurational defects are introduced, local conformational distortions must be considered so as to justify the existence of an energetically plausible defect. NET was applied for chains consisting of 200 chemical units (i.e., matrices of 3600×3600 were constructed), while IIM was restricted to shorter chains made up by 50 units (900×900).

The distribution of defects was random and the concentrations ranged from an isolated unit to 8% defects. The construction of the large matrices was made from smaller submatrices associated to each chemical unit in a given conformation or configuration. The detailed algebra and the complete expressions for such submatrices are available in [90] and can be applied to any case of polymers made up by chiral chemical repeat units.

The attention was mainly focused on the C–Cl stretching range of the supposedly mostly syndiotactic chain. Polymerization in urea clathrates provides the 'purest' sample of planar syndiotactic PVC [130]. Such a chain shows a large energy gap in the CCl stretching range ($\sim$650 to 820 cm^{-1}). It is very likely that defect modes from other structures may generate localized gap modes which may be used as useful structural probes.

While we refer to [90] for a thorough analysis of the complex structural situation, in Figures 3-33 and 3-34 we give two examples of the frequency and vibrational displacements for an isolated GG or an isolated isotactic diad defect in a fully syndiotactic planar zig-zag host chain. It must be noted that gap frequencies are different, but that the vibrational displacements associated to these gap modes involve many chemical units ($\sim$20). The complexity of the problem is shown in Figure 3-35 where the density of vibrational states of a realistic model of configurationally disordered PVC is compared with the experimental spectrum.

3.16 Structural Inhomogeneity and Raman Spectroscopy of LAM Modes

The problem raised in Section 3.14.1 requires further discussion. From longitudinal accordion motion (LAM) spectroscopy of a variety of fatty acids Vergottin et al.

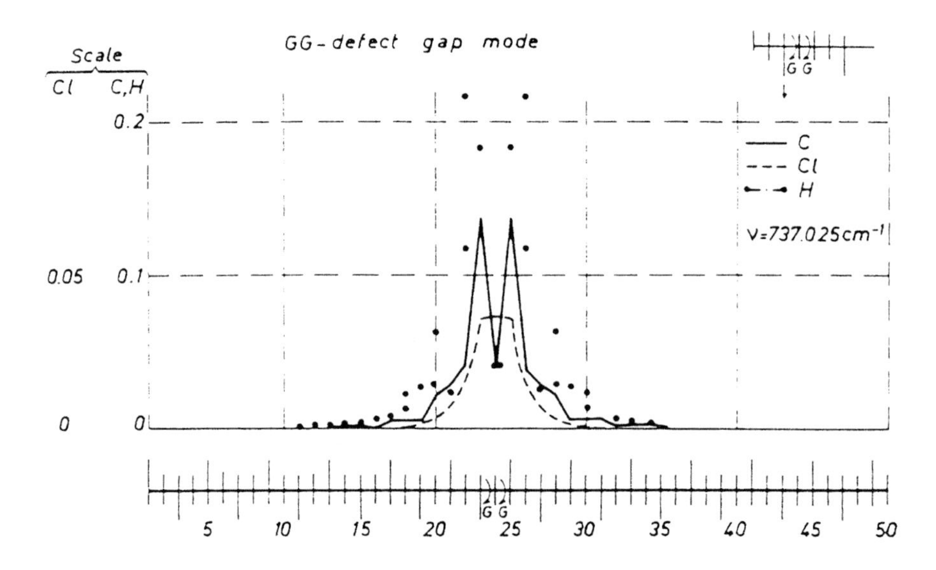

Figure 3-33. Gap mode (described in terms of total displacements of the atoms) for a GG defect in a syndiotactic PVC chain. Notice that the gap mode is localized over a large section of the chain.

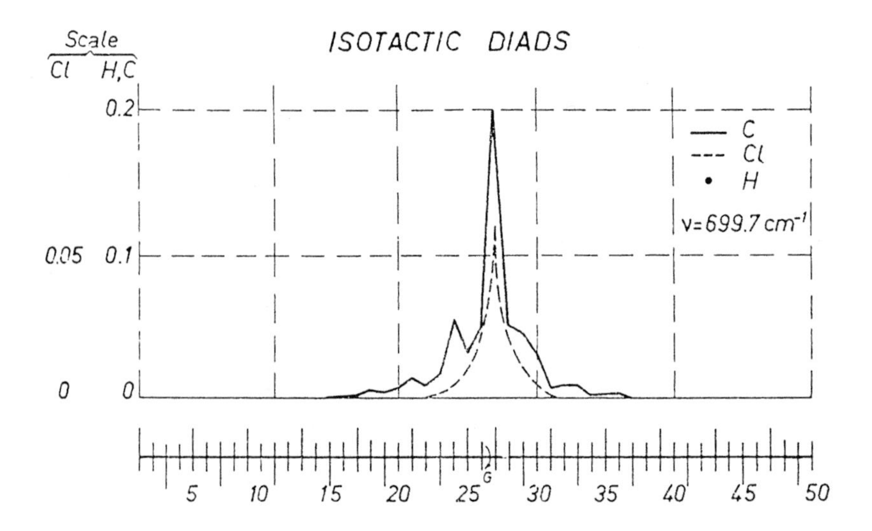

Figure 3-34. Gap mode (described in terms of total displacements of atoms) for an isotactic isolated diad. Notice the breadth of the mode which extends over at least 20 carbon atoms.

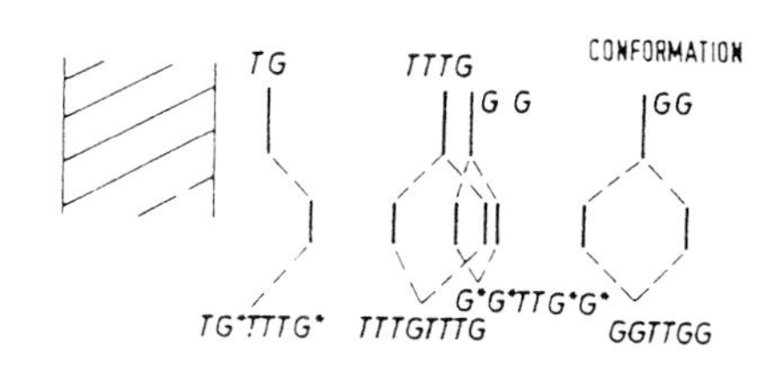

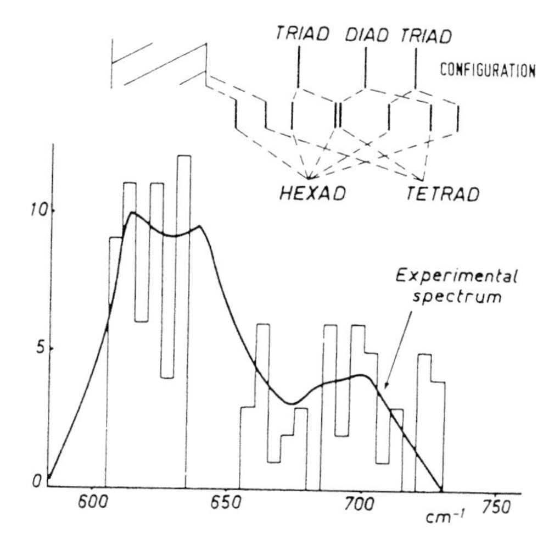

Figure 3-35. Density of vibrational states for a realistic model of configurationally disordered PVC (Bernoullian paramater $P_m = 0.47$). Comparison with the experimental spectrum (see text).

[108] suggested that the alkyl chains of fatty acids contain a few conformational defects. This conclusion was shown not to agree with the conformational mapping side-by-site obtained by defect-mode spectroscopy with the introduction of CD_2 defects in the polymethylene alkyl chain.

Such discrepancy seemed either to weaken the relevance of LAM spectroscopy for the measure of the length of the all-*trans* segment in a polyethylene chain, or to cast doubts on the reliability of defect-mode spectroscopy. Some further studies were in order.

Since dispersion curves have been clearly discussed in Sections 3.5 and 3.6, the concepts which form the grounds of LAM spectroscopy can be easily defined. Let us consider an all-*trans* polymethylene chain and focus our attention on the branch ω_8 of the phonon dispersion curve. The branch starts as mostly C–C stretching and ends as CCC bending. For a finite chain of N chemical units we expect to locate on this branch a band progression of N vibrational modes characterized by the phase coupling $\varphi_j = j\pi/N$. The 'most in-phase mode' occurs at very small frequencies; its frequency approaches zero when the chain tends to be of infinite length. Let $1, 2, 3 \ldots N$ label the components of the band progression of the modes which lie on the ω_8 branch.

Schaufele and Shimanouchi [131] have shown that for centrosymmetric all-*trans* polymethylene chains the lowest frequency mode performs an accordion-like motion (LAM) with the atomic displacements describing a wave with one node at the center of the chain [132]. Such a mode has been labeled as LAM1 mode. The higher frequencies components have been indicated analogously as LAM2, LAM3, etc. The odd modes give origin to a strong band progression in the Raman spectrum, with LAM1 extremely strong and the other components with intensity rapidly decreasing as the number of nodes increases.

Schaufele and Shimanouchi have considered the longitudinal stretching motion of an elastic rod of length L, elastic modulus E and density ρ and have shown that the frequency of its 'according' motion can be written as

$$\nu_{LAM1} = (1/2L)(E/\rho)^{1/2} \tag{3-53}$$

The same authors have shown that if E and ρ are suitably chosen for polymethylene chains, ν_{LAM1} of *trans*-planar finite polymethylene chains also follows the same law. In this case L is the length of the all-*trans* polymethylene chain which can be approximated as

$$L(\text{Å}) \approx 1.25(\text{Å})(N - 1) \tag{3-54}$$

where N is the number of CH_2 units of the chain.

LAM spectroscopy gained much popularity in polymer characterization and morphology since it has been possible to measure from Raman spectra the length of the *trans*-planar stems in polyethylene lamellae and in many other relevant cases.

As mentioned above, these concepts were applied in a straightforward manner to the case of fatty acids by the French group [108], but the results were highly controversial, thus casting some serious doubts on the applicability of LAM spectroscopy.

The dynamical problem has been re-examined with the purpose of evaluating the effects of chain ends on LAM modes. Simplified models were employed which consider chemical repeat units as point masses joined by elastic springs. Various cases of perturbations of the LAM modes were considered namely:

(i) The effect of heavier masses at one or at both ends of the polymethylene chain mimicking systems such as $X{-}(CH_2)_N{-}X$ where $X = -CH_3$, F, Cl, Br, I, etc. It transpires that ν_{LAM1} decreases when the mass of X increases [133].

(ii) The effect of joining a polymethylene chain with another chain with varying length and with different and variable spring constants. The realistic case studied was the class of emifluorinated n-alkanes $F(CF_2)_N \cdot (CH_2)_M H$ where N and M change in a broad range of values. Contrary to expectations, the two segments do not generate two independent LAM1 modes, nor can a simple heavy mass (X as (i)) at the end of a chain account for the observed spectra. Calculations show that the molecule vibrates as a whole and generates only one LAM1 mode whose frequency is critically dependent on the length of both segments. Moreover, the node of such LAM1 mode shifts along the chain depending on the length N and M of the two arms [134].

(iii) The above results opened the way to the study of dimeric systems where the two monomers are joined head-to-head by springs with varying spring constants. This model aimed at reproducing the class of dimeric fatty acids or fatty alcohols. This is the most important case both for its dynamical implications and for the consequences on structural determination as indicated at the beginning of this section. The modeling [133] considers two chains with one heavier mass at one end joined head-to-head by a spring with variable spring constants. Such a spring mimics the hydrogen bonds which link the two carboxyl groups.

The results of the calculation indicate that LAM1 extends over the whole dimer, with the node at the center of the dimer between the two carboxyl groups. The frequency of LAM1 has to be searched at much lower frequencies and the frequencies of the band progression of higher-order LAM modes are totally different from those identified in the spectra of these materials by the French group. The new band progression is easily found in the low frequency Raman spectra (Figure 3-36), thus giving further experimental support to what was shown by defect-mode spectroscopy (see Section 3.14.1), namely that fatty acids in the crystal do not contain conformational kinks and are fully *trans*-planar.

3.17 Fermi Resonances

The next step in our analysis is to focus on additional spectroscopic manifestations of molecular order and disorder which are not only intellectually interesting as physical phenomena, but also become of extreme importances as markers for the experimental routine characterization of the very many systems containing segments of polymethylene chains.

We are faced with the following experimental facts which need to be understood and possibly used for analytical and structural determinations. The Raman spectrum of polyethylene (or practically of any polymethylene chain) in the solid all-*trans* structure is very similar (or identical) to that given in Figure 3-37a; upon melting, the overall appearance of the Raman spectrum is that given in Figure 3-37b.

3.17.1 Principles

The experimental fact is that whenever a molecule contains a few CH_2 groups the Raman spectrum shows a very characteristic pattern of lines in the CH_2-stretching frequency range. Since they are strong they can be observed in a variety of materials, even under the most awkward experimental conditions. Indeed, they can be observed for monolayers of organic or biological materials spread on surfaces, in aqueous solutions, in microscopic crystals, etc. Moreover, the spectrum shows drastic changes in frequencies and intensities when the molecular structure or the environment changes. For this reason the Raman spectrum is at present a very

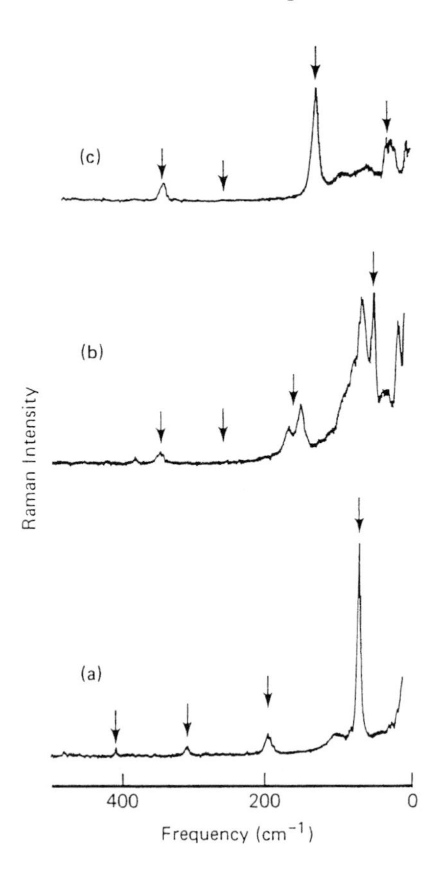

Figure 3-36. Raman spectra (0–400 cm^{-1}) of (a) hexatriacontane, n-$C_{36}H_{74}$; (b) stearic acid, n-$C_{17}H_{35}COOH$ and (c) strearyl alcohol, n-$C_{18}H_{37}OH$. Notice that since fatty acids and alcohols form head-to-head hydrogen-bonded dimers LAM modes extend throughout the whole dimeric system. Thus, the LAM progression shifts toward much lower frequencies than those expected for one isolated chain, approaching the values of hexatriacontane. The lowest frequency strong LAM1 mode has its node at the center of the molecule, inside the hydrogen-bonded dimeric structure (see text).

powerful and essential diagnostic tool on the conformationally ordered/disordered structure of complex organic, biological and industrially relevant polymethylene materials which contain polymethylene chains.

The reason of such peculiar and specific spectroscopic manifestations is that complex Fermi resonance phenomena occur in molecules containing polymethylene chains. We are concerned in this chapter on the Fermi resonances associated with the vibrations of polymethylene systems in the *trans* or *gauche* states, considered either as isolated entities and/or when they are organized in ordered and/or disordered supermolecular structures. Let us remember that Fermi resonance derives

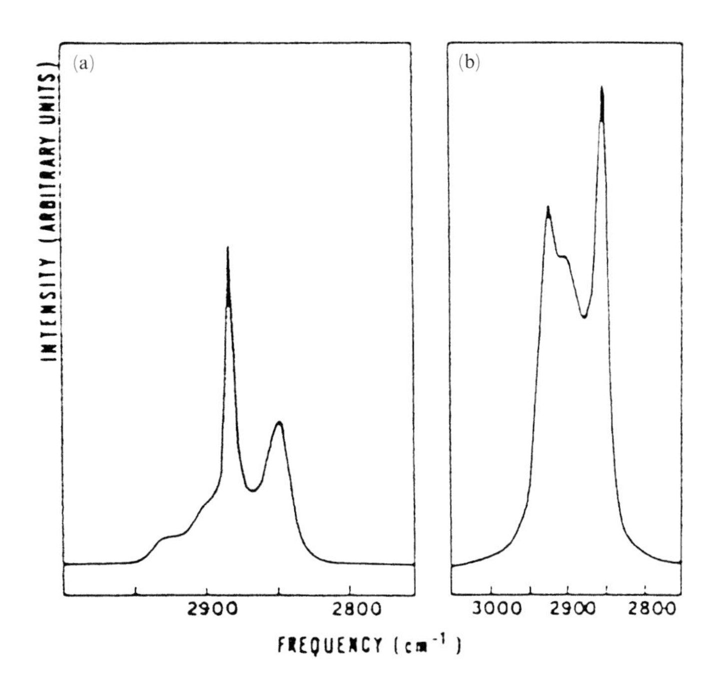

Figure 3-37. Examples of the changes with physical states of the Raman spectrum of polymethylene systems: (a) Raman spectrum of crystalline orthorhombic polyethylene in the CH stretching region; (b) Raman spectrum of polyethylene in the molten state in the same frequency range.

from the mixing, because of anharmonic terms in the potential, of two (or more) eigenstates (one fundamental and one (or more) overtone(s) or combination(s)) which belong to the same symmetry species. As a result, generally the two (many) interacting levels repel each other and the generally weak overtone(s) (or combination(s)) borrows (borrow) intensity from the much stronger fundamental [2, 3].

Fermi resonance has for a long time been a well-known and explicable fact in the case of small molecules [135]; the case of hydrocarbon chains was simply and clearly presented by Lavalley and Sheppard [136] for the molecule of hexadeuteropropane (CD_3–CH_2–CD_3), which consists of an isolated CH_2 group between two CD_3 groups with which little or no mechanical coupling is likely to occur, in a first approximation. The molecule belongs to the C_{2v} symmetry. Figure 3-38a gives the survey Raman spectrum of this molecule, while Figure 3-38b and 3-38c show an expanded section of the Raman spectrum polarized ($\|$) and depolarized ($\perp$) [137]. Normally, in the CH_2 stretching range we should reasonably expect two lines to be associated with the antisymmetric (d^-) and symmetric (d^+) CH_2 stretching of species B_2 ($\perp$) and A_1 ($\|$) respectively. Instead, we find two strong lines certainly with A_1 symmetry (polarized) and one certainly with B_2 symmetry (depolarized). The dynamical situation which justifies the observed Raman spectrum is the fol-

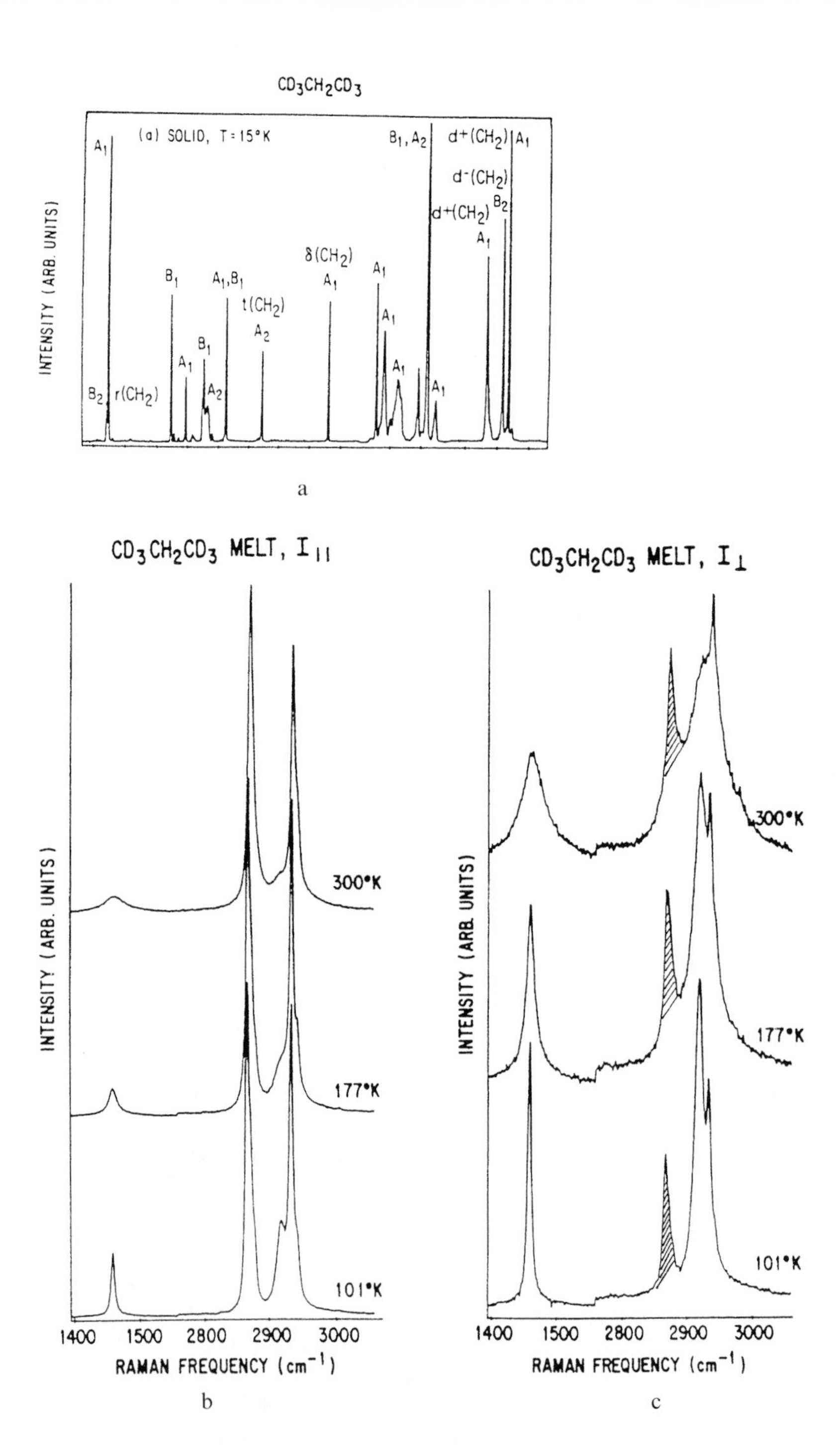

Figure 3-38. (a) Survey Raman spectrum of hexadeuteropropane (CD_3–CH_2–CD_3) in the solid state at 15 °K. The species of the vibrational transitions are indicated; (b) expanded temperature-dependent Raman spectrum in the 1400–3000 cm^{-1} range in polarized light (∥ polarization) exhibiting the transitions of A_1 species; (c) expanded temperature-dependent Raman spectrum in the 1400–3000 cm^{-1} range in polarized light (⊥) showing the transitions of B_1 species, free from Fermi resonances near 2925 cm^{-1}. The dashed scattering is due to 'bleeding' of the ∥ polarization due to experimental difficulties.

lowing. The CH_2-bending mode δ (certainly of A_1 symmetry (∥)) (Figure 3-37b) observed at 1460 cm^{-1} should have its first overtone ($2\delta : 2 \times 1460 = 2920\ cm^{-1}$, $A_1 \times A_1 = A_1$) which can enter Fermi resonance with the d^+ mode of A_1 symmetry. Eigenstates mix, frequencies are split apart with intensity borrowing from the fundamental to the overtone producing two strong lines at 2876 and 2942 cm^{-1} (Figure 3-38a–c).

It should be remembered that the first overtone 2δ should generally have vanishing intensity, but because of Fermi resonance its frequency is pushed upward and it borrows a lot of intensity from the fundamental d^+ observed at 2876 cm^{-1}.

A similar, but seemingly more complex case, occurs for polymethylene chains. Let us consider the dispersion curve of polyethylene or the corresponding vibrations of finite polymethylene chains (Figure 3-39) whose frequencies lie on these dispersion curves at finite values of the phase coupling (Section 3.5). The vibrations of the lower portion of the dispersion branch of CH_2 rocking have their first overtone levels ($\sim 720 \times 2 \approx 1440\ cm^{-1}$) close to the fundamental levels of the CH_2-bending motions. On their turn, the first overtones of the levels along the dispersion curve of the CH_2-bendings occur just in the frequency region where the fundamental CH stretching modes occur ($2 \times \sim 1440 \approx 2880\ cm^{-1}$). When symmetry allows, Fermi resonances occur and levels repel each other with drastic intensity changes. The issue becomes formally more complex since (Section 3.6) when molecules crystallize in an orthorhombic lattice (with two molecule per unit cell) the number of frequency branches doubles, thus doubling the number of levels which can generate both overtones and combinations which may produce Fermi resonances in the proper frequency ranges, as just indicated.

These Fermi resonance phenomena have found detailed mathematical treatment first tackled by Snyder et al. [138, 139] for single all-*trans* chains and later reformulated by Abbate et al. for (i) single chains, (ii) an orthorhombic lattice [140], (iii) all-*trans*, and (iv) conformationally distorted chains [141, 142]. The most interesting features to be accounted for are the Raman scattering of polymethylene chains in the frequency regions centered near 1450 and 2900 cm^{-1}.

For the sake of simplicity in the discussion which follows we label with ($^+$) and ($^-$) the symmetric and antisymmetric combinations respectively either of internal coordinates (e.g., C–H stretches within a CH_2 group) or of 'group coordinates' between two CH_2 units within the 'chemical repeat unit' made up by two CH_2 groups.

First, we discuss the case of true $\mathbf{k} = 0$ modes for an infinite chain (or the case of $\mathbf{k} \rightarrow 0$ for finite long chains). Let us, for the sake of simplicity, consider the spectrum of *trans* polyethylene both as a single chain and crystallized in an orthorhombic lattice with two molecules per unit cell. Since the spectrum of polyethylene as single chain is not available for this discussion we necessarily use the calculated spectra as simulation of reality [140]. Let us first compare the general features of the calculated Raman spectra in the CH_2-bending centered near 1435 cm^{-1} for a single chain and for crystalline polyethylene (Figures 3-40a, b) (point group D_{2h}). The single very strong line near 1435 cm^{-1} (Figure 3-39) is to be assigned to the limiting $\mathbf{k} = 0$ totally symmetric in-phase $CH_2\,\delta^+$ mode of A_g species. For the crystalline material, such a mode splits into two levels near 1420 (A_g) and 1435 cm^{-1} (B_{1g}); an additional broad scattering (with a wing towards higher frequencies) is calculated

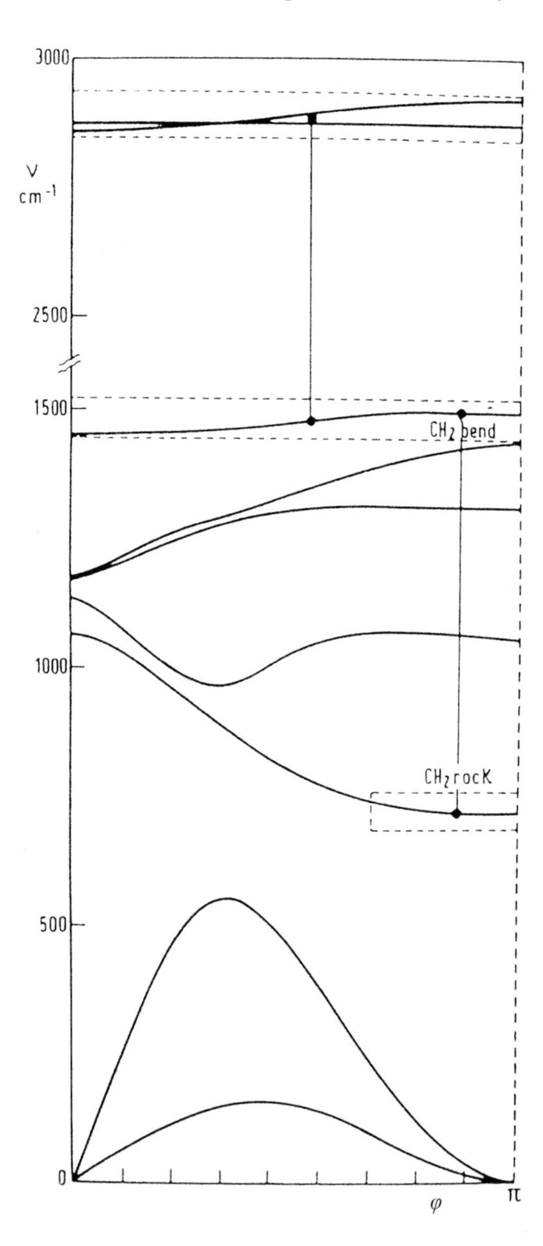

Figure 3-39. Dispersion curves of single-chain polyethylene, indicating the scheme of Fermi resonances which can occur between phonon of different branches.

near 1464 cm^{-1}. Following the previous procedure we look first at the first overtone of the infrared active CH_2 rocking P^- mode for a single chain (line group symmetry) ($2P^-$), $B_{2u} \times B_{2u} = A_g$. For the line-group, Fermi resonance can take place between $2P^-$ of A_g species and the $A_g\delta^+$ fundamental, giving rise to the weaker line near 1445 cm^{-1}. Extension of the treatment to the case of the orthorhombic lattice with two molecules per unit cell needs to consider that each level splits into two

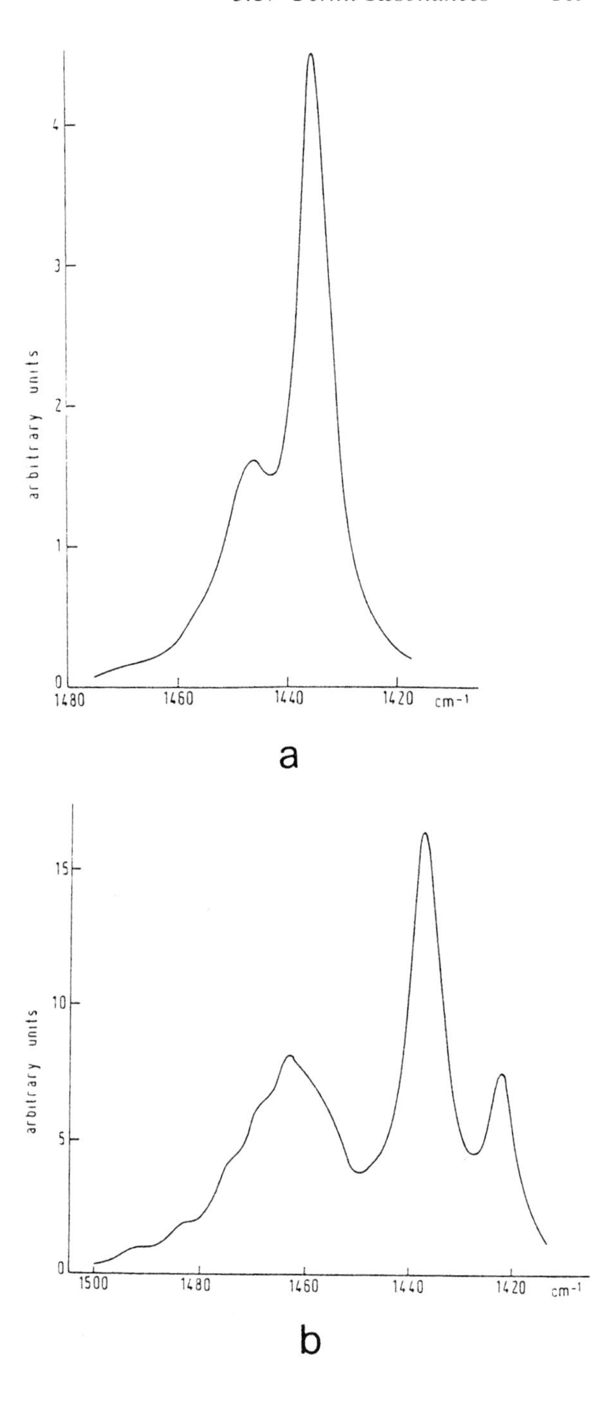

Figure 3-40. (a) Calculated Raman spectra in the CH_2-bending region of single-chain polyethylene. The BZ was divided into 72 intervals of 5°. Each Lorentzian function has $\Delta\nu_{1/2} = 5\ cm^{-1}$; (b) Calculated Raman spectrum in the CH_2-bending region of crystalline polyethylene as a convolution of Lorentzian lines as in case (a).

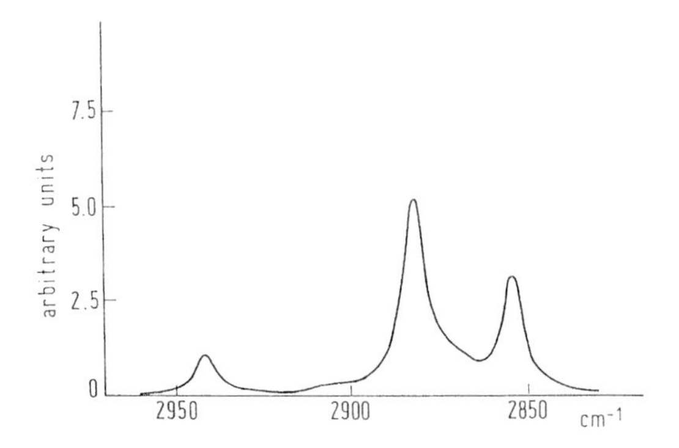

Figure 3-41. Calculated Raman spectrum for crystalline polyethylene in the CH-stretching region. The spectrum is a convolution of numerous combination and overtone levels in Fermi resonance with the nonresonant CH_2 antisymmetric stretchings. This spectrum should be compared with the experimental one of Figure 3-37.

levels; moreover, in addition to the overtone levels, we should also consider their combinations. For the space group we can use the observed experimental frequencies: $(2P^-)^-$, $B_{3u} \times B_{3u} = A_g$, $2 \times 731 = 1462\ cm^{-1}$; $2(P^-)^+$, $B_{2u} \times B_{2u} = A_g$, $2 \times 719 = 1438\ cm^{-1}$; $(P^-)^- + (P^-)^+$, $B_{3u} \times B_{2u} = B_{1g}$, $731 + 719 = 1450\ cm^{-1}$. It follows that for the space group the mode $(\delta^+)^+$ can couple with $2(P^-)^-$ and $2(P^-)^+$ while $(\delta^+)^-$ can interact with the combination $(P^-)^- + (P^-)^+$. Raman spectra of stretch-oriented (extruded rods) in various polarization directions [28, 123] show that the symmetry species of the lines at 1464 and 1440 cm^{-1} is certainly B_{1g}, while the line at 1416 cm^{-1} is certainly A_g.

The full treatment needs to consider in addition to the $\mathbf{k} \to 0$ modes also the $\mathbf{k} \neq 0$ modes (throughout the whole BZ) due to either the finiteness of the chain, or to the whole continuous dispersion curve for an infinite chain. The population of overtones and combinations becomes extremely large and many of these levels can enter Fermi resonances with a large transfer of intensity from the fundamental strong modes.

Once the bending region has been understood the same approach must be taken by considering all line-group or space-group overtones and combinations of the δ modes which are in Fermi resonance with the CH_2-stretching modes. All details of the calculations are provided in [140]. Figure 3-41 gives the result for crystalline polyethylene.

The most important and useful feature in the Raman spectra of polymethylene chains is that reported in Figure 3-37, which is easily observed every time a sequence of CH_2 units occurs in a molecule. If the system forms an orthorhombic lattice, generally one observes a broad and strong line near 2850 cm^{-1} with a broad wing which extends towards higher frequencies forming a broad background scattering that reaches $\sim$2950 cm^{-1}. Broad and sometimes ill-defined bands appear near 2930 and 2900 cm^{-1}. For the orthorhombic system, such a background scattering has been shown experimentally to have A_g symmetry [28, 123] and originates from a convolution of a large population of overtone and combination levels which are in Fermi resonance with the symmetric CH_2-stretching mode which, in principle, should occur as an unperturbed strong level near 2880 cm^{-1}, but has been pushed downward by Fermi resonances with the manyfold of overtones and combinations to which it has also lent intensity (Figure 3-41). Floating on top of such a broad A_g scattering we find the strong and sharp line associated with the CH_2 antisymmetric stretching which remains unperturbed because it does not enter in any Fermi resonance coupling.

3.17.2 Applications

We need now to translate the above theoretical results into practice. First, let us start from an orthorhombic lattice with two molecules per unit cell and lower the symmetry to a supermolecular organization in which the cell can be considered as isolated and decoupled from its neighbors. This situation can be achieved in the following case: (i) isotopic solid solutions; (ii) molecule as a guest in a chemically different host lattice (e.g., urea clathrates) [143]; (iii) liquid crystalline-like systems; and (iv) systems with conformationally distorted chain ends, but with an ordered bulk [103]. The following changes of the Raman spectrum are expected (and observed):

(i) The triplet of lines in the bending region becomes almost a singlet (δ^+), since the crystal field splitting disappears and the convolution of overtone and combination bands for finite chains (two phonon states for polyethylene) in Fermi resonance with the crystal field split fundamentals are removed (Figure 3-40).
(ii) As the number of overtone and combination states in Fermi resonance with the A_g fundamental state (CH_2 symmetric stretch) decreases we expect the lowering of the total A_g scattering and possibly a shift toward higher frequencies of the A_g fundamental. Generally, the scattering in the valley between the antisymmetric and symmetric CH_2 stretch 2980 and 2850 cm^{-1} respectively) is observed to decrease and the two fundamental lines sharpen somewhat. Attention should be paid to the fact that for chains with CH_3 end groups the symmetric CH-stretching mode shows up as a weak line within the same frequency range.

The next step is to interrupt the long *trans* chain by *gauche* defects. The following changes are expected (and observed) in the Raman spectrum [141, 142].

(iii) As both P and δ modes of CH_2 groups in *gauche* conformation occur at different frequencies the population of vibrational levels which enter Fermi resonances in the *trans* interaction scheme decreases. We expect changes of the Raman spectrum both in the CH_2-stretching and bending regions. Qualitatively, the vibrational spectrum should change from that of a full, all-*trans* single chain molecule toward that of a fully decoupled structure, thus tending to the situation described in Figure 3-38 for hexadeuteriopropane. It should be noticed that necessarily the 2850 cm^{-1} line should move towards higher frequencies approaching the unperturbed value near 2880 cm^{-1} and the 2δ line near 2930 cm^{-1} should eventually become stronger as it does not share with the manyfold of binary combinations and overtones the intensity borrowed from the fundamental originally observed near 2850 cm^{-1}.

(iv) A drastic increase in concentration of *gauche* structures occurs when the polymethylene system melts. In that case one can assume that the statistics of Flory's Rotational Isomeric State Model [48] works and we can describe the system as consisting of a distribution of *gauche* and short *trans* sequences. Upon melting, the d^+ band near 2850 cm^{-1} becomes the strongest feature in the spectrum and d^- collapses into a very broad band with its maximum shifted upward of $\sim$8–10 cm^{-1}. The overtone level near 2940 cm^{-1} increases in intensity. This is a very useful marker when the chain is conformationally collapsed (liquid like) (Figure 3-37b).

(v) There are two possible interpretations of the broadening of d^- band observed in the liquid phase. The first is that the broadening is the result of a convolution of many transitions associated to d^- mode of CH_2 groups in constrained non-equilibrium near *gauche* conformations [143, 144]. Another explanation of the broadening refers to the overall dynamics of the whole chain which may occur in the liquid (see Section 3.18) [145].

(vi) A detailed study has been carried out by Abbate et al. on the molecule CD_3–CH_2–CH_2–CD_3 [141] as a model in which (a) the possible coupling with the methyl end groups are removed by deuteration and (b) only one degree of torsional freedom between CH_2 groups is allowed. It is immediately apparent that in *gauche* conformation the symmetry of the molecule is lowered, thus changing the resonance schemes which may involve also the d^- mode which is unperturbed in the *trans* case. Calculations show that both d^- and d^+ are allowed to mix, thus reshuffling the intensities in a way that is difficult to predict.

(vii) The experimental intensity ratio $R = I_{2850}/I_{2940}$ has been proposed as an experimental way to determine the relative concentration of *trans* versus *gauche* conformers in a sample. R has been matter of strong debate (see for instance [140]). As a result of the calculations by Abbate on hexadeuteriobutane [141], because of a mixing of the modes and the unpredictable changes of the extent of Fermi resonance, we do not advise the use of R as analytical probe in liquid polymethylene systems or in partially disordered solid polyethylene samples.

3.18 Band Broadening and Conformational Flexibility

The experimental spectroscopic facts which need to be accounted for in a comprehensive way in relation to order–disorder are the following (we report here the most striking manifestations, recorded in various ways, for oligo and polymethylene systems):

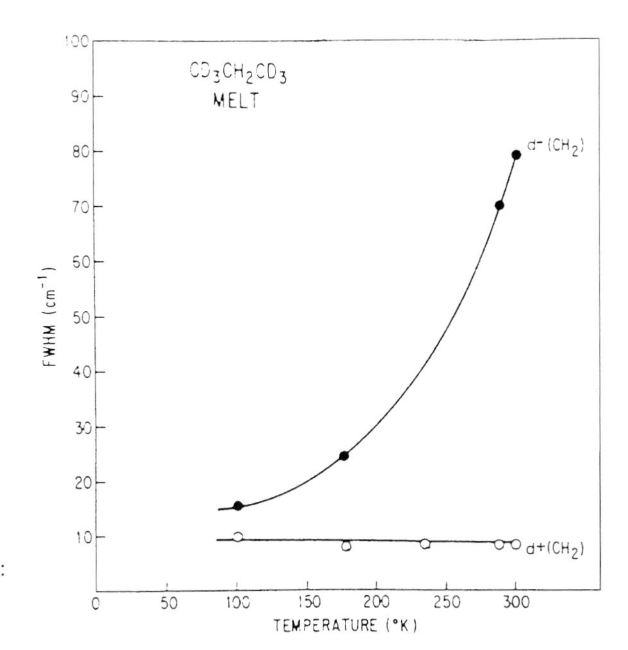

Figure 3-42. Hexadeuteropropane: $\Delta\nu_{1/2}$ versus temperature (in the melt) of d^- (•) and d^+(○).

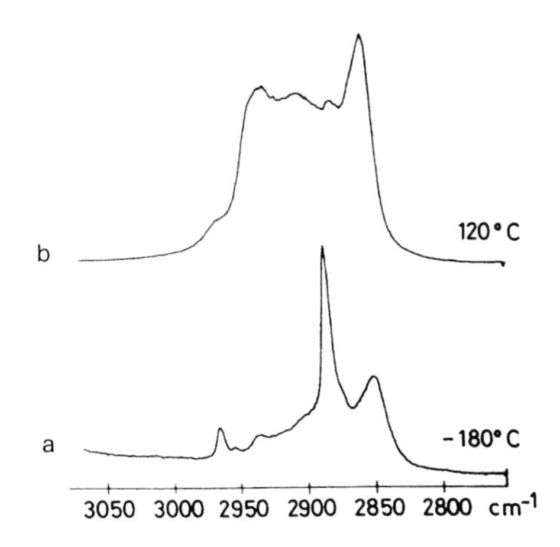

Figure 3-43. Raman spectrum in the C–H-stretching range of $C_{21}H_{44}$: (a) solid and (b) melt.

1. For short chains, the antysimmetric d^- mode broadens in the Raman spectrum by increasing temperature and sometimes shifts upward in frequency (Figure 3-42), while d^+ neither shifts substantially nor broadens. (See also Figure 3-37.)
2. For long polymethylene chains and polyethylene at melting, the strong d^- modes collapses into a featureless very broad scattering and by increasing temperature shifts its maximum towards higher frequencies (Figure 3-43).
3. For crystalline n-alkanes the width of d^- does not increase with temperature up to a few degrees below melting (Figure 3-44).
4. When n-alkanes are inserted as clathrates in urea channels, the width of d^- is

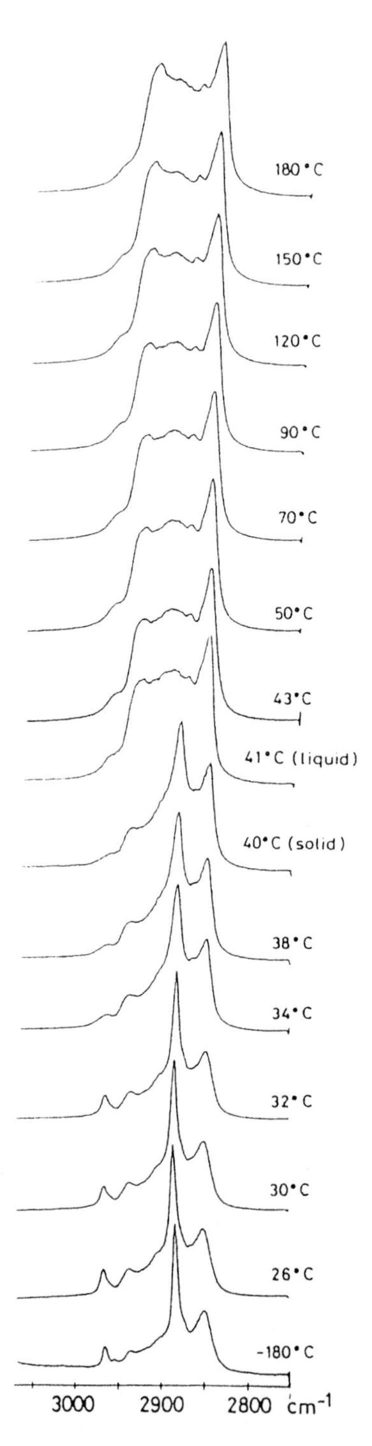

Figure 3-44. Evolution with temperature of the Raman spectrum of solid $C_{21}H_{44}$ in the CH stretching range. Notice the behavior with temperature of d^+ and d^- modes.

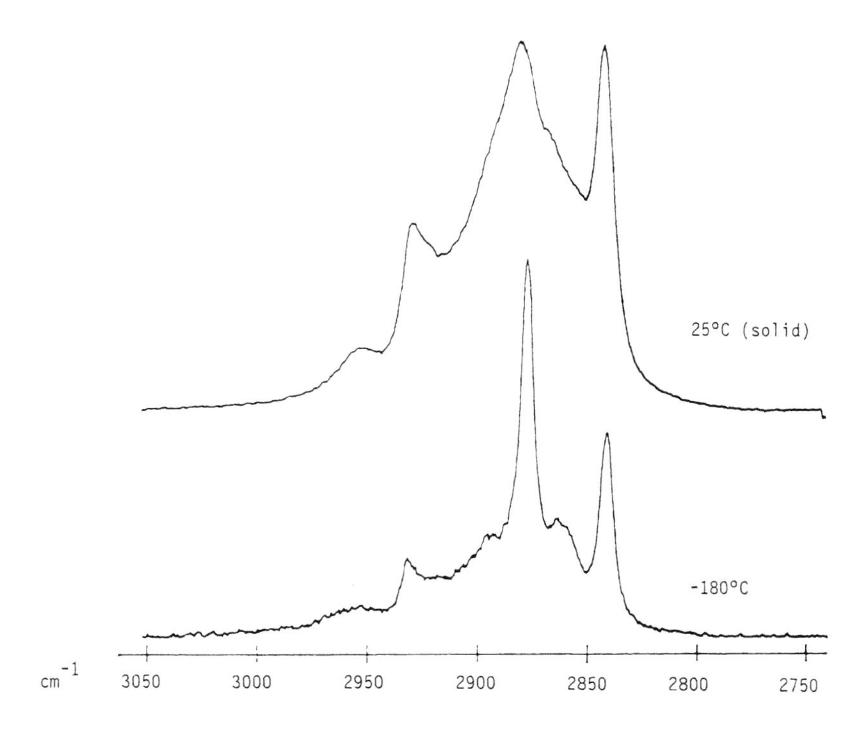

Figure 3-45. Raman spectrum of $C_{25}H_{52}$ in urea clathrate at 25 °C and −180 °C. Notice that for $d^- \Delta\nu_{1/2} = 6$ cm^{-1} at T= −180 °C and $\Delta\nu_{1/2} = 28$ cm^{-1} at T= 25 °C; $\Delta\nu_{1/2}$ of d^+ remains practically constant at the same two temperatures.

narrow at low temperatures and broadens substantially at higher temperatures (Figure 3-45).

3.18.1 Principles

In Section 3.17, experiments have been presented showing that in the case of polymethylene molecules (including polyethylene), in going from the solid all-*trans* to the liquid phases, the pattern of Raman scattering changes dramatically but in a consistent and characteristic way. We focus our attention on the temperature-dependent broadening observed for the d^- and δ modes; in contrast, the width and shape of d^+ modes are practically temperature-independent. In Section 3.17.2 it was suggested that one possible explanation of the band broadening of d^- could be a convolution of bands associated with d^- modes in different quasi *gauche* conformations existing in the liquid or 'amorphous' material.

This interpretation may be only partial since Snyder has shown that band broadening of d^- and δ modes is certainly observed also for hexadecane in urea clathrates where the molecule is believed to be (on average) in the *trans*-planar conformation [146]. Some additional or alternative processes then must exist. We wish to discuss the results of band-broadening studies which may shine more light on the conformational dynamics in systems where order–disorder may exist.

A full discussion of the theory and experiments on band-broadening effects in n-alkanes is reported in [137]. We wish to outline here a few basic concepts which, in a first approximation, are the tool to be used in the understanding of the complex dynamics in conformationally flexible systems.

Band broadening [147] both in infrared and Raman spectra of polyatomic molecules in the liquid phase may arise from vibrational relaxations (VIBR) (i.e., some kind of process dissipates the vibrational energy stored in a vibrational transition) and/or from reorientational motions of molecules (REORIENT) (i.e., the energy stored in a vibrational transition is used to generate a reorientation in space of the molecule). Vibrational relaxation can be due to vibrational energy dissipation to the rotational–translational degrees of freedom of the thermal bath, to intramolecular energy transfer between vibrational modes, or to several intermolecular phenomena such as resonance vibrational energy transfer and vibrational dephasing. Theory states that the amplitude-dependent vibrational relaxations are temperature-independent, while the angular-dependent reorientational relaxations depend strongly on temperature.

Raman scattering provides a nice way to gather information on these processes. For liquid samples, it is possible to separate experimentally the isotropic spectrum (ISOT) from the anisotropic one (ANIS), namely

$$\mathrm{I(ISOT)} = \mathrm{I}(\|) - (4/3)\mathrm{I}(\perp) \tag{3-55}$$

$$\mathrm{I(ANIS)} = \mathrm{I}(\perp) \tag{3-56}$$

From these measurable quantities, under certain approximations [137], it is also possible to give precise expressions for the relaxation times:

$$\tau(\mathrm{VIBR}) = 1/(\pi c \Delta\nu_{1/2})(\mathrm{ISOT}) \tag{3-57}$$

$$\tau(\mathrm{REORIENT}) = 1/(\pi c \Delta\nu_{1/2})(\mathrm{REORIENT}) \tag{3-58}$$

where

$$\Delta\nu_{1/2}(\mathrm{REORIENT}) = [\Delta\nu_{1/2}(\mathrm{ANIS}) - \Delta\nu_{1/2}(\mathrm{ISOT})] \tag{3-59}$$

It is commonly known that the depolarization ratio for linearly polarized light is given by

$$\rho = \mathrm{I}(\perp)/\mathrm{I}(\|) = 3\beta^2/(45\alpha^2 + 4\beta^2) \tag{3-60}$$

where α is the mean polarizability describing the isotropic component of the Raman polarizability tensor and β, the anisotropy, describes the anisotropic component.

It follows that for totally symmetric modes $\rho \approx 0$, $\mathrm{I}(\|) = \mathrm{I(ISOT)}$, and the vibrational relaxation contribution can be easily evaluated. For all antisymmetric modes $\alpha = 0$ and $\rho = \frac{3}{4}$; only the contribution from coupled vibrational–reorientational motions can then be obtained from Raman experiments. The evaluation of the pure reorientational contribution can be obtained by suitably subtracting the contribution from vibrational relaxations (Eq. (3-59)).

3.18.2 Short and long n-alkanes

The Raman spectra of a series of n-alkane suitably chosen have been studied [137] with the precise aim of extracting from d^- (antisymmetric, $\sim$ depolarized) and d^+ (symmetric $\sim$ polarized) modes of the CH_2 group information of general validity on intra- and intermolecular dynamics using the techniques discussed in Section 3.18.1.

The following molecules have been studied [137] in the solid and liquid states: propane-d_6 ($CD_3CH_2CD_3$) butane-d_6 ($CD_3CH_2CH_2CD_3$), and pentane-d_{10} ($CD_3CH_2CH_2CH_2CD_3$) and n-nonadecane. The following conclusions, which may be of general interest, have been derived:

1. For d^+ $\rho = 0$, thus allowing us to distinguish the contribution by vibrational relaxation processes; for d^- $\alpha = 0$ and $\rho = \frac{3}{4}$, thus the coupled vibrational–reorientational processes are active simultaneously. The reorientational contribution was derived by suitable deconvolution methods (Eq. (3-59)).
2. The temperature independence of the band width of d^+ modes has been verified and provides an average value of $\tau(\text{VIBR}) \sim 1.3 \times 10^{12}$ s for the three small deuterated molecules. An amplitude-dependent vibrational dephasing process (inhomogeneous broadening) best accounts for the observed temperature-independent broadening.
3. $\Delta\nu_{1/2}$ of d^- is strongly temperature-dependent and increases rapidly with increasing temperature (Figure 3-41); correspondingly, $\tau(\text{REORIENT})$ decreases by at least one order of magnitude from low to higher temperatures (see Table 3-3). It is found that, at a given temperature, identical for the three deuterated molecules, the band width increases with increasing chain length. Moreover, at a constant temperature, the reorientational relaxation time increases with increasing chain length. This trend is consistent with the view that the contribution of reorientational relaxation decreases with increasing chain length, as might be expected based on inertial and frictional effect. Indeed, it is conceivable that while the molecule of propane in the liquid phase may easily tumble in all directions about its axes, for longer molecules the end-over-end tumbling becomes progressively less likely while some sort of rotation about the long molecular axis may still occur even for longer n-alkanes.
4. For chains larger than propane (for which no torsional motion about (CH_2)–(CH_2) bonds exists) it is quite conceivable that the increasing hinderance to molecular reorientation which broadens d^- mode comes from torsional motions of the molecular skeleton, i.e., its overall torsional flexibility. In other words, when a flexible molecule tries to tumble it necessarily drives the motions of the torsional angles. We expect these contributions to increase with increasing

Table 3-3. Frequencies, bandwidths (full widths at half maximum) and reorientational–relaxation times of d^- of propane-d_6, butane-d_6 and pentane-d_{10} as a function of temperature.

	Temperature (K)	$\nu(d^-)$ (cm^{-1})	$\Delta\nu_{1/2}(d^-)$ (cm^{-1})	τ(REORIENT) d^- $\times 10^{12}$ s
$CD_3CH_2CD_3$	101	2919.0	15.7	1.3
	177	2920.5	24.8	0.58
	235			
	288	2925.0	70.0	0.17
	300	2923.5	79.0	0.15
$CD_3CH_2CH_2CD_3$	135	2896.5	15.0	1.7
	185	2896.5	24.2	0.66
	237	2899.5	40.0	0.33
	300	2898.0	65.0	0.18
$CD_3CD_2CH_2CD_2CD_3$	143	2896.5	18.1	1.2
	191	2898.0	20.1	0.93
	240	2899.5	24.8	0.67
	300	2901.0	30.9	0.48

temperature and chain length. The threshold temperature at which for a given chain length coupled rotations and torsion are able to generate conformational kinks in a more or less nonequilibrium structure (either pinned at a given site or mobile along the chain) remains an unsolved problem faced by many authors who tackled the problem by numerical simulations [148].

5. For the short-chain models it has been found that, by increasing the temperature (i.e., when the coupling between rotations and torsions increases), the frequency of d^- increases consistently by a few wavenumbers (see Table 3-3); this suggests that either the diagonal or off-diagonal force constants within each CH_2 unit change when the molecule is (even slightly) conformationally distorted or the onset of torsional distortions activates intramolecular coupling which extends at an unknown distance s at either side of the CH_2 group (see Eq. (3-45)).

3.18.3 Collective Chain Flexibility

In this section our attention is focused on attempts to describe in molecular terms the broadening and the upward shift of d^- mode as discussed in Sections 3.18.1 and 3.18.2.

Let us first consider the case of a long n-alkane molecule in the solid crystalline phase. From lattice dynamical calculations it is known that collective librational phonons (about the chain axis) exist in the lattice, some giving rise to scattering and/or absorption in infrared [70]. Let us take a n-alkane molecule with an even number of carbon atoms which belongs to the C_{2h} point group. The rigid rotation about its axis is of B_g species, the same as the Raman active d^- mode. When the molecule is embedded in a medium which transforms the free rotation into a librational motion d$^-$ and the libration can, in principle, couple because they belong to the same symmetry species. A similar observation was made by Snyder for the CH_2 rocking mode [146].

The contribution toward the understanding of chain flexibility comes from the following experiments and calculations which are described in detail in [137] and [145].

1. The first experiment has been that of recording the Raman spectra of long n-alkanes in urea clathrates [145] and in another host lattice (perhydrotriphenylene) where each guest n-alkane molecule sits in a channel, isolated from the other chains and retaining, on average, the all *trans*-planar structure. Similar experiments are also reported in [146] and [149]. By changing the temperature, the host lattice changes the average diameter of the columns where the guest molecule sits, thus allowing a varying degree of librational freedom for the guest molecule. While at low temperature the librational motion is strongly hindered, at relatively higher temperatures the lattice expands anisotropically allowing the guest molecule to perform librational motions of large amplitude.

   We expect the band of the d$^-$ mode to be narrow at low temperatures and to broaden at higher temperatures. Figure 3-45 shows that this is indeed the case. The band not only broadens but its frequency also shifts upward by a few wavenumbers. It is concluded that the almost rigid (and possibly small amplitude) collective librations about the chain axis at low temperature evolve with increasing temperature towards a collective libro-torsional (or libro-twisting) motion with nonnegligible angular distortions more or less hindered by the environment.
2. It must be pointed out that the broadening and upward frequency shift of the d$^-$ mode cannot be ascribed to the generation of *gauche* defects inside the host lattice. It has been verified experimentally that the frequencies of LAM1, LAM3, and LAM5 of the guest molecule embedded in the host lattice are identical to those observed for the same n-alkane in the crystalline state [145]. Hence, the molecule does not change its length from that of the *trans* structure [145].
3. In going from the solid to 'softer' systems, certainly the frequency of the d$^-$ mode shifts upward by a few wavenumbers. No shifting (nor broadening) is observed for crystalline n-alkanes until the so-called soft 'rotatory phase' is reached before melting. In the so-called α phase (see Section 3.19) some shifting and broadening is observed.

   Levin [150] and Gaber and Peticolas [151] have shown that the upper shift and broadening occurs also in biomembrane materials. Such a shift has been used by Gaber and Peticolas to monitor the 'lateral interactions' in phospholipids. We notice that for these phospholipid-type material the long-chain alkyl residues are fixed at one end to polar head groups and thus cannot perform rigid librational motions about their axes. Forcefully they can only perform libro-twisting or libro-torsional motions.
4. The last step is to account for the dynamical origin of the fact that when the long polymethylene chain increases its libro-torsional flexibility the frequency of d$^-$

modes shifts very selectively upward, while the frequency of d^+ mode remains unchanged. There must be some specific process which should be discovered.

With this purpose various dynamical possibilities have been considered [145]. It has been concluded that a rational explanation is that there exists a selective intramolecular coupling between d^- modes and the torsional modes of the carbon backbone. Symmetry consideration tell us that for a *trans*-planar molecule the existence of such coupling will affect d^- and not d^+ modes. In other words, within the quadratic approximations interactions of the type $f_{CH\ stretch/C\text{–}C\ torsions}$ seem to exist and are modulated by the amplitude of the collective torsional flexibility.

Another contribution to the upper shift may come from a complex scheme of Fermi resonances which are activated because of the lowering of the instantaneous symmetry during the libro-torsional motion. This difficult problem remains unsolved.

5. The final observation, relevant for the discussion which follows in Section 3.19, is that in crystalline n-alkanes the band width of d^- is relatively narrow and the frequency does not change throughout the temperature range of existence of the crystalline lattice. This suggests that libro-torsional motions are not allowed in the crystal and only small amplitude collective librational phonons seem to be allowed. Some additional observations will be added in Section 3.19.

3.19 A Worked Example: From n-Alkanes to Polyethylene Structure and Dynamics

3.19.1 Description of the Physical Phenomena

We wish to guide the reader through a realistic case worked out step-by-step using most of the concepts and methods discussed in this chapter. We consider the case of n-nonadecane (CH_3–$(CH_2)_{17}$–CH_3) as a prototypical case and try to derive from the vibrational spectra all possible information about its structure and dynamics. Both order and disorder will be considered. The methods applied and the conclusions reached form the background knowledge to be used for the understanding of: (i) the structural properties of many molecules containing long-chain alkyl derivatives; and (ii) the structural features of polyethylene or other polymers containing polymethylene segments.

The basic physics we wish to understand is the mechanism at the molecular level which generates the phase transitions of relatively short n-alkanes. The experimental nonspectroscopic facts to be accounted for are the following (for a collection of references and a full discussion of many of the data presented in the first part of this section see [103]): orthorhombic n-alkanes first undergo a solid–solid phase transition ($T_{0\rightarrow\alpha}$) to a phase variously described as ‘rotatory phase’ or ‘α’ phase whose structure has been matter of controversy in the literature. This phase characterizes a pre-melting state for n-alkanes; just a few degrees above $T_{0 \rightarrow \alpha}$ the crystal melts. The temperature range of existence for the α phase decreases with increasing the length of the chain and is also pressure-dependent. Even polyethylene seems to show a similar pre-melting state just a few degrees before melting.

The generality of the phenomenon to be understood was the reason which has convinced us of the need to perform a detailed spectroscopic study on the simplest system, n-nonadecane which shows the largest temperature range for the existence of the α phase.

We have studied the following molecules: n-nonadecane, n-nonadecane-2D_2 and n-nonadecane-10D_2 (hereafter referred to as C19, 2D2 and 10D2 respectively).

The known structural features of C19 are the following: crystalline C19 below 22.8 °C crystallizes in the orthorhombic lattice with two molecules per unit cell [152]. Chains are *trans*-planar and perpendicular to the plane formed by the methyl groups, thus being perpendicular to the lamella surface. At 22.8 °C a phase transition is observed to the α phase and melting occurs at 32 °C. The measured heat of transition is 3300 cal/mol, while the heat of melting is 10950 cal/mol [153]. The volume changes determined with a dilatometer at the solid–solid phase transition and melting are 2.5% and 10.5% respectively. The structure of the α phase below melting is reported to be 'hexagonal' with some indication of rigid rotations of the chains [154].

We add here information for the two selectively deuterated materials which were prepared in the authors' laboratory. There is no reason to think that 2D10 may not have the same structure as the parent molecule. The perturbations by the two deuterium atoms should be negligible regarding the static structure while they will affect the vibrational spectra. As a result of the 'regular' packing, the lamellae formed by these all-*trans* 10D2 molecules contain precisely in the middle a 'layer' of CD_2 groups. From the viewpoint of vibrations we expect normal crystal field splittings for the vibrations of the CD_2 since intermolecular phase coupling within the lattice may occur regularly. In contrast, 2D2 molecules in the crystal, while retaining the orthorhombic structure, are arranged at random with the heads labeled with CD_2 randomly distributed up or down within the crystal. Within Born–Oppenheimer approximation the intermolecular potential cannot distinguish between deuterated or undeuterated ends of the hydrocarbon chain. However, because of the random mechanical perturbations of the vibrations at the site of chain heads in this molecule, crystal field splitting of the CD2 vibrations is not expected (in spite of the orthorhombic packing). These predictions are indeed verified in the infrared and Raman spectra of both 10D2 and 2D2.

3.19.2 Vibrational Spectra of pure n-Nonadecane

N = 59 is the number of atoms in the C19 molecule. The symmetry point group of the single *trans*-planar chain of C19 is C_{2v} and its $59 \times 3 - 6 = 171$ vibrations can be classified in the following irreducible representations with their spectroscopic activity:

$$49A_1(\text{IR}, \text{R}) + 37A_2(\text{IR inactive}, \text{R}) + 47B_1(\text{IR}, \text{R}) + 38B_2(\text{IR}, \text{R})$$

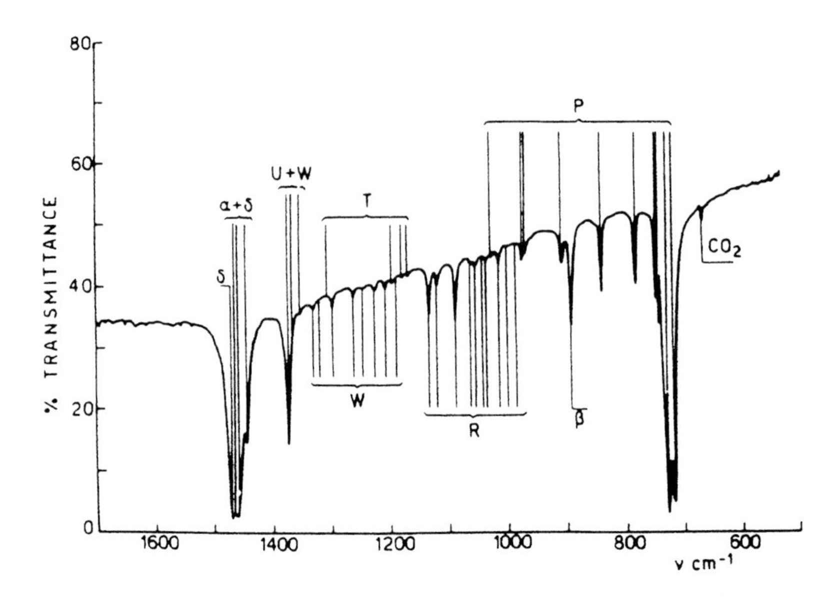

Figure 3-46. Survey infrared spectrum of n-nonadecane in the orthorhombic phase. Band progressions are indicated. For the labeling of the modes see text. R = skeletal stretching, $\beta = CH_3$ deformation.

The infrared spectrum of C19 in the solid orthorhombic phase is shown in Figure 3-46; we use this spectrum as reference in the present analysis. The Raman spectrum of the same material is given in Figure 3-47.

The identification of the large number of normal modes in the infrared and Raman spectra of these molecules becomes indeed unfeasible. Life becomes easier if we consider C19 as a segment of polyethylene made up by 17 CH_2 units and try to located the 17 normal modes lying in each frequency branch of the dispersion curves of infinite polyethylene (Figure 3-1). Each mode is then characterized by a phase shift φ_j between adjacent CH_2 units, $\varphi_j = 2\pi j/17$, (here $j = 0, \ldots, 17 - 1$). The possibility of observing all normal modes depends on their symmetry and on their intrinsic intensity in infrared or Raman, as well as on the dispersion of each phonon branch. As discussed in Section 3.9.2, we expect to observe in the infrared spectrum band progressions with decreasing intensity for branches with reasonable dispersion, while for flat branches band progression will be strongly compressed with strong overlapping.

The experimental observation predicted by theory has been discussed previously in detail by Snyder and Schachtaschneider [9, 34]. From the dispersion curves of Figure 3-1 it follows that no clear band progressions are expected for CH_2 stretching (d, overlapping) and CH_2 bending (δ, overlapping), while progressions are expected for CH_2 wagging (w, very weak in IR); in the CH_2 wagging progression we clearly locate W_5, W_7, W_9, and W_{11}; the other CH_2 wagging motions are

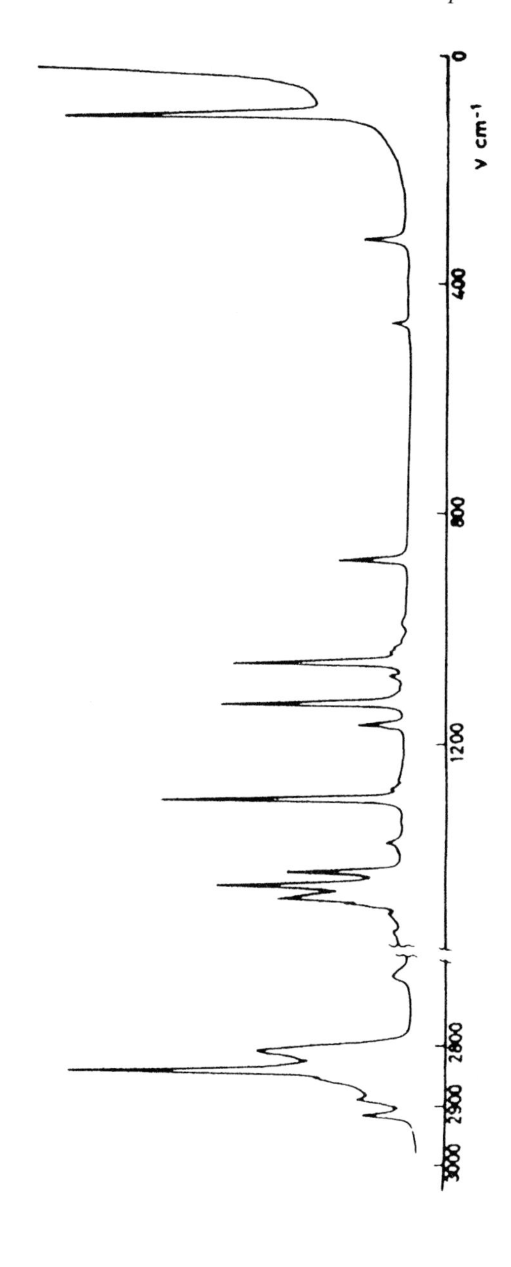

Figure 3–47. Survey Raman spectrum of orthorhombic n-nonadecane.

overlapped by the stronger CH_3 U mode near 1375 cm^{-1}. The expected progression of CH_2 twisting (t) must be very weak in IR while CH_2 rocking (P) must be strong/medium and clearly noticeable from 720 to near ~1100 cm^{-1}. C–C stretchings are weak in IR. For the band progression of the P modes we expect to observe nine components of B_2 species corresponding to quasi-phonons with odd j values. J = 1, 3, and 5 turn out to be almost overlapping and the others must be observed in the frequency range ~750–1100 cm^{-1}. The observed band progressions are clearly displayed in Figure 3-46.

The infrared bands at 1375 and 890 cm^{-1} are unquestionably assigned to the umbrella (U) and rocking (β) modes of the methyl groups respectively (Figure 3-46).

The Raman spectrum provides further information. It is generally known that in the Raman spectra only the limiting $\mathbf{k} \to 0$ modes for the infinite chain appear as strong or easily observable transitions. This is also the case for C19 where only the $\mathbf{k} \approx 0$ modes are observed.

The most important information comes from the observation of the band progression of the LAM modes (Section 3.16) with k = 1, 3, and 5 components observed at 124.5, 342, and 490 cm^{-1}. Using equations 3.53 and 3.54, the frequency of LAM1 at 124.5 cm^{-1} proves that the chain comprises exactly 19 C atoms.

The conclusions are as follows:

Conclusion no. 1

The infrared and Raman spectra provide unquestionable evidence that C19 below 22.8 °C consists of *trans*-planar molecule with 17 CH_2 units capped by two CH_3 groups. Further observation of the infrared spectrum for the orthorhombic phase shows the splitting (factor group splitting) of the δ modes and of many components of P modes into doublets with a measurable intensity ratio (Figure 3-48). Analogously, the Raman line associated to $\mathbf{k} \sim 0$ of the δ mode near 1460 cm^{-1} is split into a doublet (Figure 3-47). The number of components of the splitting is related to the number of molecules (2) in the unit cell. If the orthorhombic lattice is accepted, the intensity ratio of the components of the splitting in the infrared provides the value of the setting angle of the hydrocarbon chains in the unit cell ($\theta \approx 42°$). The shape of the Raman scattering in the 3000 and 1450 cm^{-1} range can be interpreted in terms of Fermi resonances (Section 3.17) uniquely between *trans*-planar chains.

Conclusion no. 2

IR and Raman spectra provide evidence that two *trans*-planar molecules are organized in the orthorhombic lattice, with setting angle $\theta = 42°$.

3.19.3 Temperature-Dependent Spectra

Next, we study the temperature-dependent infrared and Raman spectra of C19 through phase transition (T_α) ($T_\alpha = 22.8$ °C) and melting (T_m) ($T_m = 32$ °C). We first focus on the spectroscopy of C19. Each experimental observation is labeled and will be included in the overall analysis. In the IR spectrum:

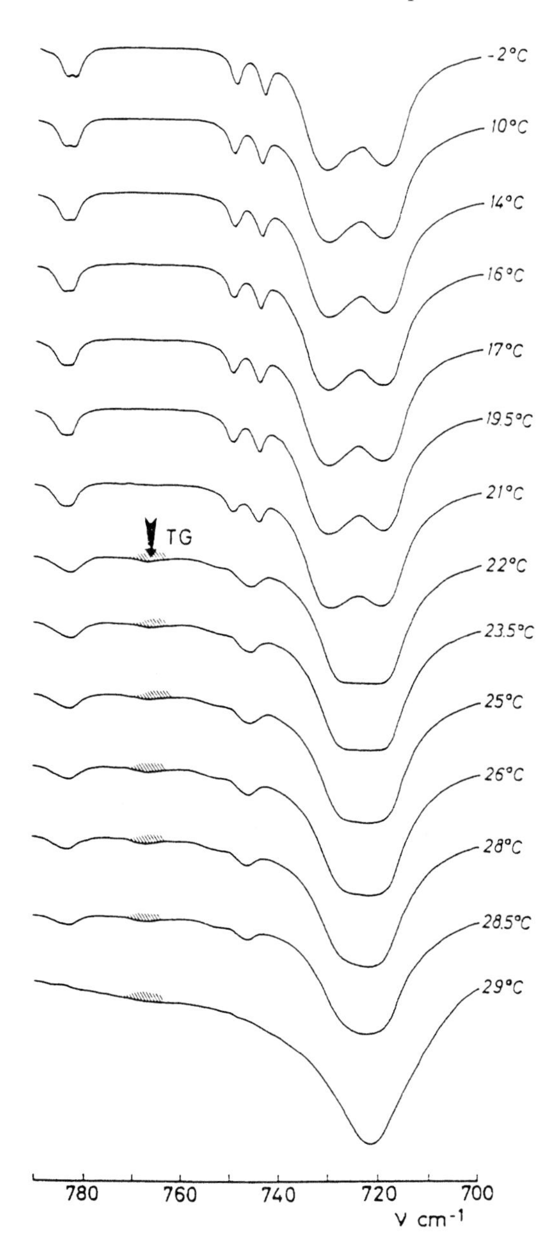

Figure 3-48. Temperature dependence of the infrared spectrum of n-nonadecane in the CH_2 rocking region showing that the crystal field doublets coalesce into singlets at the orthorhombic → α phase transitions indicating the existence of 'single chains'.

(a) At T_α the band progression of the P modes which shows factor group splitting at lower temperature loses all the splitting, but remains as prominent feature in the IR spectrum until T_m (Figure 3-48).
(b) The progression of the P modes disappears at T_m and is replaced by a broad band near 722 cm^{-1}.

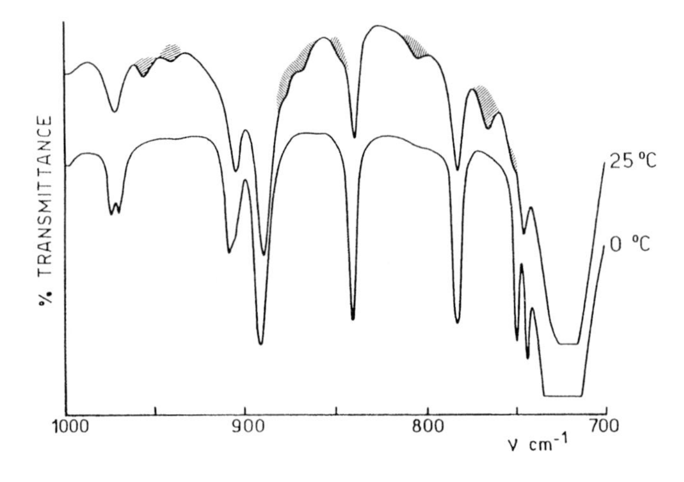

Figure 3-49. Band progression due to CH_2 rocking modes for n-nonadecane in the orthorhombic phase (at 0 °C) and in the α phase. Notice at 25 °C the appearance of a few weak bands and the disappearing of the crystal field splitting.

(c) At T_α the mode at 1375 cm^{-1} (U) shifts to 1378.5 cm^{-1}.
(d) At T_α a weak, clear and broad absorption appears at 766 cm^{-1} and remains a clear, weak and broad band in the melt (Figure 3-49).
(e) At T_α additional peaks are observed in the infrared at 766, 804, 845, 866, 877, 940, and 956 cm^{-1} (Figure 3-49). Some of these modes may be assigned to the even j values of the P band progression.
(f) At T_α in the CH_2 wagging progression we clearly locate W_5, W_7, W_9, and W_{11} as in the case of the solid, but the frequencies of these components are rigidly shifted toward higher frequencies for a few wavenumbers (Figure 3-50). The other CH_2 wagging motions are overlapped by the CH_3 U mode which shifts to ~1378 cm^{-1}.
(g) At T_α the CH_3 β mode shifts from 891 to 889 cm^{-1}.
(h) At T_α a weak, but unquestionable peak arises at 1342 cm^{-1} (Figure 3-50).

Upon melting:

(i) At T_m the band progression of the CH_2 wagging disappears (Figure 3-50).
(j) At T_m the peak at 1342 cm^{-1} gains considerable intensity (Figure 3-50).
(l) At T_m the two well-known defect modes absorptions near 1353 cm^{-1} and 1370 cm^{-1} appear together with the broad peak centered near 1306 cm^{-1} (Figure 3-50).

In the case of the Raman spectrum the most meaningful observations are the following:

(m) Above T_α the features of the $\mathbf{k} = 0$ modes of the infinite *trans*-planar molecule are still the dominant features (Figure 3-51b).
(n) At T_α the lattice band at 98 cm^{-1} disappears.
(o) At T_α the LAM1 frequency shifts from 124.5 to 123.2 cm^{-1} (Figure 3-52).

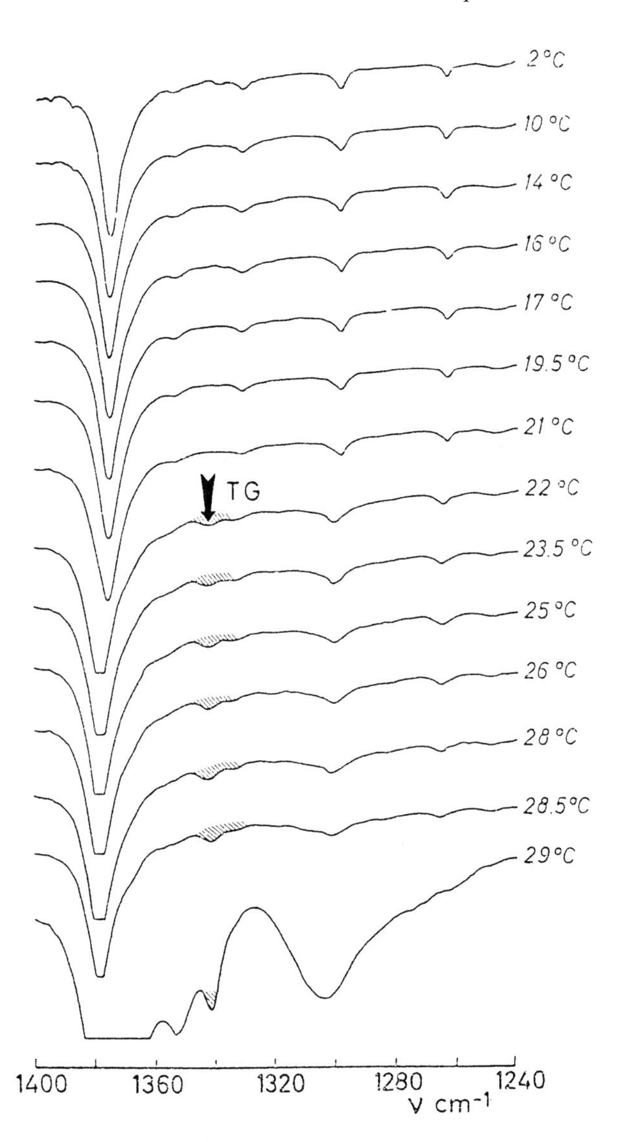

Figure 3-50. Temperature-dependent infrared spectra of n-nonadecane showing changes through the solid-α (T ~ 22 °C) and α-liquid (T ~ 29 °C) phase transitions.

(p) At T_α the A_g component of the factor group splitting of the CH_2-bending mode near 1417 cm^{-1} practically disappears, even if not completely (Figure 3-53).

(q) At T_α the valley between the A_g (2845 cm^{-1}) and B_{1g} (2882 cm^{-1}) modes of the C-H-stretching region becomes emptier and the band at 2845 cm^{-1} sharpens.

(r) At T_α a slight increase in the scattering near 1090 cm^{-1} is observed.

(s) At T_α on the lower frequency side of the well-isolated peak near 890 cm^{-1} (β mode) two weak bands are observed at 866 and 844 cm^{-1} (Figure 3-54).

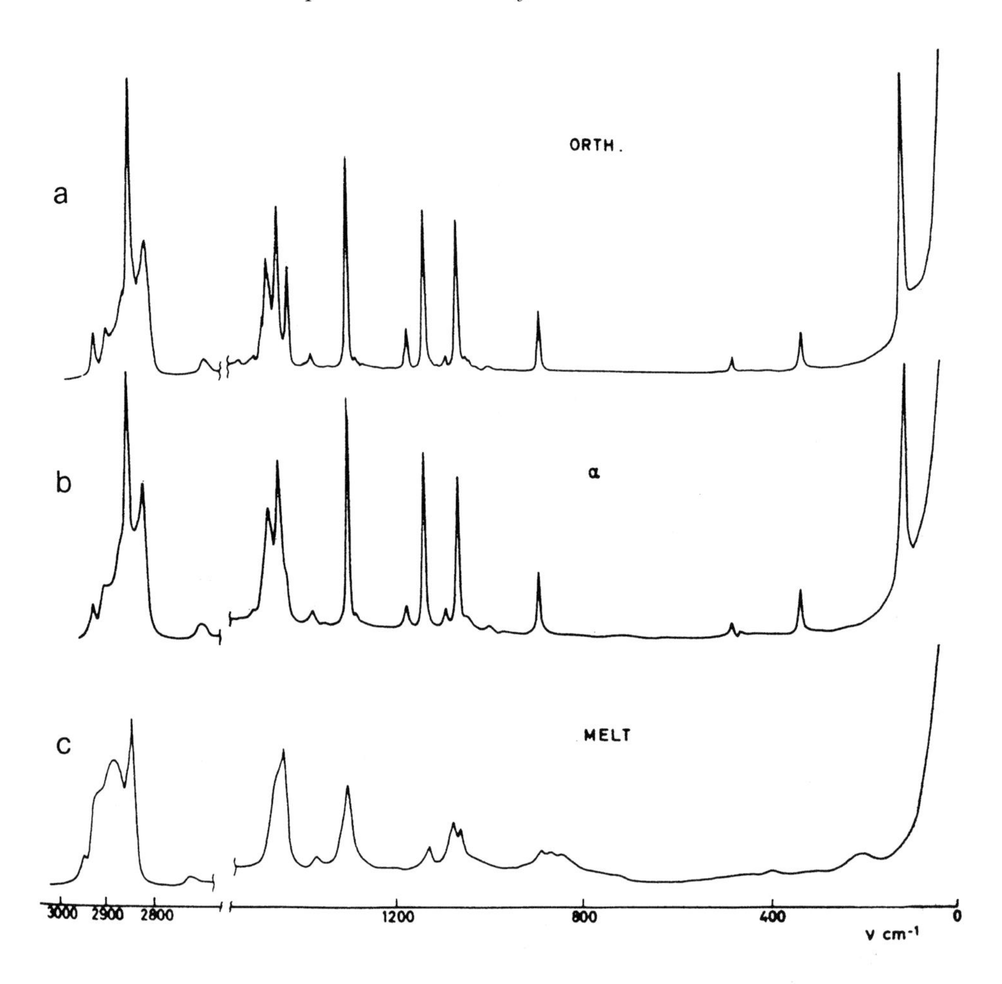

Figure 3-51. Changes of the Raman spectrum of n-nonadecane in the various phases: (a) orthorhombic, (b) α, (c) liquid.

The Raman spectrum in the melt shows the following features:

(t) At T_m the LAM1 mode disappears completely and is replaced by the so-called 'pseudo LAM', a broad and weak scattering near 220 cm^{-1} common to all liquid n-alkanes, melt polyethylene or any molecule containing long CH_2 chains in the 'melt' state (Figure 3-51c).

(u) At T_m the Raman lines at 866 and 844 cm^{-1} gain in intensity and, together with the line near 890 cm^{-1}, form a strong characteristic triplet with components of similar intensity (Figure 3-51c).

(v) At T_m the Fermi resonance line at 1461 cm^{-1} decreases while the Fermi resonance line at 2930 cm^{-1} increases.

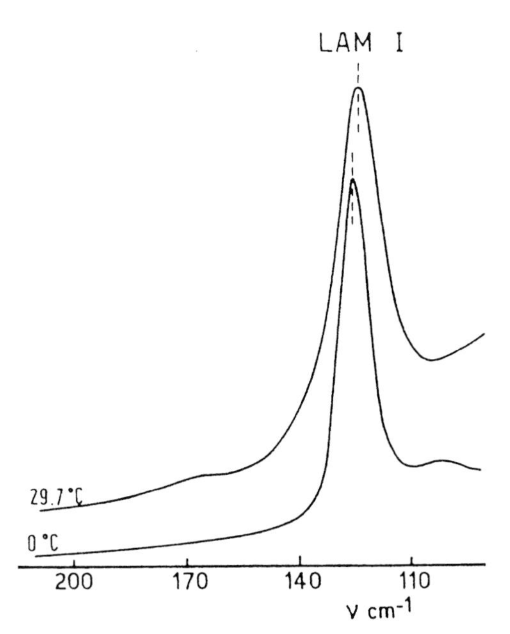

Figure 3-52. Details of the changes of the Raman spectra of n-nonadecane at 0 °C (orthorhombic) and 29.7 °C (α phase). The lowering of the frequency of LAM1 shows that interlamellar forces have weakened.

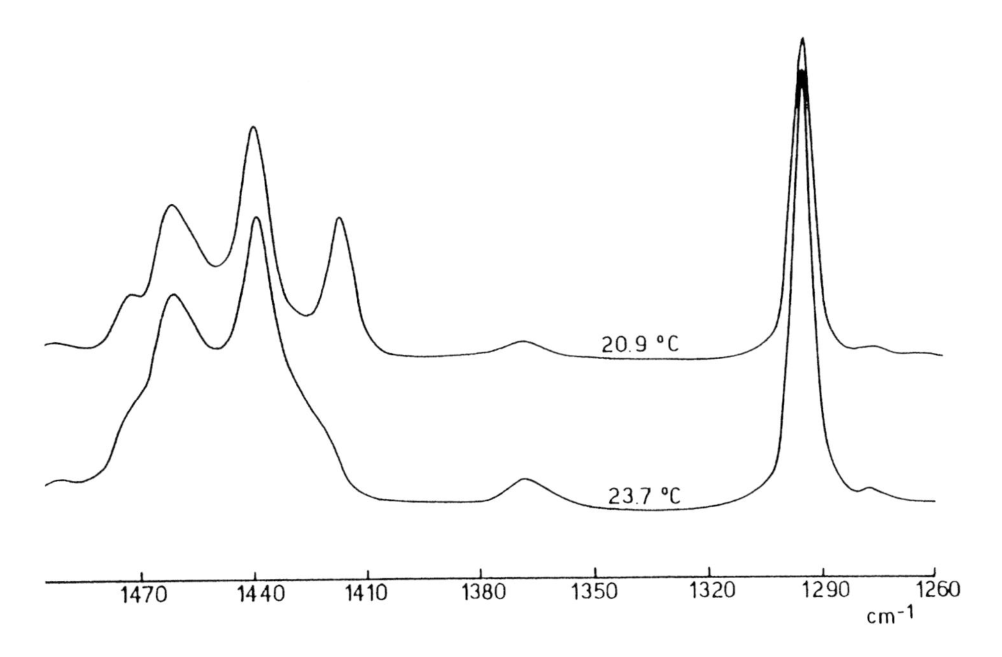

Figure 3-53. Details of the Raman spectrum of n-nonadecane just above and just below the solid $\rightarrow \alpha$ phase transition. The crystal field splitting near 1430 cm^{-1} disappears and only the single chain bending deformation remains at 1440 cm^{-1}. The Fermi resonance line near 1460 cm^{-1} is also slightly perturbed.

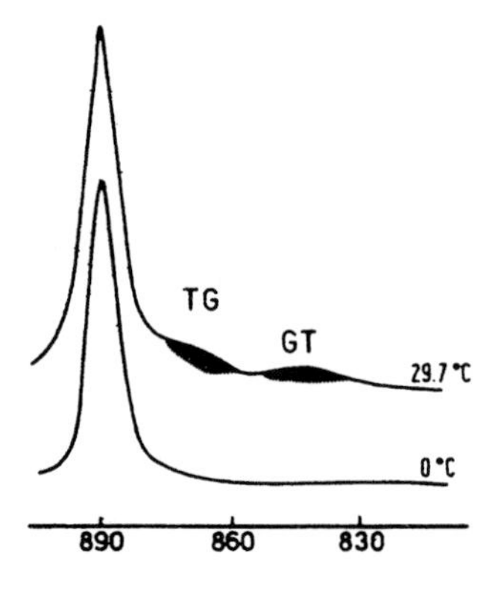

Figure 3-54. Details of the Raman spectrum of n-nonadecane in the orthorhombic (0 °C) and α phases (29.7 °C). The line near 890 cm^{-1} indicates that all chain ends are in the TT conformations. In the α phase a small population of chains has tilted their heads in TG and GT conformations.

(w) At T_m in the CH stretching range the line at 2882 cm^{-1} is submerged by the broad scattering near 3000 cm^{-1} from which two peaks emerge at 2850 (d^+) and 2875 cm^{-1} (d^-). The typical Raman spectrum of the "melt" is observed. It should be compared with Figure 3-43.
(x) At T_m near 720 cm^{-1} a clear stepwise increase of the diffuse broad background scattering is observed.
(y) At T_m the line near 1090 cm^{-1} becomes strong at the expenses of the $\mathbf{k} = 0$ singularities of the infinite *trans*-planar chain at 1135 and 1074 cm^{-1} in the orthorhombic and α phases. This line is characteristic of the stretching of the C–C bonds in *gauche* conformation (Figure 3-51c).

Conclusion no. 3

At the orthorhombic to α phase transition the disappearance of the factor group splitting (observations (a) and (n)) indicates that the orthorhombic lattice with two molecules per units cells disappears and is replaced either by a unit cell with one molecule per unit cell or by a lattice where the tridimensional order is practically lost.

Chains, however, remain mostly straight *trans*-planar (observations (f), (m), (o), (p) and (q)). Perturbations appear at chain ends (observations (c), (g) and (o)), indicating that the medium surrounding the CH_3 group has changed, in particular the inter-lamellar forces have weakened (observation (o)). The following structures appear at the end of the molecule $-CH_2(T)-CH_2(G)-CH_2-CH_3$ and $-CH_2(G)-CH_2(T)-CH_2-CH_3$ (observation (s)). This accounts for the lowering of the overall symmetry which causes the appearance of a few of the components of the P modes inactive for the orthorhombic structure (observation (e)). It seems likely that some sort of disordering at the surface of the lamella takes place with some chains having their ends tilted in gauche conformations (observation (r)). The occurrence of a few GTG, GTG′ and GG conformational defects (or unspecified gauche defects) is indicated (observations (h) and (r)); the topology of these defects along the molecules cannot, so far, be defined.

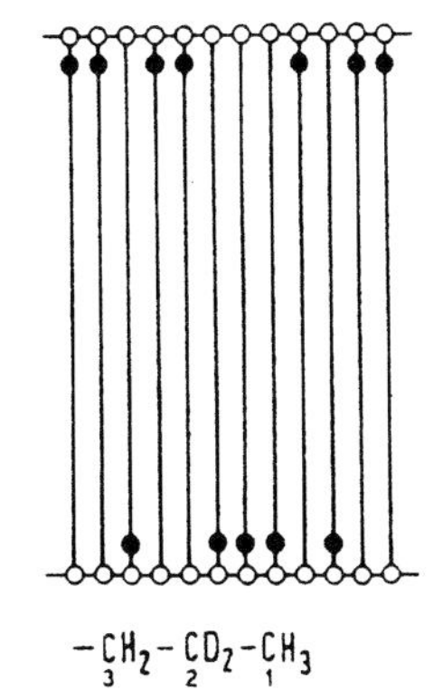

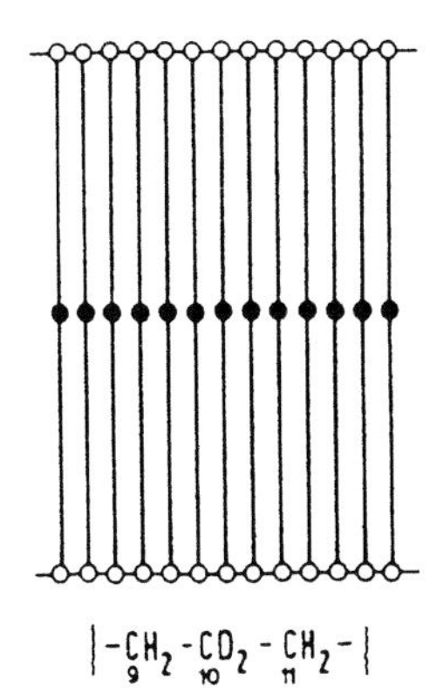

Figure 3-55. Sketch of assumed locations of selectively deuterated chains of n-nonadecane (2D2 and 10D2, see text) in the orthorhombic crystal.

Conclusion no. 4

The process of melting can be described in the following way: the *trans*-planar structure of the *trans* chains collapses (observations (i), (j) and (t)) and conformational disorder takes place accompanied by the appearance of the following conformational defects: GG, GTG, GTG′ (observations (j) and (l)), end-gauche (observation (u)) and gauche in general (observations (b), (t), (y), (v), (w), and (x)). In practice we find characteristic signals of liquid-like structures.

With the aid of the selectively deuterated species 2D2 and 10D2 further details on the processes of solid–solid phase transition and melting can be derived.

First, we consider the 2D2 molecule. Of necessity, the process of packing of *trans*-planar chains to form an orthorhombic crystalline lamella cannot distinguish deuterated and undeuterated chain ends; the resulting lamella can be described by a random distribution of 'up' or 'down' chains (Figure 3-55). With such chains we can distinguish the conformational behavior of the heads and/or the tails of the chains. It becomes thus possible to be sure of the origins of the signals coming from conformational defects.

The most direct information comes from the frequencies of the CD_2 rocking which have been calculated with normal coordinate calculations and are reported in Table 3.1.

The infrared spectrum of 2D2 in the 670–600 cm^{-1} range shows only a band near 620 cm^{-1} below T_α (Figure 3-56). The band is not split, as mentioned above, because the precise phase coupling is lost by the random distribution of 'up' and 'down' chains. At the solid–solid transition a very weak additional band appears near 655 cm^{-1} (Figure 3-56) indicating the formation of *gauche* conformations at the deuterated chain ends, thus reiterating what has been already learned from C19. From a quantitative measure of the absorption intensities of the these modes it transpires that approximately 25% of chain ends are conformationally distorted.

The infrared spectrum of 10D2 provides further information (Figure 3-57). For the orthorhombic phase the line near 620 cm^{-1} is split because in this case the phase coupling is achieved since translational symmetry is obtained within the crystal. At

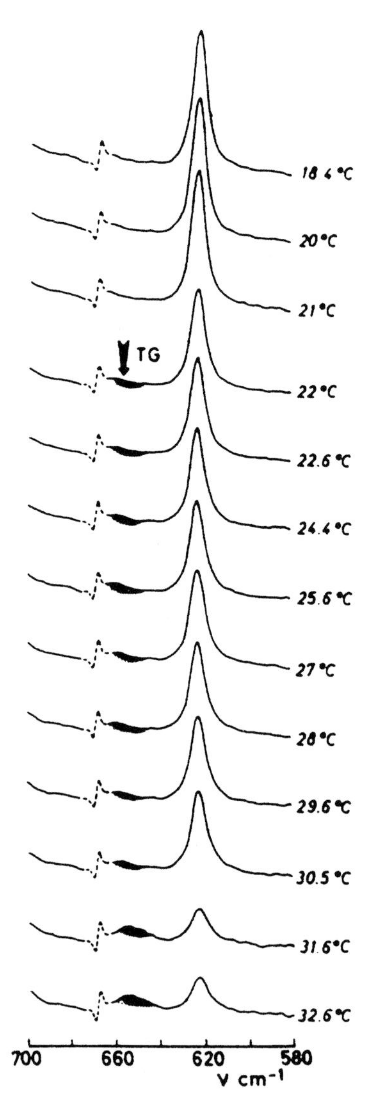

Figure 3-56. Temperature dependence of the infrared spectrum of n-nonadecane-2D_2 through orthorhombic $\rightarrow \alpha \rightarrow$ phase transitions. Notice that (i) in the α phase a small number of deuterated chain heads has taken up the TG conformation which increases dramatically in the melt, (ii) in the melt a certain number of deuterated chains do not tilt their heads but still keep the TT conformation.

T_α the doublets disappear since chains move and become uncorrelated (isolated). Since no absorption is observed near 655 and 677 cm^{-1} it means that in the middle no conformational distorsions are generated in the α phase. The liquid-like structure is, however, observed at the melting.

Conclusion no. 5

The final model emerging from this part of the study is the following: starting from a regular and ordered orthorhombic structure in which all *trans*-planar chains are

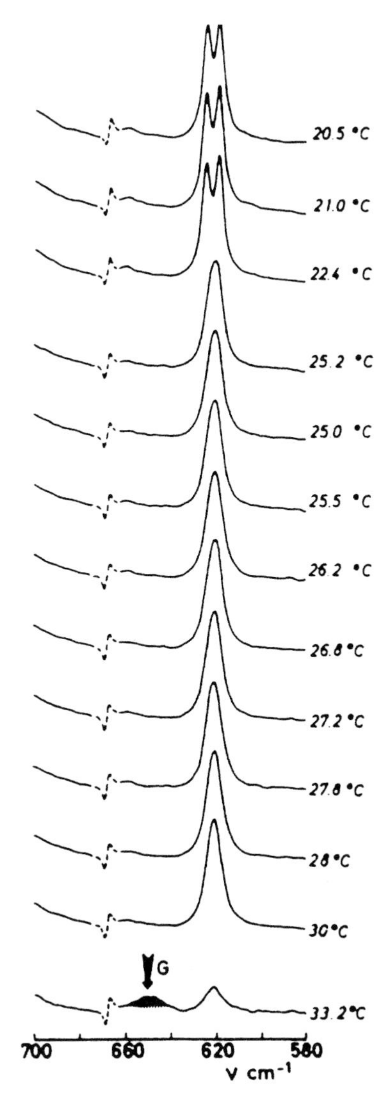

Figure 3-57. Temperature dependence of the infrared spectrum of n-nonadecane-10D_2 through orthorhombic $\rightarrow \alpha \rightarrow$ phase transitions. Notice that (i) the orthorhombic crystal disappears at 22.4 °C, (ii) in the α phase at the center of the lamella no conformational kinks are formed while they certainly appear in the melt (33.2 °C) Combination of the data of this figure and of Figure 3-56 are direct evidence of 'surface melting'.

arranged in a crystal with a setting angle of 42° at T_α the lattice expands anisotropically, interlamellar forces weaken and a conformational disordering takes place at the surface of the lamella. Such disordering is generated by the conformational deformation of approximately 25% of the chains which tilt their head with gauche conformation either in position 2–3 or 3–4. The inside of the lamella is not perturbed and all chains are still *trans*-planar (Figure 3-58).

The next question which remains to be answered is whether surface disordering is generated by the longitudinal motion of the chains. The following experiment has been made [155].

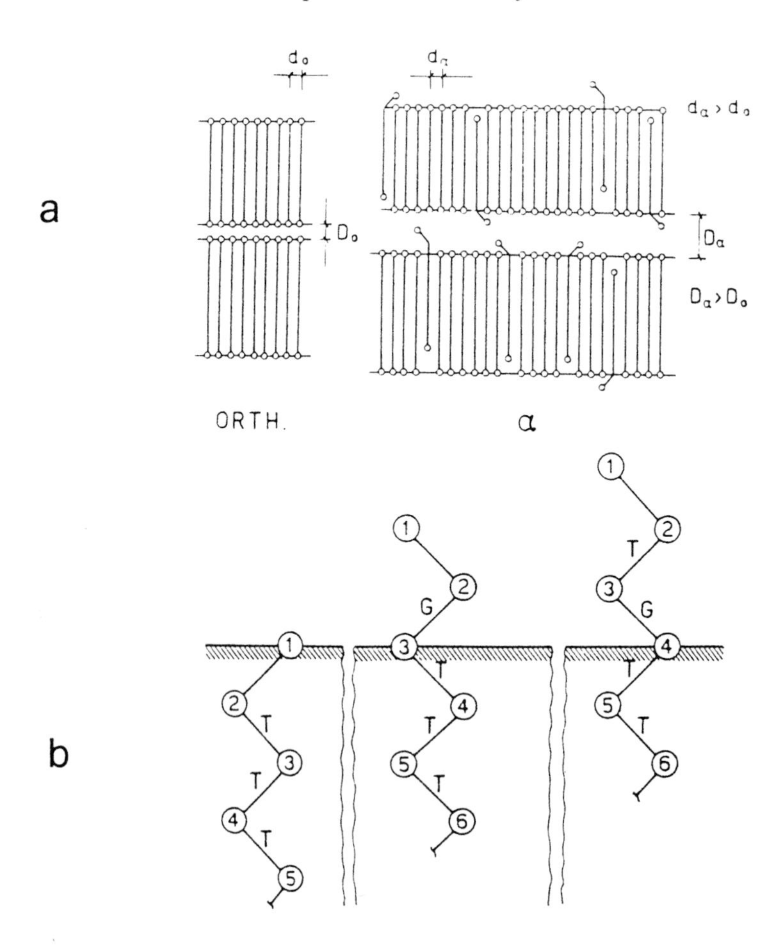

Figure 3-58. Sketch of the mechanism of phase transition of n-nonadecane based on detailed spectroscopic evidence. (a) Schematic overall view; (b) details on a molecular level.

Let us take two samples of crystalline orthorhombic n-alkanes with an odd number of carbon atoms. They show properties identical to those observed for C19. Let us make by suitable methods a mechanical mixture say of C21 and C23. The infrared spectra of the mechanical mixtures at low temperature are just the superposition of the spectra of the two independent components. Let us focus on the crystal field splitting, say of the doublets near 940 cm^{-1} (C23) and 925 cm^{-1} (C21) (Figure 3-59). If the mechanical mixture is brought near the solid–solid phase transition the infrared spectrum evolves with time. Time- and temperature-dependent spectra show that the doublets become singlets thus showing that chains have moved (Figure 3-59). The final product is a mixed crystal of C21 and C23 as also proved by parallel DSC studies. These studies were first reported by Ungar and Keller [156] and were re-examined and interpreted as reported in [155].

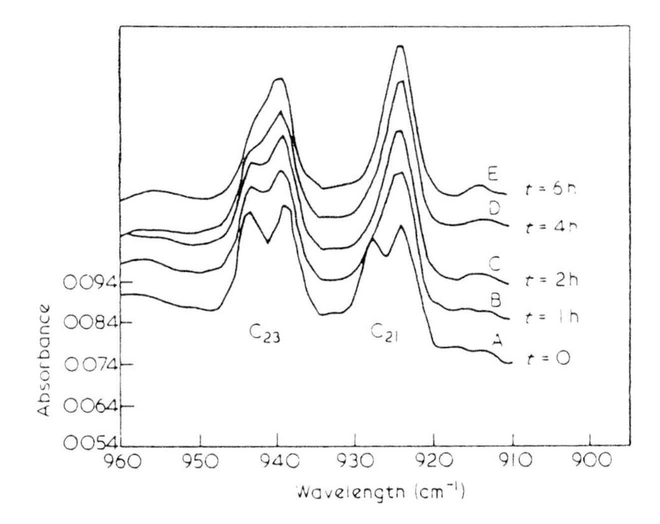

Figure 3-59. Time-dependent infrared spectrum in the 960–900 cm^{-1} range showing the P_{15} component of the CH_2 rocking progression band of n-$C_{21}H_{44}$ and the P_{17} component of n-$C_{23}H_{48}$. Evolution with annealing time. Annealing temperature 30 °C, A(t = 0); B (t = 1 hr); C (t = 2 h); D (t = 4 h); E (t = 6 h). At t = 0 the sample consists of a mechanical mixture of the two materials. At t = 0 both solids are crystalline orthorhombic as shown by the correlation field splitting doublet. By increasing the time of annealing both doublets coalesce into a singlet indicating that chains have migrated and formed a mixed crystal.

Figure 3-60 shows the kinds of kinetic studies of the mixing process which have been made.

Conclusion no. 6

In the α phase chains move longitudinally and are 'squeezed out of the lamella'; the onset of such motion occurs near the solid–solid transition temperature.

This is in agreement with recent measurements of quasi-elastic neutron scattering by Gullaume et al. on C19 at 300 °K who determined the existence of translational longitudinal diffusion with discrete steps of one or two CH_2 units with a measured diffusion coefficient $D = 6 \times 10^{-6}$ cm^2s^{-1} [157]. Statistical thermodynamics calculations give theoretical support to this kind of phenomenon [158]. These conclusions find a striking support by the calculations of molecular dynamics by Klein [159] which describe precisely all the details that we have found from spectroscopy. Moreover, recent experiments of temperature-dependent scanning tunnelling microscopy by Rabe et al. [160] give just 'photographic' support at the molecular level of the process as described by spectroscopy, namely: (i) first the chains organized in an ordered lattice; (ii) then the onset of longitudinal diffusion which generates

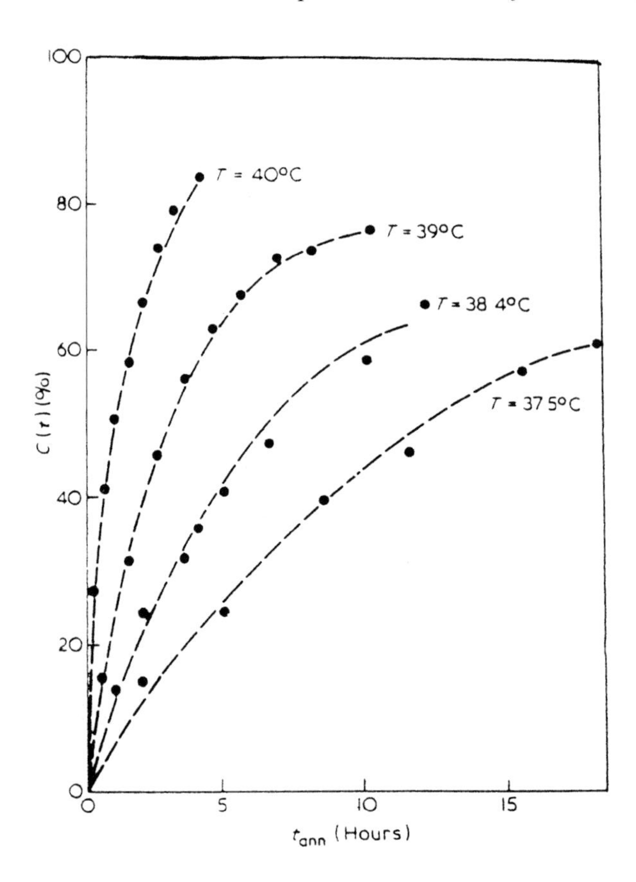

Figure 3-60. 1:1 molar mixture of n-$C_{23}H_{48}$ and n-$C_{25}H_{52}$. Kinetics of the process of diffusion at four different temperatures. Data from calorimetric measurements.

conformational deformations at chain ends; (iii) diffusion which generates the interpenetration of chains; and (iv) the full collapse towards a liquid-like structure.

Conclusion no. 7

As a consequence of such longitudinal motion, if crystallites of n-alkanes are suitably brought together thermally activated transport of matter can take place and diffusion of molecules from one crystallite to another can occur. The kinetic of diffusion is function of temperature, length of the chain, contacts between crystallites (i.e., pressure put on the sample), etc.

We can proceed further and try to understand the mechanism at the molecular level which drives the chains outside their own lamella by a thermally activated process.

We refer the reader to [103], [155] and [145] for a first analysis of the various data. We simply mention the line of thought followed by various authors on this rather complex problem.

Several theoretical models have been proposed in an attempt to account for the existence of the α phase and for the longitudinal motion. First, the general idea was that in the α phase chains rotate rigidly [152–154]. Next, models of rotational diffusion coupled with longitudinal translations of one CH_2 were proposed based on NMR data [161, 162]. Elastic neutron-scattering experiments introduced the idea of diffusional cooperative rotations or of independent translational motions with jump length of one CH_2 unit followed by 180° rotational jump [163]. A translation of two CH_2 units was proposed [164] in agreement with calculations by McCallough [165]. Most of the models proposed consider that in the solid state chains perform large-amplitude collective rototranslational motions. McCullough has presented an interesting model which shows by calculations that chain can find a favorable energy path if they librate rigidly with large amplitude librations and translate along the chain axis.

The knowledge at the molecular level of the mobility of chain molecules in the solid state is of basic importance for the understanding of the general processes of annealing of chain molecules. Attempts have been made to derive information using band shapes in the Raman spectra, especially focusing on the CH-stretching vibrations (Section 3.18).

As already stated at the end of Section 3.18, we have measured the temperature-dependent band widths and the frequency of the CH_2 d^- mode for several crystalline n-alkanes and polyethylene and found no sizable broadening or upward frequency shifts until they approach the melting point. All the theoretical models that imply large amplitude librotorsional motions are not supported by the spectroscopic criteria discussed in Section 3.18. We found evidence of large-amplitude libro-torsional oscillations only when the n-alkane chains are included as clathrates in various systems.

Conclusion no. 8

Based on the analysis of temperature-dependent Raman band shape and frequency shift of the Raman active antisymmetric CH_2-stretching modes (Section 3.18), contrary to the case of urea clathrates, in the solid crystalline state (including the α phase), n-alkane chains do not perform large-amplitude libro-torsional motions.

We must search for alternative molecular models which may account for longitudinal diffusion with transport of matter without implying large amplitude librotorsional oscillations. A conceptually and, at least, intellectually exciting model which may help in this search has been presented by Mansfield and Boyd [166] who attempted to account for molecular relaxation phenomena. We label this model as Utah-twist. The main message by these authors is that calculations suggest that an orthorhombic lattice of n-alkanes can host a chain distorted by a gentle and long twist. The long twist is achieved with a uniform succession of small torsions about the C–C bonds which is extended over approximately 12–15 CH_2 units. At the end of the gentle twist the chain is flipped by 180° with respect to the head of the twist. Such a long twist (which we label as ‘twiston’) can propagate along the polymethylene chain without finding a barrier to its propagation. Based on the atom–atom potential chosen by the two above authors the activation energy for generating a

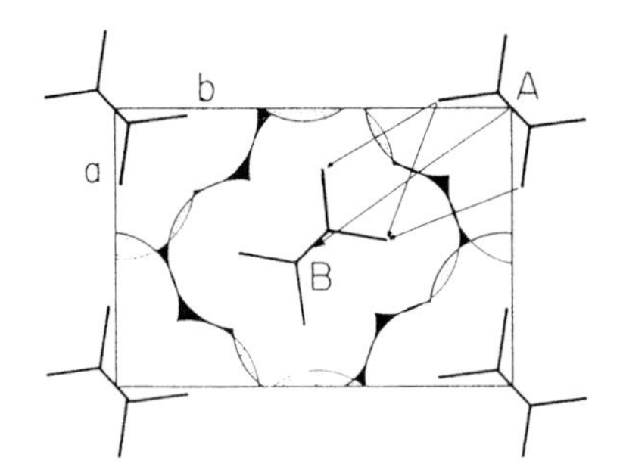

Figure 3-61. Projection onto the ab face of the structure of the orthorhombic cell of polyethylene (two molecules per unit cell, setting angle $\theta = 42°$).

twiston is estimated to be ~10 kcal/mol. The twiston model implies for a polymer: (i) the initial formation of a disordered region near the surface of the crystal; (ii) the propagation of the twiston through the crystal along the chain axis; and (iii) the disappearance of the twiston at the opposite surface leaving the chain rotated by 180° and translated by one CH_2 unit.

The idea is formally very appealing since, from the spectroscopic viewpoint, it implies a collective mobility of the polymethylene chain which accounts for the observed longitudinal mobility, surface melting, or disordering and transport of matter without requiring large-amplitude libro-torsional motions which are not detected by vibrational spectroscopy, as previously discussed.

The twiston model has been further expanded by Mansfield [167] and later by Skinner and Wolynes [168]. We shortly mention the physics behind since it may be applied to other cases in polymer physics. The whole problem of polymethylene chains is reviewed in [169].

Let us take the structure of the orthorhombic unit cell of polyethylene (Figure 3-61) and consider the CH_2 units as point masses of mass m at a distance a from the axis of the molecule. The units can rotate by an angle Θ about the molecular axis. When they rotate they are subject to the elastic response C by the torsions between two adjacent (CH_2) units. From Figure 3-61 it is apparent that when the chain rotates as a rigid body about its axis in an orthorhombic lattice it probes a twofold periodic potential $V(\Theta)$ from the crystal field.

The Hamiltonian proposed by Mansfield has the following form

$$H = \sum_i [V(\Theta_i) + (1/2)m\,a^2\dot{\Theta}_i^2 + (1/2)C(\Theta_i - \Theta_{i+1})^2] \tag{3-61}$$

In Eq. (3-61) the first term represents the contribution from the crystal potential, the second term the kinetic energy, and the third is the quadratic elastic contribution. In a gentle twist it can be assumed that Θ_i do not change much from one unit to the other. According to Mansfield one can make the continuous approximation and reach a solution of the type:

$$\Theta(x - x_0 - vt) = 4 \arctan \exp\,[\pm\gamma(x - x_0 - vt)] \tag{3-62}$$

The soliton solution gives the values of Θ_i which describes a twist in the chain which, in the absence of friction, propagates keeping the same shape and velocity v. The Utah-twist (twiston) is then approximated to a solitonic wave whose velocity of propagation is estimated by Mansfield (using some molecular parameters) to be $\sim 10^5$ cm s^{-1}.

The final concept which emerges is that in the orthorhombic crystal one can conceive the existence of conformational collective excitations thermally generated which extend over approximately 20 CH_2 units and move along the chain with a certain velocity.

The main problem is to prove the existence of such mobile twistons. The idea of using spectroscopy implies the fact that we should be able to detect somewhere in the vibrational spectrum absorption and/or scattering associated to vibrations localized on the twiston. Calculations have been made (i) of the density of vibrational states of a long polymethylene chain containing a random distribution of twistons and (ii) of the infrared spectrum (frequencies and intensities) of the molecule C19 containing a twiston. A comparison has been made with the infrared spectrum of C19 in the all-*trans* structure [169] (Figure 3-62).

In these calculations the twiston is assumed not to be mobile, but pinned at a given site of the chain. The fact that the twiston is mobile along the chain in principle does not deny the possibility, under certain conditions, to observe its vibration in the infrared/Raman spectra. The conditions are the following:

(a) Since the band width is inversely proportional to the vibrational lifetimes of each normal vibration, if the velocity of the twiston is small the vibrational lifetime is long enough to allow the formation of its own vibrational mode and can give rise to an observable IR or Raman band. If the twiston moves too fast, the vibrational lifetimes are too short and the band broadens and flattens in a broad continuum. Simple calculations show that if a band of the twiston has to be observed with $\Delta\nu_{1/2} \approx 20$ cm^{-1} it sets an upper limit to the velocity of the twiston of $\approx 10^4$ cm^{-1} which is not too different from the velocity estimated by Mansfield.
(b) The concentration of twistons has to be large enough to generate a detectable absorption in IR or scattering in the Raman. The concentration of twistons depends on their energy of formation which has to be low if twistons are to be seen spectroscopically. The energies calculated for the Utah-twist [166] are much too high and would make the twiston undetectable by spectroscopic techniques.

The results of the theoretical calculations of the spectra of twistons and a few attempts to catch spectroscopic signals possibly associated to twistons are reported in [169]. The evidence collected so far is neither compelling nor completely negative. The model needs to be supported or dismissed by other physical techniques.

The model of twistons which move as 'solitary waves' was also proposed by Dwey-Aharon et al. [170] in the case of polyvinylidene fluoride $(CH_2{-}CF_2)_n$ which upon poling transforms from the helical structure (TGTG′) (labeled either as β form or form II) into a planar all-*trans* one (α, or form I). Form I is technologically

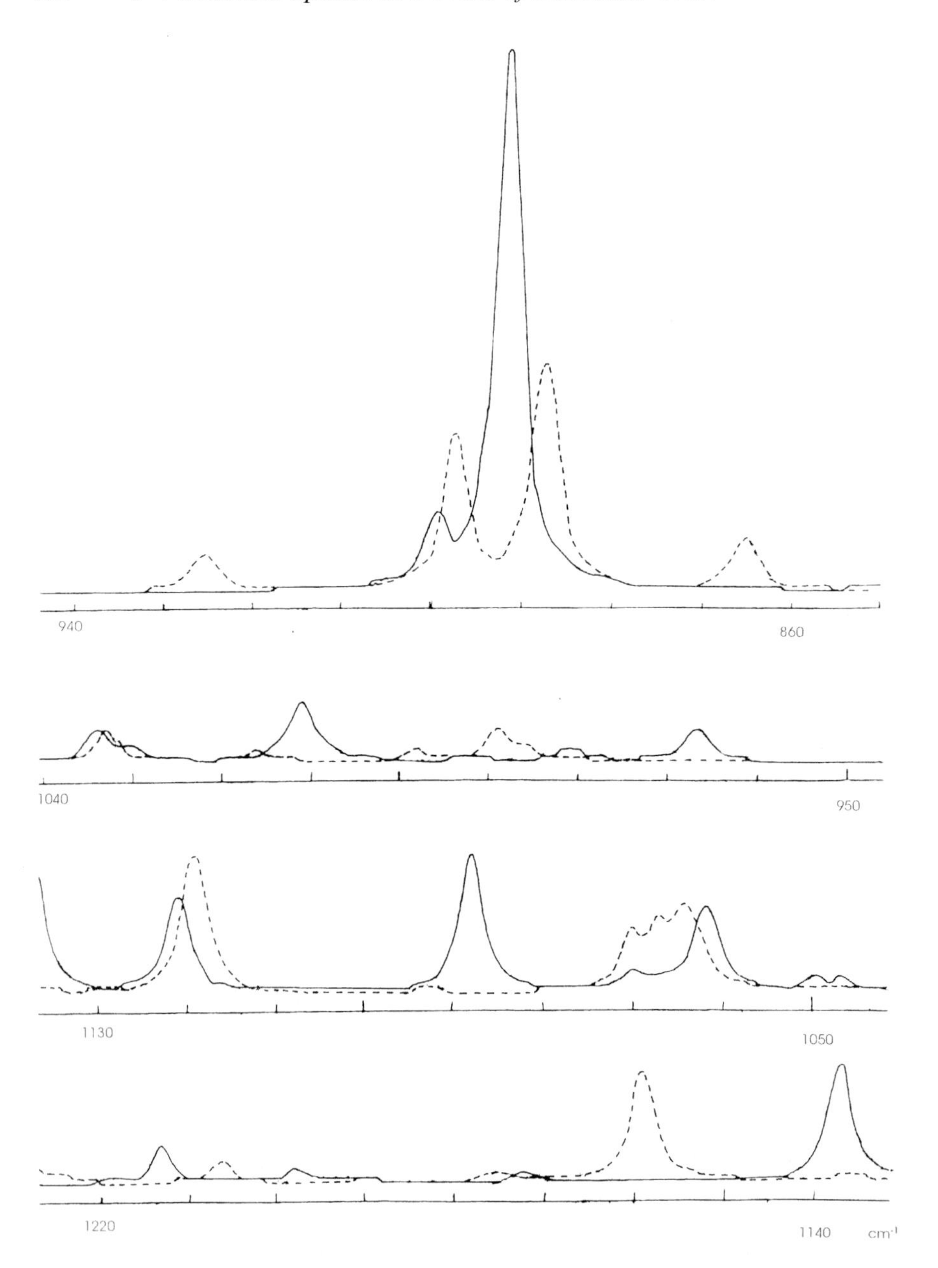

Figure 3-62. Searching for twistons: section of the calculated spectrum of n-nonadecane (—) in the all-*trans* planar conformation; (- - -) with the conformational distortion like a soliton according to Mansfield and Boyd. Both frequency and intensity were calculated.

very important, since having a strong dipole perpendicular to the chain axis shows remarkable pyro and piezoelectric properties widely used in modern technology. According to Dwey-Aharon when the conformationally disordered material existing in form II is placed in a strong electric field (poling) conformational twistons are generated which propagate and yield form I. The spectroscopy of polyvynilidene fluoride and the search for twistons has been treated in [171].

Finally, the concept of twiston has been again introduced in a spectroscopic work to account for the set of peculiar experimental data collected in the study of the phase transition in ethylene–tetrafluoroethylene alternating copolymers [172].

Acknowledgments

The contributions to knowledge presented in this chapter are the result of an enthusiastic collaboration of many researchers and students who worked in our group at the Politecnico of Milano. Our warmest thanks are extended to Dr. M. Gussoni, who laid the background for the development of this science.

3.20 References

[1] H. H. Nielsen, *Handbuch der Physik*, Springer, Berlin, Vol. 37/1, 1959.
[2] G. Herzberg, *Infrared and Raman Spectra of Polyatomic Molecules*, Van Nostrand, Princeton, USA, 1963.
[3] E. B. Wilson, J. C. Decius and P. C. Cross, *Molecular Vibrations*, McGraw Hill, New York, 1955; V. Volkenstein, M. A. Eliashevich and B. Stepanov, *Kolebaniya Molecul* II, Moscow, 1949; S. Califano, *Vibrational States*, Wiley, New York, USA, 1976.
[4] For a general discussion see: W. Person and G. Zerbi (Eds.), *Vibrational Intensities in Infrared and Raman Spectroscopy*, Elsevier, Amsterdam, The Netherlands, 1984; L. A. Gribov, *Intensity Theory for Infrared Spectra of Polyatomic Molecules*, Consulting Bureau, New York, USA, 1964.
[5] M. Gussoni in *Advances in Infrared and Raman Spectroscopy*, (Eds. R. J. H. Clark and R. H. Hester) Heyden, London, England, 1979, vol. 6 p. 96.
[6] M. Gussoni, C. Castiglioni and G. Zerbi, *J. Mol. Struct*, **1989** *198*, 475 M. Gussoni, C. Castiglioni, M. N. Ramos, M. Rui and G. Zerbi, *J. Mol. Struct*. **1990** *224*, 225 C. Decius, *J. Mol. Spectry*, *57*, 384 **1975**; A. J. Van Straten and W. Smit, *J. Mol. Spectry.*, **1976** *62*, 297 M. Gussoni, C. Castiglioni and G. Zerbi, *J. Phys. Chem.* **1984** *88*, 600.
[7] S. Trevino and H. Boutin, *J. Macromol. Sci.*, **1967** *A1*, 723.
[8] E. Curtiss, Thesis, University of Minnesota, 1958.
[9] J. H. Schachtschneider and R. G. Snyder, *Spectrochim. Acta*, **1963** *19*, 17.
[10] The most known set of original computing programs which has later originated many other programs for dynamical and force constant calculations is that written by J. H. Schachtschneider, Shell Rep, 57/65.
[11] See, for instance, J. L. Duncan, *Force Constant Calculations in Molecules*, Specialist Report, Chemical Society, London, 1975, vol. 23, p. 104; G, Fogarasi and P. Pulay, in *Vibrational Spectra and Structure*, (Ed. J. R. Durig) Elsevier, Amsterdam, 1985, Vol. 14, p. 125.

[12] M. Gussoni, C. Castiglioni and G. Zerbi, *J. Phys. Chem.*, **1984** *88*, 600.
[13] M. Gussoni, M. Ramos, C. Castiglioni and G. Zerbi, *J. Mol. Struct.*, **1988** *174*, 47.
[14] P. Pulay, X. Zhou and G. Fogarasi, in *Recent Experimental and Computational Advances in Molecular Spectroscopy*, (Ed. R: Fausto), NATO ASI Series C, Vol. 406, p. 99; W. B. Person, K. Szczesniak and J. E. Del Bene, ibidem, p. 141.
[15] K. Palmö, L.-O. Pietilä and S. Krimm, *Computers Chem.*, **1991** *15*, 249; K. Palmö, L.-O. Pietilä and S. Krimm, *J. Comp. Chem.*, **1991** *12*, 385; K. Palmö, N. G. Mirkin, L.-O. Pietilä and S. Krimm, *Macromolecules*, **1993** *26*, 6831, K. Palmö, L.-O. Pietilä and S. Krimm, *Computers Chem.*, **1993** *17*, 67; K. Palmö, L.-O. Pietilä and S. Krimm, *J. Comp. Chem.*, **1992** *13*, 1142.
[16] G. Zerbi, *Vibrational Spectra of High Polymers*, (Ed. E. G. Brame), Applied Spectroscopy Reviews, Dekker, New York, USA, 1963, vol. 2, p. 193; H. Tadokoro, *Structure of Crystalline Polymers*, Wiley, New York, 1979.
[17] P. C. Painter, M. M. Coleman and J. L. Koenig. *The Theory of Vibrational Spectroscopy and its Applications to Polymeric Materials*, Wiley, New York, USA, 1982.
[18] For a general discussion of the dynamics of simple and simplified chain molecules see: R. Zbinden, *Infrared Spectroscopy of High Polymers*, Academic, New York, USA, 1964.
[19] G. Zerbi, *Vibrational Spectroscopy of Very Large molecules*, in Advances in Infrared and Raman Spectroscopy (Eds. R. J. H. Clark and R. H. Hester) Wiley-Heyden, New York, USA, vol. 11, p. 301.
[20] J. C. Decius and R. M. Hexter, *Molecular Vibrations in Crystals*, McGraw-Hill, New York, 1977; S. Califano, V. Schettino and N. Neto, *Lattice Dynamics of Molecular Crystals*, Lecture Notes in Chemistry, Springer, Berlin, 1981.
[21] M. Gussoni, G. Dellepiane and S. Abbate, *J. Mol. Spectry.*, **1975** *57*, 323.
[22] For a discussion on the problem of redundancies see: M. Gussoni and G. Zerbi, Accademia Nazionale dei Lincei, *Rendiconti serie VIII* **1966** *40*, 843; M. Gussoni and G. Zerbi, ibid. p. 1032; M. Gussoni and G. Zerbi, *Chem. Phys. Lett.*, **1968** *2*, 145; I. M. Mills, *Chem. Phys. Lett.* **1969** *3*, 267; G. Zerbi, in *Modern Trends in Vibrational Spectroscopy*, (Eds. A. J. Barnes and W. J. Orville-Thomas) Elsevier, Amsterdam, 1977. p. 261; M. V. Volkenstein, L. A. Gribov, M. A. Eliashevich and B. Stepanov, *Kolebaniya Molekul*, Moskow, 1972, p. 203.
[23] N. B. Colthup, L. H. Daly and S. E. Wiberley, *Introduction to Infrared and Raman Spectroscopy* Academic, New York, 1975.
[24] R. N. Jones and C. Sandorfy, *Chemical Applications of Spectroscopy*, in Techniques of Organic Chemistry, (Ed. A. Weissenberger) Interscience, New York, USA, 1956, vol. IX.
[25] L. Bellamy, *The Infrared Spectra of Complex Molecules*, Wiley, New York, USA, 1958.
[26] K. Nakamoto, *Infrared Spectra of Polyatomic Molecules*, Wiley, New York, 1963.
[27] See the series of papers such as, for example, T. Shimanouchi, *Tables of Molecular Frequencies*, Part 7, *J. Physical and Chemical Reference Data*, **1973** *2*, 225.
[28] For the case of a polymer see, for instance, G. Masetti, S. Abbate, M. Gussoni and G. Zerbi, *J. Chem. Phys.* **1980** *73*, 4671.
[29] M. Gussoni and G. Zerbi. *J. Mol. Spectry.*, **1968** *26*, 485.
[30] G. Dellepiane, M. Gussoni and G. Zerbi, *J. Chem. Phys.* **1970** *53*, 3450; M. Gussoni, G. Dellepiane and G. Zerbi, in *Molecular Structures and Vibrations* (Ed. S. J. Cyvin) Elsevier, Amsterdam, 1972, p. 101.
[31] P. Torkington, *J. Chem. Phys.*, **1949** *17*, 347.
[32] Y. Morino and K. Kuchitsu, *J. Chem. Phys.*, **1952** *20*, 1809.
[33] J. H. Schachtschneider and R. G. Snyder, *Spectrochim. Acta* **1963**, *19*, 17.
[34] R. G. Snyder and J. H. Schachtschneider, *Spectrochim. Acta.* **1963** *19*, 117.
[35] J. H. Schachtschneider and R. G. Snyder, *Spectrochim. Acta*, **1965** *21*, 1527.
[36] R. G. Snyder, *J. Chem. Phys.*, **1967** *47*, 1316.
[37] R. G. Snyder and G. Zerbi, *Spectrochim. Acta*, **1967** *23 A*, 391.
[38] W. T. King and B. L. Crawford, *J. Mol. Spectry.*, **1960** *5*, 429.
[39] J. Overend and J. R. Scherer, *Spectrochim. Acta*, **1960** *16*, 773.
[40] *Handbook of Conducting Polymers*, (Ed. T. A. Skotheim) Dekker, New York, 1986, Vols. 1 and 2.

[41] M. Gussoni, C. Castiglioni and G. Zerbi, in *Spectroscopy of Advanced Materials* (Eds. R. J. H. Clark and R. E. Hester) Wiley, New York, 1991, p. 251.
[42] G. Zerbi, M. Gussoni and G. Castiglioni, in *Conjugated Polymers*, (Eds. J. L. Brédas and R. Silbey) Kluwer, Netherland, 1991, p. 435.
[43] C. Castiglioni, M. Del Zoppo and G. Zerbi, *J. Raman Spectry.*, **1993** *24*, 485.
[44] We consider as almost unique example of the existence of long range interactions in covalent systems the superlinear increase with conjugation length of the second order hyperpolarizability γ in long chain polyenes as reported by: I. D. W. Samuel, I. Ledoux, C. Dhenaut, J. Zyss, H. M. Fox, R. R. Scrock and R. J. Silbey, *Science*, **1994** *256*, 1070.
[45] G. Natta and P. Corradini, *Nuovo Cimento, Suppl.*, **1960** *15*, 40.
[46] B. Wunderlich, *Macromolecular Physics*, Vols 1 and 2, Academic, New York, 1976.
[47] See, for instance, P. De Sanctis, E. Giglio, A. Liquori and A. Ripamonti, *J. Polym. Sci.*, **1963** *A1*, 1383.
[48] P. Flory, *Statistical Mechanics of Chain Molecules*, Interscience, New York, 1969.
[49] M. Tasumi and T. Shimanouchi, *J. Chem. Phys.*, **1965** *43*, 1245; M. Tasumi and S. Krimm, *J. Chem. Phys.*, **1967** *46*, 755.
[50] L. Piseri and G. Zerbi, *J. Mol. Spectry.*, **1968** *26*, 254.
[51] A. A. Maradudin, E. W. Montroll and G. H. Weiss, *Solid State Phys., Suppl.*, **1963** *3*, 1.
[52] L. Brillouin, *Wave Propagation in Periodic Structures*, Dover, New York, 1953.
[53] G. Zerbi, in *Advances in Infrared and Raman Spectroscopy*, (Eds. R. J. H. Clark and R. E. Hester) Wiley, 1984, p. 301.
[54] G. Zerbi and M. Sacchi, *Macromolecules*, **1973** *6*, 692.
[55] P. Bosi, R. Tubino and G. Zerbi, *J. Chem. Phys.*, **1973** *59*, 4578.
[56] L. Piseri and G. Zerbi, *J. Chem. Phys.*, **1968** *48*, 3561; G. Zerbi and L. Piseri, *J. Chem. Phys.*, **1968** *49*, 3840.
[57] For Calculations of dispersion curves of Hydrogen bonded systems in one dimension: R. Tubino and G. Zerbi, *J. Chem. Phys.*, **1969** *51*, 4509; A. B. Dempster and G. Zerbi, *J. Chem. Phys.*, **1971** *54*, 3600; R. Tubino and G. Zerbi, *J. Chem. Phys.* **1970** *53*, 1428.
[58] G. Zerbi, C. Castiglioni and M. Del Zoppo, in *Conducting Polymer: The Oligomer Approach (K. Mullen and G. Wegner Eds.)* VCH, Keinheim (1998).
[59] C. Castiglioni, M, Gussoni and G. Zerbi, *J. Chem. Phys.*, **1991** *95*, 7144.
[60] R. Tubino, L. Piseri and G. Zerbi, *J. Chem. Phys.*, **1972** *56*, 1022.
[61] P. W. Higgs, *Proc. Roy. Soc.*, London, **1953** *A220*, 472.
[62] T. Miyazawa, *J. Chem. Phys.*, **1961** *35*, 693.
[63] G. Masetti, S, Abbate, M. Gussoni and G. Zerbi, *J. Chem. Phys.*, **1980** *73*, 4671.
[64] B. Orel, R. Tubino and G. Zerbi, *Molecular Phys.*, **1975** *30*, 37.
[65] G. Zerbi, F. Ciampelli and V. Zamboni, *J. Polym. Sci., Part C*, **1964** 141.
[66] G. Zerbi, *Probing the Real Structure of Chain Molecules by Vibrational Spectroscopy*, American Chemical Society, *Specialist Reports*, **1983** *203*, 487.
[67] T. Miyazawa and T. Kitagawa, *J. Polym. Sci.*, **1965** *B2*, 395; S. Enomoto and M. Asahina, *Polym. Sci.*, **1964** *A2*, 3523.
[68] R. G. Snyder, *J. Mol. Spectry.*, **1961** *7*, 116; R. G. Snyder, in *Methods of Experimental Physics* (Ed. R. A. Fava), Academic, New York, Vol. 16, part A, p. 73.
[69] S. Krimm, *Advances in Polymer Science*, **1960** *2*, 51.
[70] D. Dows, in *Physics and Chemistry of Organic Solid State*, (Eds. I. Fox, M. M. Labes and A. Weissberger) Wiley-Interscience, New York, 1963, Vol. 1; W. Vedder and D. F. Hornig, *Advances in Spectroscopy*, **1962** *2*, 189; M. I. Bank and S. Krimm, *J. Appl. Phys.*, **1969** *10*, 248; S. Krimm and T. C. Cheam, *Faraday Disc. Chem. Soc.* **1979** *68*, 244.
[71] V. Zamboni and G. Zerbi, *J. Polym. Sci.*, **1964** *C7*, 153; G. Zerbi and G. Masetti, *J. Mol. Spectry*, **1967** *22*, 284; G. Zerbi and J. P. Hendra, *J. Mol. Spectry.*, **1968** *27*, 17.
[72] For Polytetrafluoroethylene: F. J. Boerio and J. L. Koenig, *J. Chem. Phys.*, **1971** *54*, 3667 and C. J. Peacock, P. J. Hendra, H. A. Willis and M. E. A. Cudby, *J. Chem. Soc. A*, **1970** 2943.
[73] For Isotactic Polypropylene, ref. 66, p. 501, and G. Masetti, unpublished work.
[74] M. Peraldo, *Gazz. Chim. Ital.*, **1959** *89*, 798; M. Peraldo and M. Farina, *Chim. Ind* (Milan), **1960** *42*, 1349.

[75] G. Zerbi, M. Gussoni and F. Ciampelli, *Spectrochim. Acta*, **1967** *23*, 301.
[76] G. Masetti, F. Cabassi and G. Zerbi, *Polymer*, **1980** *21*, 143.
[77] G. Natta and P. Corradini, Atti Accad. Nazle. Lincei, *Mem. Classe Sci. Fis. Mat. Nat.*, **1955** *4*, 73; G. Natta, P. Corradini and M. Cesari, *ibid.*, **1960** *21*, 365.
[78] G. Zerbi, *Specialist Reports* **1983**, *203*, 501.
[79] H. Jeffreys and B. S. Jeffreys, *Methods of Mathematical Physics*, Cambridge, New York, 1950, p. 40; P. Dean and J. L. Martin, *Proc. Roy. Soc.* (London), **1960** *A259*, 409; P. Dean, *Rev. Mod. Phys.*, **1972** *44*, 127.
[80] A. Rubcic and G. Zerbi, *Macromolecules*, **1974** *7*, 754.
[81] W. Myers, J. L. Donovan and J. S. King, *J. Chem. Phys.*, **1965** *42*, 4299; G. C. Summerfield, *J. Chem. Phys.*, **1965** *43*, 1079.
[82] A. Sjolander, *Arkiv. Fysik*, **1958** *14*, 315.
[83] L. V. Tarasov, *Soviet Phys.-Solid State, English Transl.*, **1961** *3*, 1039.
[84] S. Trevino and H. Boutin, *J. Macromol. Sci.*, **1967** *A1*, 723.
[85] W. Myers, G. C. Summerfield and J. S. King, *J. Chem. Phys.*, **1986** *44*, 184.
[86] S. F. Parker J. Chem. Soc., *Faraday Trans.*, **1996** *92*, 1941.
[87] Faraday Discussions of the Chemical Society *Organization of Macromolecules in the Condensed Phase*, No. 68, Roy. Soc. Chem. 1979.
[88] G. Zerbi in P*honons*, (M. A. Nusimovici Ed.), Proceedings of the international conference, Rennes, France 1971, p. 248.
[89] G. Zerbi and G. Cortili, *Chem. Comm.*, **1965** *13*, 295; G. Cortili and G. Zerbi, *Spectrochim. Acta*, **1967** *23A*, 285.
[90] A. Rubcic and G. Zerbi, *Macromolecules*, **1974** *7*, 759.
[91] M. Tasumi and G. Zerbi, *J. Chem. Phys.* **1968** *48*, 3813.
[92] B. Fanconi, *Ann. Rev. Phys. Chem.* **1980** *31*, 265.
[93] C. Schmid and K. Holzl, *J. Polym. Sci., Phys. Ed.*, **1972** *10*, 1881; C. Schmid and K. Holzl, *J. Phys. C*, **1973** *6*, 2401; V. N. Kozyrenko, L. Kumpanenko and V. Mikhailov, *J. Polym. Sci., Polym. Phys. Ed.*, **1977** *15*, 1721.
[94] G. Zannoni and G. Zerbi, *J. Mol. Struct.*, **1983** *1000*, 385. See also J. J. Ladik, *Quantum Theory of Solids*, Plenum, New York, 1987, p. 140.
[95] G. Zerbi, L. Piseri and F. Cabassi, *Mol. Phys.*, **1971** *22*, 241.
[96] J. H. Wilkinson, *Computer J.* **1958** *1*, 90.
[97] M. Gussoni and G. Zerbi, *J. Chem. Phys.*, **1974** *60*, 4862.
[98] I. F. Chang and S. S. Mitra, *Adv. Phys.*, **1971** *20*, 359.
[99] G. Zerbi and M. Gussoni, *Polymer*, **1980** *21*, 1129.
[100] A. Rubcic and G. Zerbi, *Macromolecules*, **1974** *7*, 759.
[101] M. Tasumi and G. Zerbi, *J. Chem. Phys.*, **1968** *48*, 3813.
[102] R. G. Snyder and M. Poore, *Macromolecules*, **1973** *6*, 709.
[103] G. Zerbi, R. Magni, M. Gussoni, K. Holland-Moritz, A, Bigotto and S. Dirlikov, *J. Chem. Phys.*, **1981** *75*, 3175.
[104] For the works by the school of R. G. Snyder in this field see, for instance, M. Maroncelli, S. P. Qi, H. L. Strauss and R. G. Snyder, *J. Am. Chem. Soc.*, **1982** *104*, 6237.
[105] A. Keller, *Faraday Disc. Chem. Soc.* **1979** *68*, 244.
[106] G. Bucci and T. Simonazzi, *J. Polym. Sci., part C, Polymer Symposia*, **1964** *7*, 203.
[107] G. Zerbi, G. Minoni and A. P. Tulloch, *J. Chem. Phys.*, **1983** *78*, 5853.
[108] G. Vergottin, G. Fleury and I. Moschetto, *Advances in Infrared and Raman Spectroscopy*, (Ed. R. J. H. Clark and R. E. Hester) Heyden, London, 1978, vol. 4, p. 195.
[109] G. Zerbi, G. Conti, G. Minoni, S. Pison and A. Bigotto, *J. Phys. Chem.*, **1987** *91*, 2386.
[110] G. Minoni and G. Zerbi, *J. Phys. Chem.*, **1982** *96*, 4791.
[111] M. Del Zoppo and G. Zerbi, *Polymer*, **1990** *31*, 658.
[112] E. Galbiati and G. Zerbi, *J. Chem. Phys.* **1986** *84*, 3500.
[113] E. Galbiati and G. Zerbi, *J. Chem. Phys.*, **1987** *87*, 3653.
[114] E. Galbiati, G. Zerbi, E. Benedetti and E. Chiellini, *Polymer*, **1991** *32*, 1555.
[115] S. J. Spells, S. J. Organ, A. Keller and G. Zerbi, *Polymer*, **1987** *28*, 697.
[116] H. Kay and B. A. Newman, *Acta Crystallogr.* **1968** *B24*, 615.

[117] P. Jona, M. Gussoni and G. Zerbi, *J. Appl. Phys.*, **1985** *57*, 834.
[118] M. Gussoni, P Jona and G. Zerbi, *J. Mol. Struct., Theochem*, **1985** *119*, 329; ibid. p. 347.
[119] T. Shimanouchi, *J. Phys. Chem. Ref. Data*, **1973** *2*, 225.
[120] G. Zerbi, G. Gallino, N. Del Fanti and L. Baini, *Polymer*, **1989** *30*, 2324.
[121] G. Avitabile, R. Napolitano, R. Pinozzi, K. D. Rouse, W. Thomas and B. T. M. Willis, *Polym. Lett.*, **1975** *13*, 371.
[122] S. Abbate, M. Gussoni and G. Zerbi, *J. Chem. Phys.*, **1979** *70*, 3577.
[123] E. Agosti, G. Zerbi and I. M. Ward, *Polymer* **1992** *33*, 4219.
[124] G. Zerbi, G. Conti, G. Minoni, S. Pison and A. Bigotto, *J. Phys. Chem.* **1987** *91*, 2386.
[125] M. Del Zoppo and G. Zerbi, *Polymer*, **1990** *31*, 658.
[126] D. Chapmann, *The Structure of Lipids*, Methuen, London, 1965.
[127] C. Almirante, G. Minoni and G. Zerbi, *J. Phys., Chem.*, **1986** *90*, 852.
[128] L. Basini, A. Raffaelli and G. Zerbi, *Chem. Mater.* **1990** *2*, 679.
[129] G. Ferrari and G. Zerbi, *J. Raman Spectroscopy*, **1994** *25*, 713.
[130] For a discussion on the structure of PVC see, for instance, F. A. Bovey, F. P. Hood, E. W. Anderson and R. L. Kornegay, *J. Phys. Chem.* **1967** *71*, 312; L. Cavalli, G. C. Borsini, G. Carraro and G. Confalonieri, *J. Polym. Sci., Part A-1*, **1970** *8*, 801.
[131] R. F. Schaufele and T. Shimanouchi, *J. Chem. Phys.* **1967** *47*, 3605.
[132] T. Shimanouchi, *Pure and Appl. Chem.*, **1973** *36*, 93.
[133] G. Minoni and G. Zerbi, *J. Phys. Chem.*, **1982** *86*, 4791.
[134] G. Minoni and G. Zerbi, *J. Polym. Sci. Polym. Lett.*, **1984** *22*, 533.
[135] E. Fermi, *Z. Physik*, **1931** *71*, 250.
[136] J. Lavalley and N. Sheppard, *Spectrochim. Acta, Part A*, **1972** *28*, 845.
[137] S. Wunder, M. Bell and G. Zerbi, *J. Chem. Phys.*, **1986** *85*, 3827.
[138] R. G. Snyder, S. L. Hsu and S. Krimm, *Spectrochim. Acta, Part A*, **1978** *34*, 395.
[139] R. G. Snyder and J. R. Scherer, *J. Chem. Phys.*, **1979** *71*, 3221.
[140] S. Abbate, G. Zerbi and S. L. Wunder, *J. Phys. Chem.*, **1982** *86*, 3140.
[141] S. Abbate, S. L. Wunder and G. Zerbi, *J. Phys. Chem.*, **1984** *88*, 593.
[142] L. Ricard, S. Abbate and G. Zerbi, *J. Phys. Chem.*, **1985** *89*, 4793.
[143] G. Zerbi and S. Abbate, *Chem. Phys. Lett.*, **1981** *80*, 455.
[144] M. Del Zoppo and G. Zerbi, *Polymer*, **1991** *32*, 1581.
[145] G. Zerbi, P. Roncone, G. Longhi and S. L. Wunder, *J. Chem. Phys.*, **1988** *89*, 166.
[146] R. G. Snyder, J. R. Scherer and P. Gaber, *Biochim. Biophys. Acta*, **1980** *601*, 47.
[147] W. G. Rotschild, *Dynamics of Molecular Liquids*, Wiley, New York, 1984.
[148] B. G. Sumpter, D. W. Noid and B. Wunderlich, *J. Chem. Phys.* **1990** *93*, 6875; G. L. Liang, D. W. Noid, B. G. Sumpter and B. Wunderlich, *Acta Polymer.*, **1993** *44*, 219.
[149] M. Kobayashi, T. Kobayashi, Y. Cho and F. Kaneko, *J. Chem. Phys.* **1986** *84*, 4636.
[150] I. W. Levin, in *Advances in IR and Raman Spectroscopy*, Wiley, New York, 1984, Vol 11.
[151] B. P. Gaber and W. Peticolas, *Biochim. Biophys. Acta* **1977** *465*, 260.
[152] A. Muller, *Proc. Royal Soc., London Ser.* **1930** *A127*, 417.
[153] M. Broadhurst, *J. Res. Natl. Bur. Stand. Sect* **1962** *A66*, 241.
[154] K. Larsson, *Nature* (London) **1967** *213*, 383.
[155] G. Zerbi, R. Piazza and K. Holland-Moriytz, *Polymer*, **1982** *23*, 1921.
[156] G. Ungar and A. Keller, *Colloid Polym. Sci.*, **1979** *257*, 90.
[157] F. Guillaume, C. Sourisseau and A. J. Dianoux, *J. Chem. Phys.*, **1990** *93*, 3536.
[158] P. Jona, B. Bassetti, V. Benza and G. Zerbi, *J. Chem. Phys.*, **1987** *86*, 1561.
[159] M. L. Klein, *J. Chem. Soc. Faraday Trans.*, **1992** *13*, 1701.
[160] L. Askadskaya and J. P. Rabe, *Phys. Rev. Lett.*, **1992** *69*, 1395.
[161] For review on model calculations see D. W. McClure, *J. Chem. Phys.*, **1968** *49*, 1830.
[162] H. G. Olf and A. Peterlin, *J. Poly. Sci.*, **1970** *A2*, 791.
[163] B. Ewen and D. Richter, *J. Chem. Phys.* **1978** *69*, 2954.
[164] J. D. Barnes, *J. Chem. Phys.*, **1973** *58*, 5193.
[165] R. L. McCullough, *J. Macromol. Sci., Phys.*, **1974** *9*, 97.
[166] M. Mansfield and R. H. Boyd, *J. Polym. Sci., Polym. Phys. Ed.*, **1978** *16*, 1227.
[167] M. L. Mansfield, *Chem. Phys. Lett.*, **1980** *69*, 383.

[168] J. L. Skinner and P. J. Wolynes, *Chem. Phys*, **1980** *73*, 4015.
[169] G. Zerbi and M. Del Zoppo, *J. Chem. Soc. Faraday Trans.*, **1992** *88*, 1835.
[170] H. Dwey-Aharon, T. S. Slucking, P. L. Taylor and A. J. Hopfinger, *Phys. Rev. Ser. B*, **1980** 21.
[171] G. Borionetti, G. Zannoni and G. Zerbi, *J. Mol. Struct.* **1990** *224*, 425.
[172] S. Radice, N. Del Fanti and G. Zerbi, *Polymer*, **1997** *38*, 2753.

4 Vibrational Spectroscopy of Intact and Doped Conjugated Polymers and Their Models

Y. Furukawa and M. Tasumi

4.1 Introduction

Most organic polymers are electrical insulators, a property which distinguishes them from metals. However, the development of a new class of organic polymers with quasimetallic electrical conductivities has been actively pursued during the past 18 years, following the discovery in 1977 of high electrical conductivities for doped polyacetylenes [1, 2]. The novel concept of a conducting organic polymer has aroused the interest of a large number of researchers in various areas such as chemistry, physics, electrical engineering, material science, etc. Detailed discussions in each area can be found in many review articles [3–17]. In particular, a new field of physics has been opened for the purpose of understanding their electrical properties.

A goal of basic research on conducting polymers is to understand the metallic properties of such polymers. Since conducting organic polymers have conjugated π-electrons in common, chemists intuitively believe that studies on the physical and chemical properties of conjugated systems would lead to an elucidation of the mechanism of charge transport. On the other hand, new concepts, *solitons* [18, 19], *polarons* [20–22], and *bipolarons* [21–23] have been proposed by solid-state physicists as elementary excitations in conjugated polymers, in order to explain the physical properties of these polymers. They are collectively called *self-localized excitations*. These concepts and terminologies are unfamiliar to molecular spectroscopists. Thus, it seems desirable to build a bridge between solid-state physicists and molecular spectroscopists.

The structures and properties of conducting polymers have been studied by various spectroscopic techniques. Among them, vibrational (Raman and infrared) spectroscopy is a powerful tool for elucidating the molecular and electronic structures of conducting polymers as described in previous reviews [13–15]. In spite of numerous spectroscopic studies, discussions on polarons, bipolarons, and solitons were not based upon reliable evidence from vibrational spectroscopy until very recently. Thus, the aims of this review are: (i) to provide a general introduction to the concepts of polarons, bipolarons, and solitons from the standpoint of molecular spectroscopists; (ii) to describe the methodology of Raman studies on these self-localized excitations; (iii) to review the results of studies on poly(*p*-phenylene) and other polymers; and (iv) to discuss the mechanism of charge transport in conducting polymers.

4.2 Materials

The conductivity of iodine-doped polyacetylene first reported by Shirakawa et al. [1] in 1977 was 30 S cm^{-1}. Since then, the conductivity reported for doped polyacetylene has kept increasing, the highest conductivity obtained so far for an iodine-doped stretched polyacetylene film [17] being $>10^5$ S cm^{-1}, a value comparable with that of copper (6×10^5 S cm^{-1}).

A film of intact polyacetylene usually shows a conductivity lower than 10^{-5} S cm^{-1}. However, the conductivity increases dramatically when the film is exposed to oxidizing agents (electron acceptors) such as iodine, AsF_5, H_2SO_4, etc. or reducing agents (electron donors) such as alkali metals. This process is referred to as *doping*, by analogy with the doping of inorganic semiconductors. The polymers that are not doped are referred to as *intact* polymers in this review. The main process of doping is a redox reaction between the polymer chains and acceptors (or donors). Upon doping, an ionic complex consisting of positively (or negatively) charged polymer chains and counter ions such as I_3^-, AsF_6^-, etc. (or Na^+, K^+, etc.) is formed. Counter anions or cations are generated by reduction of acceptors or oxidation of donors, respectively. The use of an acceptor causes the p-type doping, and that of a donor the n-type doping. The electrical conductivity can be controlled by the content of a dopant. A sharp increase in conductivity is observed when the dopant content is <1 mole% per C_2H_2 unit. After this sharp increase, the conductivity becomes gradually higher with further increase in the dopant content. At low doping levels, polyacetylene does not exhibit metallic properties, whereas its conductivity is high. When the dopant content is more than about 13 mole% per C_2H_2 unit, polyacetylene shows metallic properties such as a Pauli susceptibility [24, 25], a linear temperature dependence of the thermoelectric power [26], and a high reflectivity in the infrared region [27], though the temperature dependence of conductivity is not like that of a metal. The origin of such properties of heavily doped polyacetylene is not yet fully understood.

Following polyacetylene, a large number of conducting polymers have been reported. The chemical structures of typical conducting polymers are depicted in Figure 4-1. These conducting polymers have conjugated π-electrons in common,

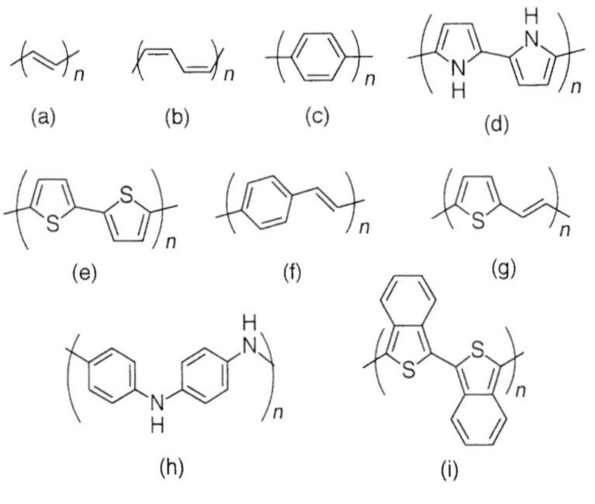

Figure 4-1. Chemical structures of conducting polymers. (a) *Trans*-polyacetylene; (b) *cis*-polyacetylene; (c) poly(*p*-phenylene); (d) polypyrrole; (e) polythiophene; (f) poly(*p*-phenylenevinylene); (g) poly(2,5-thienylenevinylene); (h) polyaniline (leucoemeraldine base form); (i) polyisothianaphthene.

and all are either insulators or semiconductors in their intact states, and they become conductors upon doping. Metallic properties are observed for poly(*p*-phenylene) [28], polythiophene [29–33], polypyrrole [31], and polyaniline [34, 35] at heavy doping levels, although reported data depend on samples and preparation methods. The origin of the metallic states of heavily doped polymers is one of the major unresolved problem in the field of conducting polymers.

4.3 Geometry of Intact Polymers

The question of whether the C–C bonds in an infinite polyene chain are equal or alternate in length has been discussed by quantum chemists since the 1950s [36, 37]. Various experimental results on oligoenes have shown the existence of alternating single and double bonds. The bond alternation in *trans*-polyacetylene has been confirmed by X-ray [38] and nutation NMR [39] studies. However, it is difficult to determine exact structure parameters for conducting polymers from X-ray diffraction studies, because single crystals of the polymers are unavailable. It is then useful to examine the geometries of the polymers and model oligomers by the molecular orbital (MO) method, especially at *ab initio* Hartree–Fock levels or in higher approximations.

Conducting polymers can be divided into two types, *degenerate* and *nondegenerate*, according to the structure of the ground-states of intact polymers. The total energy curve of a degenerate system in the ground state is shown schematically as a function of a structural deformation coordinate, R, in Figure 4-2a, and that for a nondegenerate system in Figure 4-2b. Let us consider an infinite *trans*-polyacetylene chain as a prototype of the degenerate ground-state polymers. There are two degrees of freedom in bond lengths: r_{C-C} and $r_{C=C}$, which correspond, respectively, to the lengths of the alternating single and double CC bonds. The coordinate for structural deformation is simply expressed by the following equation.

$$R = \frac{1}{\sqrt{2}}(r_{C-C} - r_{C=C}) \tag{4-1}$$

This coordinate reflects the degree of bond alternation. This coordinate is used in the effective-conjugation-coordinate model proposed by Zerbi et al. [15] for the purpose

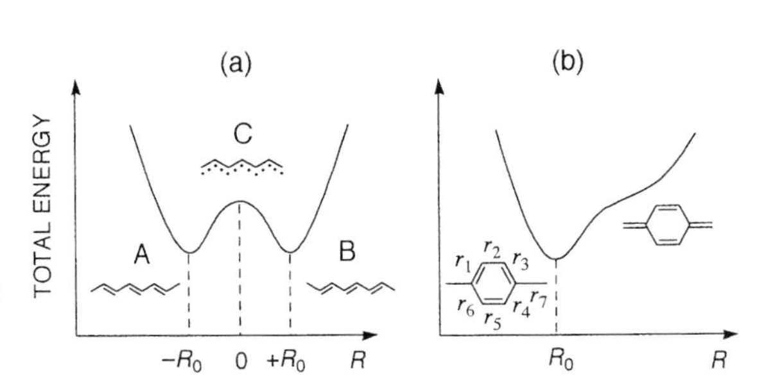

Figure 4-2. Total energy as a function of a structural deformation coordinate. (a) *Trans*-polyacetylene (degenerate ground-state system); (b) poly(*p*-phenylene) (nondegenerate ground-state system).

of explaining the Raman spectra of intact polymers and doping-induced infrared bands. (This model will be mentioned again in Section 4.5.) There are two stable structures (A and B in Figure 4-2a) with alternating C–C and C=C bonds, while the structure with equal C–C bond lengths (C in Figure 4-2a) is unstable. These two structures are identical to each other and have the same total energy; in other words, they are degenerate. According to a nutation NMR study [39], the lengths of the C–C and C=C bonds in *trans*-polyacetylene are 1.44 and 1.36 Å, respectively. An ab initio MO calculation at the Hartree–Fock level with the 6-31G basis set for a model compound, $C_{22}H_{24}$, has shown that the C–C and C=C bond lengths of the central unit are 1.450 and 1.338 Å, respectively [40]. Similar results have been obtained from calculations which take into account the effects of electron correlation [41]. From these experimental and theoretical results, R_0 (equilibrium value of R, see Figure 4-2a) is calculated to be 0.06 and 0.079 Å, respectively.

A nondegenerate polymer has no two identical structures in the ground state. Let us consider poly(*p*-phenylene) as an example of nondegenerate polymers. In order to express the structural deformation in this polymer, the following coordinate may be chosen, as proposed by Castiglioni et al. [42].

$$R = \frac{1}{\sqrt{7}}(r_1 + r_3 + r_4 + r_6 - r_2 - r_5 - r_7) \tag{4-2}$$

where r_is are shown in Figure 4-2b. The total energy of this polymer has only one minimum at R_0 as shown in Figure 4-2b. If we use the bond lengths, $r_1 = r_3 = r_4 = r_6 = 1.388$ Å, $r_2 = r_5 = 1.382$ Å, and $r_7 = 1.492$ Å, of the central unit of neutral *p*-terphenyl obtained by an ab initio MO calculation at the Hartree–Fock level with the double-zeta quality 3-21G basis set [43], R_0 is calculated to be 0.490 Å. As shown in Figure 4-2b, an increase in R means the structural deformation toward a quinoid structure from a benzenoid structure.

4.4 Geometric Changes Induced by Doping

The main process of doping is a charge-transfer reaction between an organic polymer and a dopant. When charges are removed from (or added to) a polymer chain upon chemical doping, geometric changes occur over several repeating units and the charge is localized over this region. Structure parameters such as bond lengths, bond angles, etc. are changed in this region. For example, the bond lengths [44] of neutral *p*-terphenyl, its radical anion, and its dianion calculated by the PM3 method are shown in Table 4-1. From this table, we can see readily that the bond lengths of the neutral and charged species are different from each other. The phenyl and phenylene rings in neutral *p*-terphenyl have benzenoid structures. In the radical anion and dianion, an increase in R occurs as the result of shortening of the r_{23}, r_{45}, and r_{67} bonds and lengthening of the r_{12}, r_{34}, and r_{56} bonds; in other words, geometric changes from benzenoid to quinoid occur upon ionization. The geometric

Table 4-1. Bond lengths (in Å) calculated by the PM3 method for *p*-terphenyl, its radical anion, and its dianion.

Species	Bond length[a]						Coordinate[b]
	r_{12}	r_{23}	r_{34}	r_{45}	r_{56}	r_{67}	R
Neutral	1.391	1.390	1.397	1.469	1.397	1.389	0.507
Radical anion	1.397	1.385	1.417	1.426	1.423	1.366	0.580
Dianion	1.400	1.371	1.438	1.388	1.446	1.351	0.640

[a] Numbering of carbon atoms is shown below. Dihedral angles between neighboring benzene rings are 48°, 0°, and 0° for *p*-terphenyl, its radical anion, and its dianion, respectively. The data are taken from [44].

[b] The structural deformation coordinate defined in Eq. 4-2 is calculated by using the equation, $R = 1/\sqrt{7}(4r_{56} - 2r_{67} - r_{45})$.

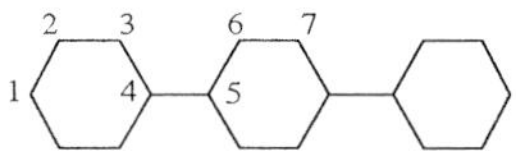

changes from the neutral species are larger for the dianion than for the radical anion. Since the localized charges, which are accompanied by geometric changes, can move along the polymer chain, they are regarded as charge carriers in conducting polymers. These quasiparticles are classified into polarons, bipolarons, and solitons according to their charge and spin.

4.4.1 Polarons, Bipolarons, and Solitons

Self-localized excitations and corresponding chemical terminologies are listed in Table 4-2. Schematic structures of the self-localized excitations in poly(*p*-phenylene) and *trans*-polyacetylene are depicted in Figure 4-3. In these illustrations, the charge and spin are localized on one carbon atom. In real polymers, however, they are considered to be localized over several repeating units with geometric changes.

When an electron is removed from an infinite polymer chain, charge $+e$ and spin 1/2 are localized over several repeating units with geometric changes. This is a self-localized excitation called a positive polaron (Figure 4-3a, e). Since a positive polaron has charge $+e$ and spin 1/2, it is a radical cation in chemical terminology. When an electron is added to a polymer chain, a negative polaron, which is a radical anion, is formed (Figure 4-3b, f).

When another electron is removed from a positive polaron, charge $+2e$ is localized over several repeating units. The charge $+2e$ is considered to be localized in a region narrower than that of a positive polaron. This species is called a positive bipolaron (Figure 4-3c). Since a positive bipolaron has charge $+2e$ and no spin, it is a dication in chemical terminology. A negative bipolaron corresponding to a dianion can also be considered (Figure 4-3d).

Table 4-2. Self-localized excitations and chemical terminologies.

Self-localized excitation	Chemical term	Charge	Spin
positive polaron	radical cation	$+e$	1/2
negative polaron	radical anion	$-e$	1/2
positive bipolaron	dication	$+2e$	0
negative bipolaron	dianion	$-2e$	0
neutral soliton	neutral radical	0	1/2
positive soliton	cation	$+e$	0
negative soliton	anion	$-e$	0

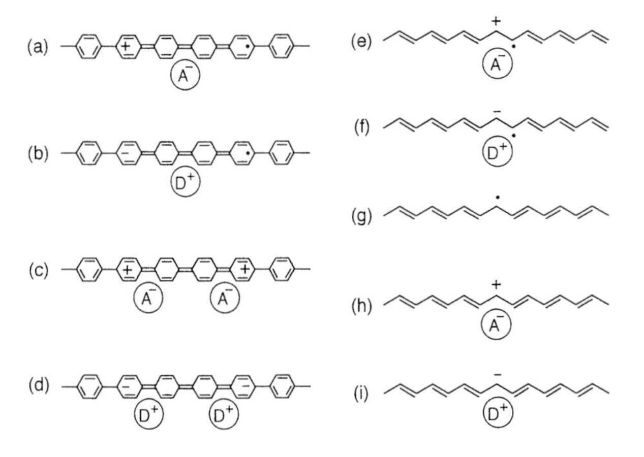

Figure 4-3. Schematic structures of self-localized excitations in poly(p-phenylene) and *trans*-polyacetylene. (a) Positive polaron; (b) negative polaron; (c) positive bipolaron; (d) negative bipolaron; (e) positive polaron; (f) negative polaron; (g) neutral soliton; (h) positive soliton; (i) negative soliton;. D, donor; A, acceptor; +, positive charge; −, negative charge; •, unpaired electron.

In *trans*-polyacetylene having a degenerate ground-state structure, soliton excitations can be formed between the A and B phases (see Figure 4-2a). Solitons are classified into neutral, positive, and negative types according to their charge. A neutral soliton has no charge and spin 1/2 (Figure 4-3g). A positive (or a negative) soliton has charge $+e$ (or $-e$) and no spin (Figure 4-3h, i). A neutral soliton, a positive soliton, and a negative soliton correspond, respectively, to a neutral radical, a cation, and an anion of a linear *trans*-oligoene with an odd number of carbon atoms, C_nH_{n+2} (where n is an odd number).

Bond-alternation defects or misfits in polyenes were studied using the Hückel MO method in the early 1960s [36, 45, 46], though these are nothing but solitons proposed by Su, Schrieffer, and Heeger [18, 19] about two decades after. These workers have studied solitons in *trans*-polyacetylene by using a simple tight-binding Hamil-

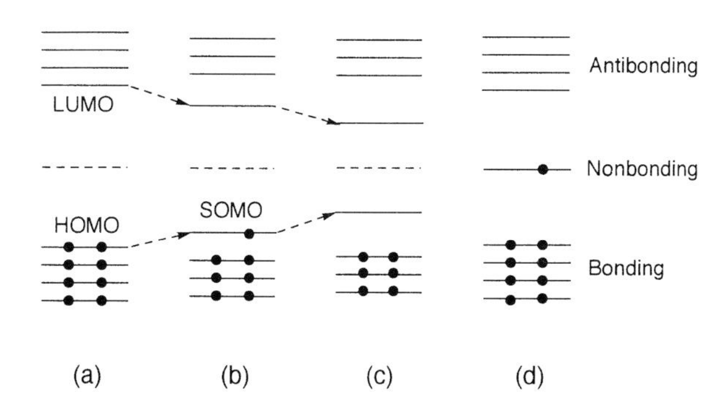

Figure 4-4. Schematic molecular orbital energy levels. (a) Neutral conjugated system; (b) its radical cation; (c) its dication; (d) neutral radical of an odd oligoene. HOMO, highest occupied molecular orbital; LUMO, lowest unoccupied molecular orbital; SOMO, singly occupied molecular orbital; •, electron.

tonian with an electron–lattice interaction term (SSH model), their model being essentially the same as the Hückel approximation. The continuum versions of the SSH model are very useful for describing polarons and bipolarons as well as solitons, because analytic solutions can be obtained for physical parameters (creation energy, electronic absorption, etc.) relating to the excitations [21, 47–49].

4.4.2 Electronic States of Polarons, Bipolarons, and Solitons

We will first discuss the electronic states of linear conjugated molecules at the Hückel level with electron–lattice interaction [7], as models of self-localized excitations. The MO energy levels of these systems are schematically shown in Figure 4-4. In a neutral conjugated molecule (e.g., *p*-terphenyl), the bonding and antibonding levels are formed as the result of interaction between π-electron levels (Figure 4-4a). In its radical cation, geometric changes lead to an upward shift of the highest occupied molecular orbital (HOMO) and a downward shift of the lowest unoccupied molecular orbital (LUMO), and one electron is removed from the HOMO level (Figure 4-4b); the singly occupied molecular orbital (SOMO) is then formed. In its radical anion, the HOMO level is occupied by two electrons and the LUMO has one electron. Since geometric changes in the dication, where another electron is removed from the HOMO level (Figure 4-4c), are larger than those in the radical cation, the HOMO and LUMO levels shift further. In the dianion, each of the HOMO and LUMO levels is occupied by two electrons. In the MO energy levels of an odd *trans*-oligoene radical, $C_nH_{n+2}^{\cdot}$ (where n is an odd number), a nonbonding level, which is occupied by one electron, is formed in the center of the band gap as shown in Figure 4-4d [36]. In its cation and anion, the nonbonding level is occupied by null and two electrons, respectively.

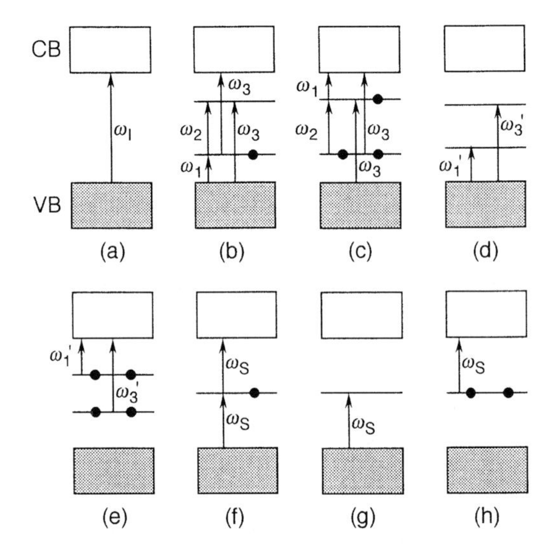

Figure 4-5. Schematic electronic band structures. (a) Neutral polymer; (b) positive polaron; (c) negative polaron; (d) positive bipolaron; (e) negative bipolaron; (f) neutral soliton; (g) positive soliton; (h) negative soliton. CB, conduction band; VB, valence band; •, electron; arrow, electronic transition.

Next, we discuss the electronic transitions due to polarons, bipolarons, and solitons in a polymer chain on the basis of a theoretical study reported by Fesser et al. [48] on a continuum electron–phonon-coupled model. The electronic energy levels of a neutral infinite polymer and those of polarons, bipolarons, and solitons are shown schematically in Figure 4-5.

In an infinite neutral polymer, interaction between repeating units leads to the formation of electronic energy bands. The bonding and antibonding levels constitute, respectively, the valence band (VB) and the conduction band (CB) as shown in Figure 5-5a. The band gaps (ω_I in Figure 5-5a) of most intact polymers are in the range between 35000 and 8000 cm^{-1} (4.3 and 1.0 eV, 1 eV = 8066 cm^{-1}).

For a positive polaron (Figure 5-5b), two localized electronic levels, bonding and antibonding, are formed within the band gap. One electron is removed from the polaron bonding level. Thus, a positive polaron is expected to have the following three intragap transitions:

- ω_1, polaron bonding level $\leftarrow$ valence band
- ω_2, polaron antibonding level $\leftarrow$ polaron bonding level
- ω_3, polaron antibonding level $\leftarrow$ valence band, and conduction band $\leftarrow$ polaron bonding level.

When a negative polaron is formed, one electron is added to the polaron antibonding level. In this case also, three transitions are expected (Figure 4-5c).

The electronic structure of a positive bipolaron is shown in Figure 4-5d. Since the geometric changes for a bipolaron are larger than those for a polaron, localized electronic levels appearing in the band gap for a bipolaron are farther away from the band edges than for a polaron. Two electrons are removed from the bipolaron bonding level. Thus, a positive bipolaron is expected to have the following two intragap transitions:

- ω_1', bipolaron bonding level $\leftarrow$ valence band
- ω_3', bipolaron antibonding level $\leftarrow$ valence band.

When a negative bipolaron is formed, two electrons are added to the bipolaron antibonding level (Figure 4-5e) and thus two transitions are again expected. The intensities of the electronic transitions for polarons and bipolarons will be discussed in Section 4.7.1.2.

When a soliton is formed, a nonbonding electronic level is formed at the center of the band gap (Figure 4-5f–h). For a neutral soliton, the nonbonding electronic level is occupied by one electron, while for a positive soliton and a negative soliton, the level is occupied by null and two electrons, respectively. Thus, all the neutral, positive, and negative solitons are expected to have only one intragap transition, ω_S, as shown in Figure 4-5f–h.

Polarons, bipolarons, and solitons are different from each other as described above. It is expected that we can detect these excitations separately by electronic absorption, ESR, and vibrational spectroscopies. In particular, geometric changes induced by doping can be studied by vibrational (Raman and infrared) spectroscopy.

4.5 Methodology of Raman Studies of Polarons, Bipolarons, and Solitons

Doping induces strong infrared absorptions that are attributed to the vibrational modes arising from charged domains generated by doping [10, 14, 15, 50]. In the case of *trans*-polyacetylene, Horovitz et al. [51–53] have proposed the amplitude-mode model for explaining doping-induced infrared absorptions. According to their theory, the normal vibrations that induce oscillations in bond alternation are strongly observed in the infrared spectra of doped polyacetylene. Castiglioni et al. [54] have reformulated the amplitude-mode theory in terms of the GF matrix method [55] used in molecular spectroscopy. In their treatment, they have proposed an effective conjugation coordinate (expressed by Eq. 4-1) which reflects bond alternation. They have also explained the doping-induced infrared absorptions of other conducting polymers such as polypyrrole, polythiophene, poly(*p*-phenylenevinylene), etc., by using effective conjugation coordinates [50]. However, the types of self-localized excitations (polarons, bipolarons, and solitons) existing in these doped polymers have not been identified.

Raman spectroscopy of conducting polymers has mainly treated the structures of intact polymers (geometry of polymer chains, conjugation length, force field, etc.) [13–15]. Although the Raman spectra of doped polymers have been reported [13–15], the observed Raman bands have not been correlated to the types of self-localized excitations created by doping. Zerbi et al. [15] have drawn a conclusion from the effective-conjugation-coordinate model that the Raman bands arising from the charged domains are extremely weak and would never be observed. However, a new approach to Raman identification of the self-localized excitations will be described in this review.

Most of intact polymers show electronic absorptions in the region from ultraviolet to visible. Upon doping, new absorptions with several peaks appear in the region from visible to infrared. These new absorptions are associated with self-localized excitations created by doping. Thus, we can observe vibrational spectra arising from the charged domains, by using resonance Raman spectroscopy with a wide range of excitation wavelengths from visible to infrared [56, 57]. So far, although most of the Raman spectra of doped polymers have been measured by using visible laser lines for excitation, it is also desirable to use laser lines with longer wavelengths.

A useful method for analyzing the vibrational and electronic spectra of polymers is to compare them with the spectra of oligomers. In chemical terminology, polarons, bipolarons, and charged solitons are nothing but radical ions, divalent ions, and ions of oligomers, respectively (see Section 4.4.1). If these charged oligomers have geometries similar to those of self-localized excitations, these compounds would give rise to the electronic and Raman spectra similar to those of the self-localized excitations. We can identify self-localized excitations on the basis of the spectroscopic data for charged oligomers (charged oligomer approach) [56, 57].

4.6 Near-Infrared (NIR) Raman Spectrometry

Since the 1970s, Raman spectrometry has been using visible laser lines as the most common excitation sources. It was generally believed that NIR laser lines were not suitable for obtaining high-quality Raman spectra, though since Hirshfeld and Chase [58] and Fujiwara et al. [59] reported successful NIR Raman spectrometry measurements in 1986, this technique has advanced dramatically and NIR Raman spectrophotometers are now commercially available.

We will describe two NIR Raman spectrophotometers, one with a diffraction grating, and the other with an interferometer. A Ti:sapphire laser (Coherent Radiation 890) pumped with an argon ion laser (Coherent Radiation Innova 90-6) is used as an NIR source for Raman excitation. This laser covers the wavelength range between 680 and 1000 nm. Since the output intensity of the Ti:sapphire laser is very stable for a long time in comparison with a NIR dye laser, it is a good excitation source for NIR Raman spectrometry. Continuous-wave lasers are used for measuring the Raman spectra of doped polymers, because the doped samples are liable to damage due to high peak powers of pulsed lasers. Raman spectra are usually measured on a single polychromator (Spex 1870) equipped with a charge-coupled-device (CCD) detector (Princeton Instruments LN/CCD-1024TKBS, back-illuminated type, 1024 × 1024 pixels) operated at −120 °C. The CCD detector functions with an ST-135 controller (Princeton Instruments) linked with an NEC PC-9801FX computer through an IEEE-488 interface. The spectroscopic response of the CCD detector extends to about 1000 nm. Optical filters (holographic notch filters) are used to eliminate the Rayleigh scattering. The advantage of this system lies in its high optical throughput due to the use of the single polychromator and optical filter.

We have modified an NIR Fourier-transform spectrophotometer (JEOL JIR-5500) for the measurements of Raman spectra excited with the 1064-nm line of a continuous-wave Nd:YAG laser (CVI YAG-MAX C-92) [60]. Raman-scattered radiation is collected with a 90° off-axis parabolic mirror in a back-scattering geometry. Collected radiation is passed through two or three long-wavelength-pass filters to reduce the Rayleigh scattering.

4.7 Poly(*p*-phenylene)

As a typical example, we will first discuss the results of the electronic absorption and Raman spectra of Na-doped poly(*p*-phenylene). Poly(*p*-phenylene) consists of benzene rings linked at the para positions as shown in Figure 4-1c. Since poly(*p*-phenylene) has a nondegenerate structure, polarons and/or bipolarons are expected to be formed by chemical doping. As model compounds of poly(*p*-phenylene), *p*-oligophenyls (abbreviated as PP*n*, *n* being the number of phenylene and phenyl rings) such as biphenyl (PP2), *p*-terphenyl (PP3), *p*-quaterphenyl (PP4), *p*-quinquephenyl (PP5), and *p*-sexiphenyl (PP6) are used. It has been shown by X-ray diffraction studies that the two phenyl rings of PP2 [61, 62] in crystal at room temperature are coplanar, and PP3 [63] and PP4 [64] in similar conditions deviate slightly from coplanarity. However, these deviations are not significant in analyzing their vibrational spectra. Accordingly, the phenylene and phenyl rings of PP5 and PP6 in solids, and poly(*p*-phenylene) are assumed to be coplanar.

4.7.1 Electronic Absorption Spectra

4.7.1.1 Intact and Doped Poly(*p*-phenylene)

The electronic absorption spectra of undoped and electrochemically Bu_4N^+(n-type)-doped poly(*p*-phenylene) films are shown in Figure 4-6a, b, respectively [65, 66]. The undoped poly(*p*-phenylene) film shows the electronic absorption band at 27400 cm^{-1} (365 nm), which is associated with the interband transition from the valence band to the conduction band. The value of the band gap (ω_I in Figure 4-5a) is estimated to be 24200 cm^{-1} from the onset of the electronic absorption [66]. The electronic transition energies of PP*n* [67, 68] are plotted in Figure 4-7. The transition energy decreases as the chain length becomes longer, an observation explained by the fact that increased conjugation occurs for longer chains. The observed tran-

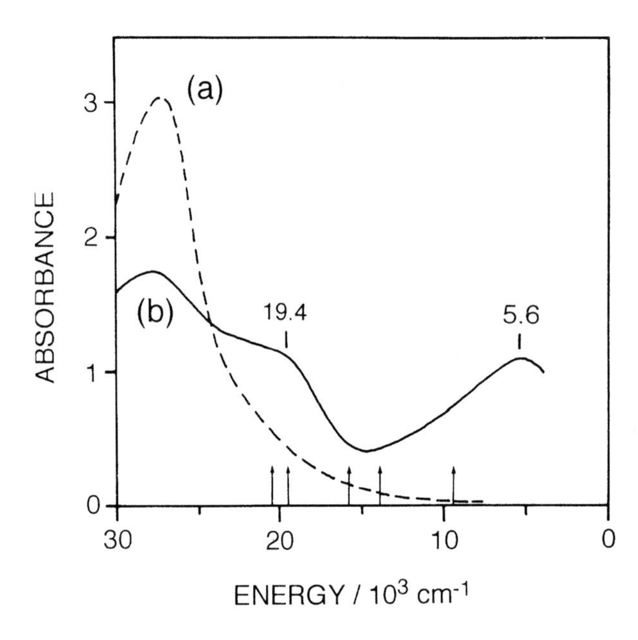

Figure 4-6. Absorption spectra of (a) undoped and (b) electrochemically Bu_4N^+(n-type)-doped poly(*p*-phenylene) films. Arrows indicate the positions of excitation wavelengths (488.0, 514.5, 632.8, 720, and 1064 nm) used for Raman measurements [66].

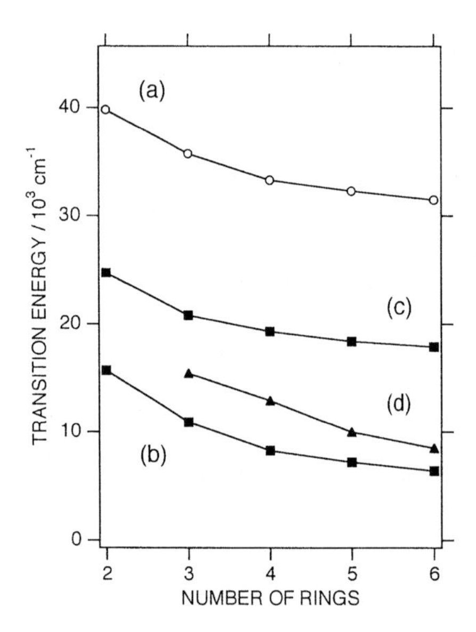

Figure 4-7. Observed electronic transition energies of *p*-oligophenyls. (a) Neutral species; (b) band I of the radical anions; (c) band II of the radical anions; (d) band I′ of the dianions. The data of neutral *p*-oligophenyls are taken from [67]. The data of the radical anions and dianions are taken from [69, 70].

sition energy of poly(*p*-phenylene), 27400 cm^{-1}, is lower than that of PP6, 31 500 cm^{-1}, indicating that the conjugation length of poly(*p*-phenylene) is much longer than 6. Upon Bu_4N^+ doping, two broad absorptions appear at about 5600 cm^{-1} (1800 nm) and 19400 cm^{-1} (515 nm). The assignments of these bands will be discussed in the next section on the basis of the spectra of the radical anions ($PPn^{\cdot -}$)

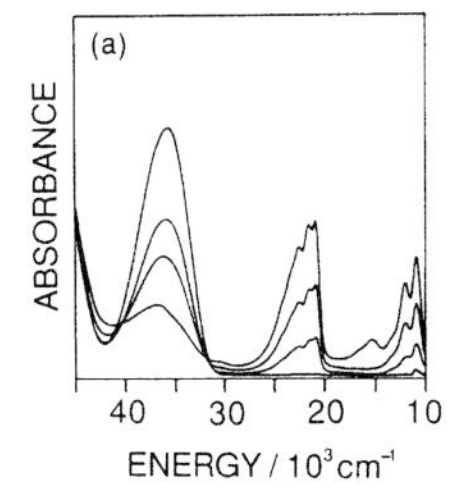

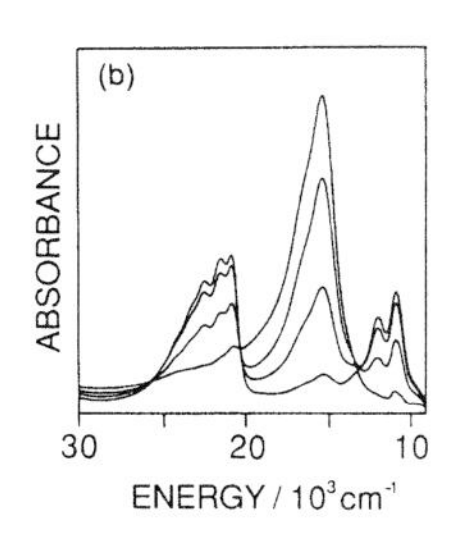

Figure 4-8. Absorption spectra of a THF solution of *p*-terphenyl at (a) early and (b) later stages of reduction. In (a), the 35 600 cm^{-1} band decreases and the 20 800 and 10 900 cm^{-1} bands increase in intensity. In (b), the 20 800 and 10 900 cm^{-1} bands decrease and the 15 400 cm^{-1} band increases in intensity.

and dianions (PPn^{2-}) of *p*-oligophenyls, which are models of negative polarons and bipolarons, respectively.

4.7.1.2 Radical Anions and Dianions of *p*-Oligophenyls

The electronic absorption spectra of $PPn^{\cdot -}$ and PPn^{2-} have been systematically studied [69, 70]. When a tetrahydrofuran (THF) solution of PP3 is exposed to a sodium mirror, its absorption spectrum gives new bands (Figure 4-8). At an early stage, the band observed at 35 700 cm^{-1} (280 nm) arising from neutral PP3 decreases and the bands observed at 20 800 and 10 900 cm^{-1} (481 and 916 nm) increase in intensity. These new bands are attributed to the radical anion of PP3 ($PP3^{\cdot -}$). In the course of this spectral change, an isosbestic point is observed at about 31 500 cm^{-1}, indicating that PP3 is quantitatively converted into $PP3^{\cdot -}$. When the solution is further exposed to the sodium mirror, the 20 800 and 10 900 cm^{-1} bands decrease and the band observed at 15 400 cm^{-1} (650 nm) increases in intensity. The 15 400 cm^{-1} band is ascribed to the dianion of PP3 ($PP3^{2-}$). During this reaction, three isosbestic points are observed at about 25 800, 20 400, and 13 400 cm^{-1}. The THF solutions of PP*n* ($n = 4, 5$, and 6) show similar reduction reactions which may be described as

$$PPn + Na \rightarrow PPn^{\cdot -} + Na^{+}$$

$$PPn^{\cdot -} + Na \rightarrow PPn^{2-} + Na^{+}$$

Since $PPn^{\cdot -}$ is stable for a long time, we can conclude that the following disproportionation reaction does not occur.

$$2PPn^{\cdot -} \rightarrow PPn + PPn^{2-}$$

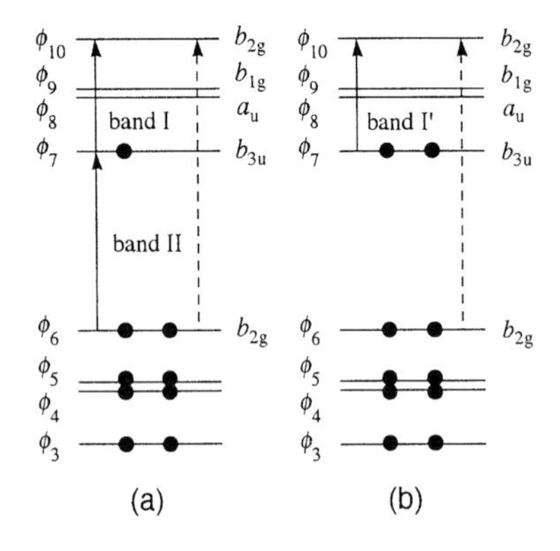

Figure 4-9. Schematic energy level diagrams. (a) The radical anion of biphenyl ($PP2^{\cdot -}$); (b) the dianion of biphenyl ($PP2^{2-}$). •, electron; arrow, electronic transition. The molecular orbital levels are taken from [77].

This reaction is similar to the process of the intramolecular formation of a bipolaron from two polarons.

The electronic transition energies [69, 70] of the radical anions and dianions of *p*-oligophenyls are plotted in Figure 4-7. Two strong bands are observed for the radical anions and are called bands I and II (curves b and c in Figure 4-7, respectively). On the other hand, one intense band is observed for the dianions, which is called band I′ (curve d in Figure 4-7). In the electronic absorption spectra of the radical cations and dications of α-oligothiophenes [71–73], two strong bands and one strong band are commonly observed for the radical cations and dications, respectively. In the case of the radical cations of oligophenylenevinylenes, two bands are observed [74, 75], although different results are also reported [76].

The absorption spectra of conducting polymers including poly(*p*-phenylene) have been discussed within the frame work of the Hückel approximation [7, 10]. In order to correlate the spectra of the radical anions and dianions of *p*-oligophenyls with those of doped poly(*p*-phenylene), we will discuss the observed absorption spectra of the charged *p*-oligophenyls on a similar theoretical basis. The energy diagram of $PP2^{\cdot -}$ is shown schematically in Figure 4-9a. Molecular orbital levels in this figure are taken from the results calculated by the Pariser–Parr–Pople–SCF–MO method for PP2 (D_{2h} symmetry) [77]. A Pariser–Parr–Pople–SCF–MO–CI calculation for $PP2^{\cdot -}$ [78] shows that band I is assigned to the transition mostly attributable to the electronic excitation of $\phi_{10} \leftarrow \phi_7$ and band II to $\phi_7 \leftarrow \phi_6$. These transitions are polarized along the long molecular axis. The ϕ_8 and ϕ_9 levels are nonbonding with respect to the inter-ring CC bond. It is reasonable to consider that these assignments hold also for $PPn^{\cdot -}$ ($n = 3$–6). Moreover, band I′ of $PP2^{2-}$ is assigned to the transition mostly attributable to the electronic excitation $\phi_{10} \leftarrow \phi_7$ (Figure 4-9b). This assignment would also be applicable to PPn^{2-} ($n = 3$–6). The $\phi_{10} \leftarrow \phi_6$ transitions of $PP2^{\cdot -}$ and $PP2^{2-}$ are symmetry forbidden under D_{2h} symmetry.

Since the radical anions and dianions of *p*-oligophenyls correspond, respectively, to negative polarons and bipolarons in the polymer, bands I and II of the radical anions are associated, respectively, with the ω_1 and ω_2 transitions of polarons (Figure 4-5c), and band I′ of the dianions with the ω_1' transition of bipolarons (Figure 4-5e). The $\phi_{10} \leftarrow \phi_6$ transitions correspond to the ω_3 and ω_3' transitions. It is worthwhile pointing out that the ω_3 and ω_3' transitions are symmetry forbidden. According to a continuum electron–phonon-coupled model reported by Fesser et al. [48], the ω_1 and ω_2 transitions are dominant among the three expected for a polaron, and the ω_1' transition is more intense between the two expected for a bipolaron. Finally, it is expected that *a polaron has two intense intragap transitions and a bipolaron one intense transition* [79].

The absorption spectrum of Bu_4N^+-doped poly(*p*-phenylene) can now be explained as follows. Two broad bands centered at about 5600 and 19400 cm^{-1} (Figure 4-6) are attributable to the ω_1 and ω_2 transitions of negative polarons, respectively. These assignments will be discussed again in Section 4.7.3 on the basis of the Raman results.

4.7.2 Raman Spectra

4.7.2.1 Intact Poly(*p*-phenylene) and *p*-Oligophenyls

The observed infrared and Raman spectra of intact poly(*p*-phenylene) [80–84] have been analyzed by normal coordinate calculations [43, 84–87]. The factor group of a coplanar polymer is isomorphous with the point group D_{2h}. When the yz plane is taken in the phenylene-ring plane (z axis along the polymer chain) and the x axis perpendicular to the phenylene-ring plane, the irreducible representation at the zone center ($k = 0$) is as follows:

$$\Gamma^{\text{vib}}{}_{\text{in-plane}} = 5a_g(\text{R}) + 4b_{1u}(\text{IR}) + 4b_{2u}(\text{IR}) + 5b_{3g}(\text{R})$$

$$\Gamma^{\text{vib}}{}_{\text{out-of-plane}} = 2a_u + 1b_{1g}(\text{R}) + 3b_{2g}(\text{R}) + 2b_{3u}(\text{IR})$$

where R and IR denote Raman- and infrared-active vibrations, respectively. The 1064-nm excited Raman spectrum and infrared absorption spectrum of intact poly(*p*-phenylene) prepared by dehalogenation polycondensation of *p*-dihalogeno-benzene are shown in Figure 4-10a, b, respectively. The observed and calculated vibrational frequencies of poly(*p*-phenylene), ^{13}C-substituted analog, and perdeuterated analog are listed in Table 4-3, except for CH (CD) stretches [$\nu_1(a_g)$, $\nu_9(b_{1u})$, $\nu_{16}(b_{2u})$, and $\nu_{20}(b_{3g})$]. Major Raman and infrared bands have been assigned.

In Figure 4-10b, the 809 cm^{-1} infrared band is assigned to the CH out-of-plane bend of the phenylene rings, and the 765 and 694 cm^{-1} bands are assigned to the CH out-of-plane bends of the terminal phenyl rings.

The Raman spectrum of poly(*p*-phenylene) has been compared with those of *p*-oligophenyls. The 1064-nm excited Raman spectra of *p*-oligophenyls in the solid state [88] are shown in Figure 4-11. The bands at 1605–1593, 1278–1275, 1221–1220, and 795–774 cm^{-1} correspond to those at 1595, 1282, 1222, and 798 cm^{-1} of intact poly(*p*-phenylene), respectively. These bands are assigned to the a_g vibrations (ν_2–ν_5). The atomic displacements of these modes obtained by normal coordinate calculations based on the PM3 method [84] are depicted in Figure 4-12a–d. The 1595 cm^{-1} mode (Figure 4-12a) is contributed mainly by the CC stretch of the phenylene ring. The 1282 cm^{-1} mode (Figure 4-12b) is mainly contributed by the

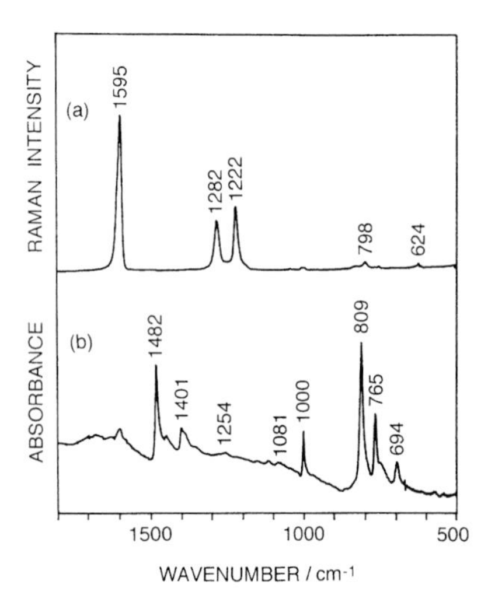

Figure 4-10. Vibrational spectra of intact poly(*p*-phenylene). (a) Raman spectrum taken with the 1064-nm line in powder sample; (b) infrared spectrum in a KBr disk [83].

inter-ring CC stretch. The 1222 cm^{-1} mode (Figure 4-12c) is a mixture of CC stretch and CH in-plane bend. The 798 cm^{-1} mode (Figure 4-12d) is a mixture of CC stretch and CCC deformation. The intensity of the 1278–1275 cm^{-1} band relative to that of the 1221–1220 cm^{-1} band decreases as the chain length becomes longer. Thus, this ratio is a marker of the length of conjugated segments [81, 82, 88]. The wavenumbers of the Raman bands of poly(*p*-phenylene) and *p*-oligophenyls are almost independent on the excitation wavelength [81]. It is well known that the wavenumbers of the two strong Raman bands of *trans*-polyacetylene depend greatly on the wavelength of the excitation laser. These large dispersions are explained by the existence of segments of various conjugation lengths that give rise to the Raman bands at different wavenumbers [13, 14]. However, the Raman spectra of other conducting polymers do not show such dependence on the excitation wavelength.

The (resonance) Raman intensities of conjugated molecules have been analyzed by the classical vibronic theory due to Tang and Albrecht [89]. Inagaki et al. [90] have shown that resonance Raman intensities of totally symmetric modes of β-carotene (an oligoene with eleven conjugated C=C bonds) arise from the A term (Franck–Condon term) of the Albrecht theory. Probably, the Franck–Condon factor plays an important role in determining the resonance Raman intensities of almost all conjugated molecules. The resonance Raman intensity of each totally symmetric mode is proportional to the square of the shift of the potential minimum in the resonant excited electronic state from that in the ground electronic state along the normal mode giving rise to the resonance Raman band, when the shift is small [91–94]. The effective conjugation coordinate represents approximately the change of equilibrium geometry between the ground electronic state and the first dipole-allowed excited electronic state [15]. Thus, the resonance Raman intensity of a normal mode is determined by the contribution of the effective conjugation coordinate to the mode. Details are described in [15]. It has been demonstrated that the

Table 4-3. Vibrational frequencies (in cm^{-1}) of neutral poly(*p*-phenylene) and its ^{13}C-substituted and perdeuterated analogs.

Symmetric species	No.	Normal species					^{13}C-substituted analog		Perdeuterated analog		
		Observed	Calculated				Observed	Calculated	Observed	Calculated	
		[83]	[84]	[43]	[86]	[87]	[83]	[84]	[84]	[84]	[87]
a_g	ν_2	1595	1599	1626	1661	1601	1543	1547	1563	1563	1572
	ν_3	1282	1295	1282	1289	1290	1239	1246	1255	1281	1253
	ν_4	1222	1226	1233	1127	1184	1209	1216	892	887	863
	ν_5	798	776	792	846	802	768	747	764	751	776
a_u	ν_6	–	969	951	961	–	–	960	–	775	–
	ν_7	–	427	367	401	–	–	415	–	378	–
b_{1g}	ν_8	–	831	823	834	–	–	825	–	647	–
b_{1u}	ν_{10}	1482	1477	1487	1510	1490	1452	1445	1354	1353	1364
	ν_{11}	1048	1060	1045	1051	1044	–	1035	817	812	811
	ν_{12}	1000	991	984	968	1004	965	958	978	984	977
b_{2g}	ν_{13}	–	945	968	945	–	–	936	–	824	–
	ν_{14}	–	768	754	760	–	–	742	–	663	–
	ν_{15}	–	441	410	402	–	–	426	–	414	–
b_{2u}	ν_{17}	1401	1407	1399	1440	1412	1361	1375	1338	1334	1347
	ν_{18}	1254	1232	1275	1268	1343	–	1193	–	1185	1268
	ν_{19}	1081	1119	1162	1075	1118	–	1092	–	854	855
b_{3g}	ν_{21}	–	1603	1601	1654	1652	–	1546	–	1579	1620
	ν_{22}	–	1290	1325	1328	1343	–	1277	–	1016	1044
	ν_{23}	624	633	625	602	616	602	610	606	600	593
	ν_{24}	–	487	488	460	418	–	470	–	450	383
b_{3u}	ν_{25}	809	798	797	790	–	801	789	650	661	–
	ν_{26}	–	497	457	458	–	–	481	–	435	–

Raman spectrum of neutral poly(*p*-phenylene) is well explained by the effective-conjugation-coordinate model [95].

4.7.2.2 Doped Poly(*p*-phenylene) and the Radical Anions and Dianions of *p*-Oligophenyls

The Raman spectra of heavily Na-doped poly(*p*-phenylene) measured with the 488.0, 514.5, 632.8, 720, and 1064 nm laser lines are shown in Figure 4-13 [70]. The Raman spectra of Na-doped poly(*p*-phenylene) are apparently different from that of intact poly(*p*-phenylene), indicating that these bands arise from charged domains (i.e., negative polarons and/or negative bipolarons) created by Na-doping. The assignments of these bands will be discussed on the basis of the data for the radical anions and dianions of *p*-oligophenyls.

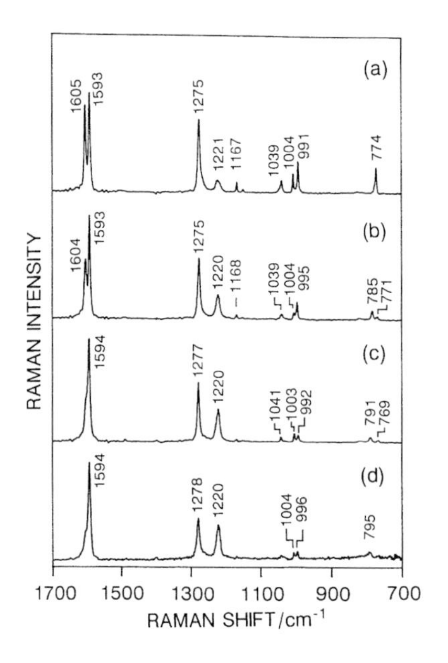

Figure 4-11. Raman spectra of *p*-oligophenyls in solid. (a) *p*-Terphenyl (PP3); (b) *p*-quaterphenyl (PP4); (c) *p*-quinquephenyl (PP5); (d) *p*-sexiphenyl (PP6). Excitation wavelength is 1064 nm [88].

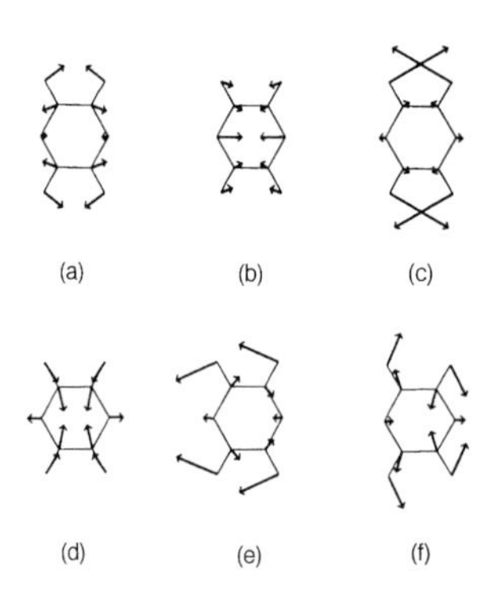

Figure 4-12. Atomic displacements of major vibrational modes ($k = 0$) of poly(*p*-phenylene). (a) 1599 cm^{-1} [$\nu_2(a_g)$]; (b) 1295 cm^{-1} [$\nu_3(a_g)$]; (c) 1226 cm^{-1} [$\nu_4(a_g)$]; (d) 776 cm^{-1} [$\nu_5(a_g)$]; (e) 1477 cm^{-1} [$\nu_{10}(b_{1u})$]; (f) 991 cm^{-1} [$\nu_{12}(b_{1u})$] [84].

The Raman spectra of the radical anions of PP2 [92, 96, 97], PP3 [70, 98], PP4 [70, 98], PP5 [70] and PP6 [70], and the dianions of PP3 [70, 98], PP4 [70, 98], PP5 [70], and PP6 [70] have been reported by several authors. The Raman spectra of the radical anions ($PPn^{\cdot-}$) and dianions (PPn^{2-}) in THF solutions are shown in Figures 4-14 and 4-15, respectively. The excitation wavelengths used are rigorously or nearly resonant with band II of the radical anions and band I′ of the dianions. The Raman spectra of the radical anions are similar to each other, and so are the Raman spectra of the dianions. However, the two groups of the Raman spectra are different from each other and from those of neutral *p*-oligophenyls. These results suggest that the negative polarons and bipolarons, corresponding respectively to the radical anions and dianions, can be identified by Raman spectroscopy.

The Raman spectra of the radical anions in Figure 4-14 show small changes with

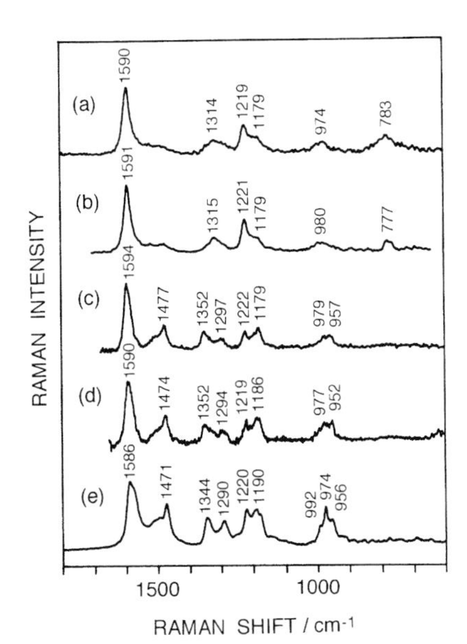

Figure 4-13. Raman spectra of Na-doped poly(-*p*-phenylene) taken with various wavelengths of excitation lasers. (a) 488.0; (b) 514.5; (c) 632.8; (d) 720, (e) 1064 nm [70].

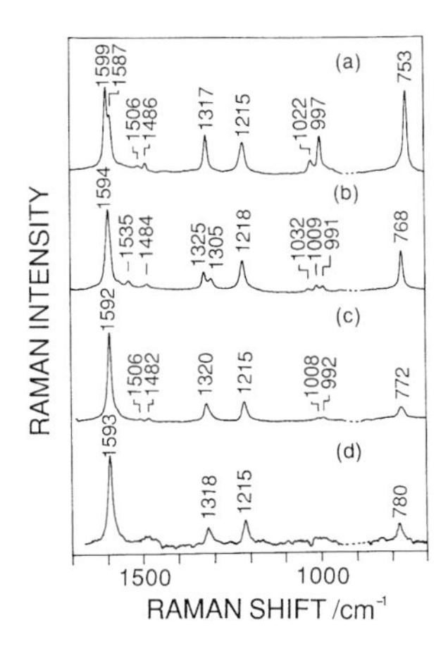

Figure 4-14. Resonance Raman spectra of the radical anions of *p*-oligophenyls in THF solutions. (a) *p*-Terphenyl; (b) *p*-quaterphenyl; (c) *p*-quinquephenyl; (d) *p*-sexiphenyl. Excitation wavelengths are 488.0 and 514.5 nm for (a) and (b–d), respectively. The bands of the solvent are subtracted [70].

the chain length. The intensities of the bands observed in the 1032–991 and 780–753 cm^{-1} regions decrease as the chain length becomes longer, indicating that these bands are associated with the terminal rings. The bands observed in the 780–753 cm^{-1} region show an upshift in wavenumber, as the chain length becomes longer. This result indicates that the wavenumbers of these bands can be used as a measure of the localization length of polarons. The bands observed in the 1325–1305 cm^{-1} region are assigned to the inter-ring CC stretch [70, 92, 98]. The corresponding bands in neutral PPn are observed in the range between 1278 and 1275 cm^{-1}. About 40-cm^{-1} upshifts are explained by the increased π-bond orders of the inter-ring CC bonds, i.e., shortening of the inter-ring CC bond lengths. In PP3, the calculated

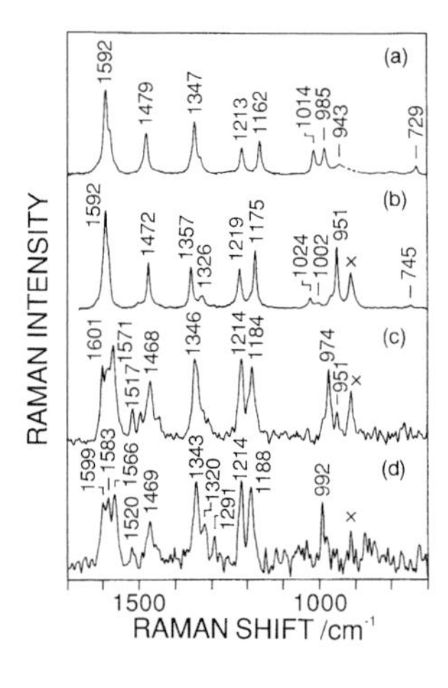

Figure 4-15. Resonance Raman spectra of the dianions of *p*-oligophenyls in THF solutions. (a) *p*-Terphenyl; (b) *p*-quaterphenyl; (c) *p*-quinquephenyl; (d) *p*-sexiphenyl. Excitation wavelengths are (a) 632.8, (b) 720, (c) 1064, and (d) 1064 nm, respectively. The bands with × are due to the solvent. Fluorescence background is subtracted in all the spectra [70].

length of the inter-ring CC bond (r_{45}) shrinks from 1.469 to 1.426 Å in going from neutral PP3 to $PP3^{\cdot -}$ (see Table 4-1).

In the Raman spectra of the dianions in Figure 4-15, the bands in the 1479–1468 cm^{-1} and 1188–1162 cm^{-1} regions are strongly observed. Corresponding bands are weakly observed or are not observed in the spectra of the radical anions. This reflects structural differences between the electronic ground state and the excited state associated with band II of the radical anions and band I′ of the dianions. The bands observed in the 1357–1320 cm^{-1} region are assigned to the inter-ring CC stretch [70, 98]. About 70 cm^{-1} upshifts in going from neutral PP*n* to PPn^{2-} are also explained by the shortening of the inter-ring CC bond lengths. These upshifts of PPn^{2-} are larger than those of $PPn^{\cdot -}$. This observation suggests that the degrees of shrinkage of the inter-ring CC bonds in PPn^{2-} are larger than those in $PPn^{\cdot -}$. This is consistent with the results of calculation for PP3 given in Table 4-1: 1.469 Å (PP3) and 1.388 Å ($PP3^{2-}$). It is worth pointing out that the frequency of inter-ring CC stretch increases with the value of structural deformation coordinate *R* (column 8 in Table 4-1) for PP3, $PP3^{\cdot -}$, and $PP3^{2-}$. In the 1000–940 cm^{-1} region, there is a series of bands at 943 ($PP3^{2-}$), 951 ($PP4^{2-}$), 974 ($PP5^{2-}$), and 992 cm^{-1} ($PP6^{2-}$). This band series may be useful in estimating the localization length of bipolarons.

Next, we will analyze the Raman spectra of Na-doped poly(*p*-phenylene) (Figure 4-13) on the basis of those of the radical anions and dianions of *p*-oligophenyls. The Raman spectra of Na-doped poly(*p*-phenylene) taken with the 488.0- and 514.5-nm lines (Figure 4-13a, b) are different from those with the other three laser lines (Figure 4-13c–e), and are similar to those of the radical anions of *p*-oligophenyls (Figure 4-14). This observation can be understood by considering that the Raman bands arising from charged domains formed by Na-doping in the polymer chains are quite similar to those of the radical anions. Since the radical anions are viewed as negative polarons in the polymer, negative polarons exist in Na-doped poly(*p*-phenylene) [70]. These Raman spectra of Na-doped poly(*p*-phenylene) are observed under resonant conditions. Thus, the results described above indicate that the

Table 4-4. Observed Raman frequencies (in cm^{-1}) of negative polarons and negative bipolarons in Na-doped poly(*p*-phenylene).

No.	Negative polaron[a]	Negative bipolaron[a]	Assignment[b]
A	1591–1590	1594–1586	sym. in-phase mode consisting of $\nu_2(a_g)$
B		1477–1471	sym., $\nu_{10}(b_{1u})$
C	1315–1314	1352–1344	sym., in-phase, $\nu_3(a_g)$
D		1297–1290	sym., out-of-phase, $\nu_3(a_g)$
E	1221–1219	1222–1219	sym., in-phase, $\nu_4(a_g)$
F	1179	1190–1179	sym., out-of-phase, $\nu_4(a_g)$
G	980–974	992–952	sym., $\nu_{12}(b_{1u})$
H	783–777		sym., in-phase, $\nu_5(a_g)$

[a] Data are taken from [83].
[b] Numbering and symmetry correspond to those of neutral poly(*p*-phenylene). Sym. denotes a symmetric mode.

absorption around 488.0 and 514.5 nm (20 500 and 19 400 cm^{-1}) is attributed to polarons.

The Raman spectra excited with laser lines in the 632.8–1064 nm region (Figure 4-13c–e) are quite similar to those of the dianions of *p*-oligophenyls (Figure 4-15), which correspond to negative bipolarons in the polymer. Accordingly, negative bipolarons also exist in Na-doped poly(*p*-phenylene), and the absorption in the 632.8–1064 nm (15 800–9400 cm^{-1}) region arises from bipolarons [70]. In the 1000–950 cm^{-1} region of the Raman spectra, a few bands are observed at 979 and 957 cm^{-1} (632.8-nm excitation), at 977 and 952 cm^{-1} (720 nm), and at 992, 974, and 956 cm^{-1} (1064 nm). By comparing these bands with those of the Raman spectra of the dianions, it may be concluded that negative bipolarons localized over four, five, and six phenylene rings coexist in Na-doped poly(*p*-phenylene).

The Raman bands arising from negative polarons and bipolarons, together with the tentative mode assignments based on the wavenumber shifts upon ^{13}C-substitution [83], are listed in Table 4-4. The observed bands are called bands A–H as shown in Table 4-4. Bands A, C, E, and H correspond to the a_g modes (ν_i, $i = 2$–5) of neutral poly(*p*-phenylene), respectively. The 1282 cm^{-1} Raman band of neutral poly(*p*-phenylene) upshifts to 1315–1314 cm^{-1} in the spectra of polarons and 1352–1344 cm^{-1} in the spectra of bipolarons. According to normal coordinate calculations [43, 84, 86, 87], the 1282-cm^{-1} mode is mainly attributed to the inter-ring CC stretch. The observed upshifts upon Na-doping reflect the increasing π-bond order of the inter-ring CC bond, i.e., structural changes from benzenoid to quinoid [70]. The upshifts for bipolarons (70–62 cm^{-1}) are larger than those for polarons (33–32 cm^{-1}), indicating that the geometric changes in bipolarons are larger than those in polarons. On the other hand, bands A and E show little shifts in wavenumber and band H shows a moderate shift.

In the Raman spectra of negative bipolarons (Figure 4-13c–e), bands B, D, F, and G are strongly observed. Band B is assignable to a symmetric mode consisting of the $\nu_{10}(b_{1u})$ mode (Figure 4-12e) of each benzene ring in the charged domain. Band G is also attributed to a symmetric mode consisting of the $\nu_{12}(b_{1u})$ mode (Figure 4-12f). Bands D and F are assignable to out-of-phase symmetric modes consisting of the $\nu_3(a_g)$ and $\nu_4(a_g)$ modes (Figure 4-12b, c) of each benzene ring, respectively. The appearance of bands B, D, F, and G indicates the loss of translational symmetry of neutral poly(*p*-phenylene) chain, which is consistent with the formation of localized charged domains upon doping. Bands B, D, F, and G are not observed or weakly observed for negative polarons, but strongly observed for negative bipolarons. These observations seem to reflect different geometric changes occurring between polarons and bipolarons on going from their respective ground electronic states to their respective excited electronic states.

4.7.3 Assignments of Electronic Absorption Spectra of Doped Poly(*p*-phenylene)

In Section 4.7.1, two broad bands centered at about 5600 and 19400 cm^{-1} in the electronic absorption spectrum of Bu_4N^+-doped poly(*p*-phenylene) have been attributed to the ω_1 and ω_2 transitions of negative polarons, respectively, on the basis of the data of the radical anions and dianions of *p*-oligophenyls. Raman spectroscopy provides information on the electronic levels associated with polarons and bipolarons as well as on their geometries. The Raman results indicate the co-existence of polarons and bipolarons in Na-doped poly(*p*-phenylene). The Raman bands due to polarons are resonantly enhanced by the use of the 488.0 and 514.5 nm laser lines, which are located within the electronic absorption around 19400 cm^{-1} (515 nm). These observations confirm the proposed assignment that the 19400 cm^{-1} band is due to the ω_2 transition of polarons. The Raman bands due to bipolarons are resonantly enhanced by the use of the 632.8, 720, and 1064 nm laser lines. However, in the absorption spectrum of Bu_4N^+-doped poly(*p*-phenylene), no peaks are observed in the 15800–9400 cm^{-1} (632.8–1064 nm) region. This fact can be explained by considering that the content of the bipolaron is very small and the ω_1' absorption due to the bipolaron is overlapped by the strong ω_1 absorption band due to polarons.

4.8 Other Polymers

The Raman spectra of doped states of polyaniline [99], poly(*p*-phenylenevinylene) [75, 100–102], and polythiophene [72, 73] have been also analyzed on the basis of the data of charged oligomers (charged oligomer approach). The detected self-localized excitations are summarized in Table 4-5. Some important conclusions can be drawn from these Raman studies.

Table 4-5. Types of self-localized excitations detected by Raman spectroscopy.

Polymer	Dopant			Species	Reference
	Type	Reagent	Content		
poly(*p*-phenylene)	n	Na	heavy	polaron, bipolaron	[70]
poly(*p*-phenylenevinylene)	n	Na	heavy	polaron, bipolaron	[100]
	n	Na	light	polaron, bipolaron	[101]
	n	K	heavy	polaron, bipolaron	[101]
	p	H_2SO_4	heavy	polaron	[75]
	p	SO_3	heavy	polaron	[102]
	p	SO_3	light	polaron	[102]
polythiophene	p	BF_4^-	heavy	polaron	[72, 73]
polyaniline (emeraldine base form, 2A)[a]	p	HCl	heavy	polaron	[99]
polyaniline (leucoemeraldine, 1A)	p	BF_4^-	heavy	polaron	[99]

[a] HCl-doped emeraldine base form is called emeraldine acid form or 2S form.

1. Only polarons are detected for p-type doping. The Raman spectra of the polymers doped with acceptors do not show large changes with various excitation wavelengths [72, 73, 75]. The polymer-chain structures of p-type-doped polymers are more homogeneous, and more regular arrays of polarons are formed.
2. Polarons and bipolarons coexist in n-type doped polymers in contrast to the results of p-type doping. The Raman spectra of the polymers doped with donors depend on the excitation wavelength [70, 100]. These observations are explained by the existence of various localization lengths of polarons and/or bipolarons [70, 100]. The polymer-chain structures of the n-type-doped polymers are inhomogeneous. Possibly, Na-doped polymers are easily hydrogenated, even in the presence of a small amount of water during the sample preparation, and negative polarons and/or bipolarons are formed on the short conjugated segments.

The Raman spectra of doped polyacetylene have been reported with the excitation of visible laser lines [103–110]. There have been some discussions about the question of whether these bands arise from charged domains or undoped parts remaining in the doped films. Recently, the Raman spectra of maximally Na-doped polyacetylene excited with NIR lasers have been reported [57, 60, 111] and compared with those of the radical anions of α, ω-diphenyloligoenes and α, ω-dithienyloligoenes [111]. It has been concluded that the observed bands arise from charged domains (charged solitons and/or polarons), because the observed Raman spectra of the doped polymer are similar to those of the radical anions of α, ω-diphenyloligoenes. However, further studies are required for characterizing doped polyacetylene in more detail.

4.9 Electronic Absorption and ESR Spectroscopies and Theory

It is widely accepted [7, 10] that the major species generated by chemical doping in nondegenerate polymers is bipolarons, except for polyaniline. However, the results obtained from Raman spectroscopy indicate the existence of polarons at heavily doped polymers. These results are inconsistent with the established view. The experimental basis for claiming the existence of bipolarons has been electronic absorption and ESR spectra [7, 9, 10]. We will comment on these experimental results.

The electronic absorption spectra of p-type-doped polypyrrole have been interpreted in terms of polarons and bipolarons for the first time [112]. Electronic absorption spectra of doped polypyrrole in the range from visible to NIR show three bands due to intragap transitions at low dopant contents and two bands at high dopant contents [113]. The three bands are attributed to polarons and the two bands to bipolarons [112], because a polaron is expected to have three intragap transitions and a bipolaron is expected to have two intragap transitions (see Section 4.4.2). A quantitative ESR study [114] has confirmed the interpretation of the electronic absorption experiments. This rule of thumb for band assignment has so far also been applied to other conducting polymers.

However, the two electronic absorption bands of doped poly(*p*-phenylene) are not explained by bipolarons, as demonstrated in Section 4.7.3. Furthermore, p-type-doped polythiophene shows two absorptions at about 12000 and 5200 cm^{-1} [72, 73, 115, 116] and p-type-doped poly(*p*-phenylenevinylene) shows two broad absorptions at about 19000 and 7300 cm^{-1} [75, 117, 118]. In these two systems, only positive polarons are detected by Raman spectroscopy (see Table 4-5). These results of Raman spectroscopy suggest that the two-band pattern in the electronic absorption spectrum is attributed to polarons. As discussed in Section 4.7.1.2, it is expected that a polaron has two intense intragap transitions and a bipolaron one intense transition. From these findings, we propose a new rule: *a two-band pattern corresponds to polarons and one-band pattern to bipolarons* [79], when these species do not coexist. When polarons and bipolarons coexist, and their electronic absorptions are overlapped, a more careful examination of the observed spectrum is of course necessary.

Recently, Hill et al. [119] have proposed the existence of a singlet radical-cation dimer (i.e., polaron dimer) as an alternative to a spinless bipolaron to explain weak ESR signals from doped polymers. Their proposal is based on the observed dimerization of the radical cation of 2,5″-dimethylterthiophene. They have found by electronic absorption and ESR measurements that the above radical cation is dimerized at low temperature even at a low concentration (4×10^{-4} mol/l). Singlet intrachain polaron pairs and interchain polaron pairs give no ESR signals. Thus, the absence of ESR signals or observation of weak ESR signals does not directly lead to the conclusion that the doped polymer contains only bipolarons.

The stability of polarons, bipolarons, and solitons has been studied theoretically under various models. According to the SSH model and similar approximations with electron–lattice coupling for a single polymer chain [7, 10], the creation energy of one bipolaron in a nondegenerate system (or two charged solitons in a degenerate system) is lower than that of two polarons. Thus, it is considered that two separate polarons are always unstable, and changed into a bipolaron (or two charged solitons). For doped polyacetylene, the effect of the electron–electron interaction and the dopant potential, which are not taken into account in the SSH model, have been studied [120]. On the other hand, the existence of polarons has been theoretically proposed by several authors. Kivelson and Heeger [121] have proposed a polaron lattice (regular infinite array of polarons) for explaining the metallic properties of heavily doped polyacetylene. Mizes and Conwell [122] have reported that polarons are stabilized in short conjugated segments for *trans*-polyacetylene and poly(*p*-phenylenevinylene) from the results obtained by a three-dimensional tight-binding calculation. Shimoi and Abe [123] have studied the stability of polarons and bipolarons in nondegenerate systems by using the Pariser–Parr–Pople method combined with the SSH model. A polaron lattice is stabilized by the electron–electron interaction at dopant contents lower than a critical concentration, whereas a bipolaron lattice is stabilized at dopant contents higher than the critical concentration. Thus, importance of the electron–electron interaction and three-dimensional interaction as well as the electron–lattice coupling has been demonstrated. However, the nature of the metallic states of heavily doped polymers remains an unresolved theoretical problem.

4.10 Mechanism of Charge Transport

Overall electrical conduction in doped polymers is considered to have contributions from various processes such as intrachain, interchain, interdomain, interfibril transport, etc. The contributions of individual processes would depend on the solid-state structure and morphology [8]. Charge transport has been discussed in terms of polarons, bipolarons, charged solitons, and their lattices [6, 8, 10]. Even when the dopant content is very small, electrical conductivity increases dramatically. At this doping level, discrete localized electronic levels associated with self-localized excitations are formed within the band gap, and the temperature dependencies of d.c. conductivity have been explained by the hopping between localized electronic states [8]. It is conceivable that the energy gap between the lower (or upper) localized state and the valence (or conduction) band for a positive (or a negative) polaron is smaller than that of a positive (or a negative) bipolaron. Then, it is expected that polarons play a more important role in hopping conduction than bipolarons.

Heavily doped polymers are of special interest, because these polymers show metallic properties such as a Pauli susceptibility, a linear temperature dependence of the thermoelectric power, a high reflectivity in the infrared region, etc., as described in Section 4.2. The temperature dependence of d.c. conductivity in heavily doped polymers is not like a metal, whereas metallic properties are observed. These results

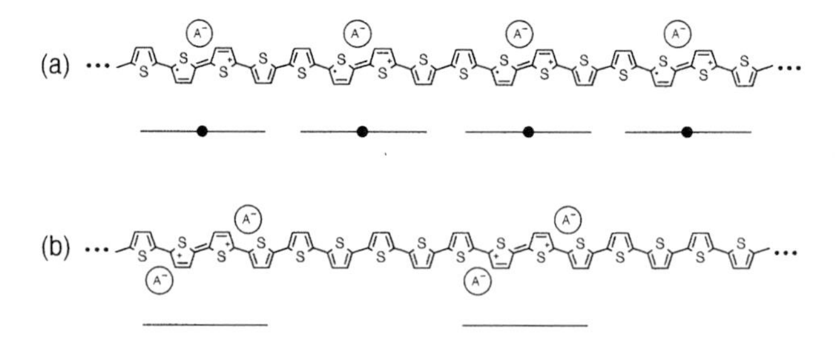

Figure 4-16. Schematic structures of polythiophene chains doped with electron acceptors (dopant content, 25 mole% per thiophene ring) and bonding electronic levels of positive polarons and bipolarons. (a) Polaron lattice; (b) bipolaron lattice. A, acceptor; +, positive charge; –, negative charge; •, electron; —, electronic energy level.

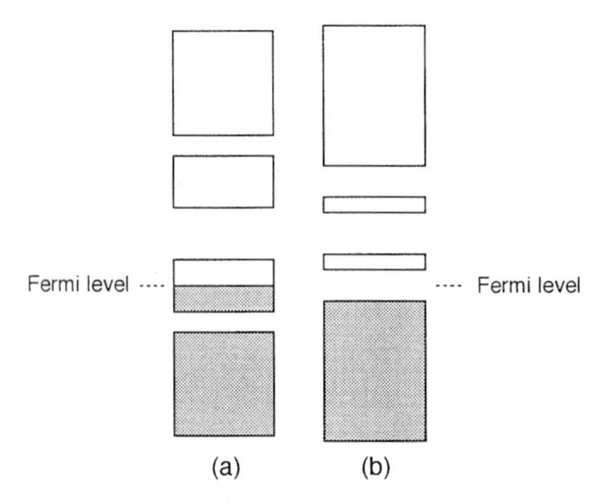

Figure 4-17. Schematic band structures of an infinite polymer chain doped with acceptors (dopant content, 25 mole% per thiophene ring). (a) Polaron lattice; (b) bipolaron lattice. Shadowed areas are filled with electrons [126].

can be explained as follows. Although a single chain itself behaves like a metal, macroscopic electrical conduction is limited by interchain, interdomain, or interfibril charge transport. Here, we will discuss the intrachain metallic charge transport.

Let us suppose an infinite nondegenerate polymer chain (e.g., polythiophene) doped heavily with electron acceptors. At a high dopant content, the polymer-chain structure and electronic structure of the doped polymer are radically different from those of the intact polymer. As typical cases, we will describe two kinds of lattice structures of doped polythiophene (dopant content, 25 mole% per thiophene ring): a polaron lattice and a bipolaron lattice. They are the regular infinite arrays of polarons and bipolarons. The schematic polymer-chain structures are shown in Figure 4-16. Band-structure calculations have been performed for polaron and/or bipolaron lattices of poly(*p*-phenylene) [124], polypyrrole [124], polyaniline [125], polythiophene [124, 126], and poly(*p*-phenylenevinylene) [127], with the valence-effective Hamiltonian pseudopotential method on the basis of geometries obtained by MO methods. The schematic electronic band structures shown in Figure 4-17 are based on the results calculated with this method for a polaron and a bipolaron lattice of p-type-doped polythiophene (dopant content, 25 mole% per thiophene ring) [126].

A polaron has a localized electronic level occupied by one electron (i.e., SOMO). In a polaron lattice formed at a heavy doping level, electronic wave functions of neighboring polarons are overlapped and interact with each other. As a result, a half-filled metallic band is formed from the polaron bonding levels occupied by one electron (Figure 4-17a). The Fermi level is located at the center of the polaron bonding band. The electronic states of the polaron lattice are not localized, whereas the electronic states of a polaron are localized. The calculated bandwidths of the polaron bands are 8900, 7900, and 5600 cm^{-1} (1.1, 0.98, and 0.69 eV) for heavily doped polyaniline [125], polythiophene [126], and poly(*p*-phenylenevinylene) [127], respectively. If the conjugation length of the polymer chain is not sufficiently long, the metallic band is not formed. In this case, discrete levels are formed and hopping conduction is expected to take place. In contrast to the polaron lattice, no metallic band is formed for the bipolaron lattice, even if the conjugation length is long enough (Figure 4-17b). This is because a bipolaron has no singly occupied electronic level, whereas a polaron has a singly occupied level. In addition, the differences in the electronic band structure between the polaron and bipolaron lattices suggest that the bipolaron lattice is the result of a dimerization of the polaron lattice like the Peierls transition [126].

In the case of *trans*-polyacetylene (a degenerate polymer), band-structure calculations [126] indicate that a polaron lattice has a metallic band whereas a charged soliton lattice has no metallic band. An alternating polaron-charged soliton lattice [128] has been proposed for explaining the metallic properties of heavily doped states.

We will discuss polymer-chain structures obtained by Raman spectroscopy and metallic properties. The Raman measurements indicate the existence of positive polarons for p-type-doped polythiophene, poly(*p*-phenylenevinylene), and polyaniline at heavy doping levels (Table 4-5). These results indicate that positive polarons are formed on the polymer chains upon acceptor doping. If conjugation length is sufficiently long and the dopant content sufficiently high, a large number of polarons are formed and interact with each other. Then, a polaron lattice is formed. Therefore, the results of p-type-doped polythiophene, poly(*p*-phenylenevinylene), and polyaniline are consistent with the formation of polaron lattices at heavy doping levels. *The metallic properties observed for doped polymers probably originate from the polaron lattice.* From the results obtained so far, we can point out some important conditions for obtaining metallic polymers: (i) a long conjugation length; (ii) a high dopant content (large degree of charge transfer); and (iii) formation of polarons. In the case of n-type doping, both negative polarons and bipolarons are detected for doped poly(*p*-phenylene) and poly(*p*-phenylenevinylene) (Table 4-5). Polarons and bipolarons with various localization widths are detected, and electronic states are probably localized, even at heavy doping levels. It is considered that such an inhomogeneous structure does not lead to formation of a metallic band.

It is well known that physical properties of conducting polymers depend on synthetic conditions, doping conditions, and other treatments of samples. Thus, in order to confirm the proposal that the metallic properties originate from the polaron lattice, it is requisite to use the same polymer film or the films from the same batch for measuring various physical properties and Raman spectra.

4.11 Summary

It has been demonstrated that Raman spectroscopy is a powerful tool for studying self-localized excitations such as polarons and bipolarons in doped conducting polymers. Resonance Raman spectroscopy by using exciting laser lines in a wide range between visible and near-infrared gives valuable information on self-localized excitations; especially, near-infrared Raman spectroscopy developed recently is important. Spectroscopic data on the radical ions and divalent ions of oligomers are useful for analyzing the electronic and vibrational spectra of doped polymers. New assignments of the electronic absorption spectra of polarons and bipolarons are proposed: a two-band pattern is attributed to polarons, and a one-band pattern to bipolarons. Polarons and/or bipolarons have been detected for doped poly(*p*-phenylene), polythiophene, poly(*p*-phenylenevinylene), and polyaniline by Raman spectroscopy. On the basis of the Raman results, metallic properties at heavily doped polymers is attributed to formation of a polaron lattice with a half-filled metallic band.

Postscript: After the manuscript of this Chapter was completed, a review book treating the polaron/bipolaron in conjugated polymers was published [129].

4.12 References

[1] Shirakawa, H., Louis, E. J., MacDiarmid, A. G., Chiang, C. K., Heeger, A. J., *J. Chem. Soc., Chem. Commun.* **1977**, 578–580.
[2] Chiang, C. K., Fincher, Jr., C. R., Park, Y. W., Heeger, A. J., Shirakawa, H., Louis, E. J., Gau, S. C., MacDiarmid, A.G., *Phys. Rev. Lett.* **1977**, *39*, 1098–1101.
[3] Wegner, G., *Angew. Chem. Int. Ed. Engl.* **1981**, *20*, 361–381.
[4] Etemad, S., Heeger, A. J., MacDiarmid, A. G., *Ann. Rev. Phys. Chem.* **1982**, *33*, 443–469.
[5] Baughman, R. H., Brédas, J. L., Chance, R. R., Elsenbaumer, R. L., Shacklette, L. W., *Chem. Rev.* **1982**, *82*, 209–222.
[6] Chien, J. C. W., *Polyacetylene–Chemistry, Physics, and Material Science*. New York: Academic Press, 1984.
[7] Brédas, J. L., Street, G. B. *Acc. Chem. Res.* **1985**, *18*, 309–315.
[8] Roth, S., Bleier, H., *Adv. Phys.* **1987**, *36*, 385–462.
[9] Patil, A. O., Heeger, A. J., Wudl, F., *Chem. Rev.* **1988**, *88*, 183–200.
[10] Heeger, A. J., Kivelson, S., Schrieffer, J. R., Su, W.-P., *Rev. Mod. Phys.* **1988**, *60*, 781–850.
[11] Kuzmany, H., Mehring, M., Roth, S. (Eds.) *Springer Ser. Solid-State Sci.* **1987**, 76; **1990**, 91 ; **1992**, 107.
[12] Lu, Y., *Solitons and Polarons*. Singapore: World Scientific, 1988.

[13] Kuzmany, H., *Makromol. Chem., Macromol. Symp.* **1990**, *37*, 81–97.
[14] Harada, I., Furukawa, Y., in: *Vibrational Spectra and Structure*: Durig, J. R. (Ed.). Amsterdam: Elsevier, 1991; Vol. 19, pp. 369–469.
[15] Gussoni, M., Castiglioni, C., Zerbi, G., in: *Advances in Spectroscopy*: Clark, R. J. H., Hester, R. E. (Eds.). Chichester: John Wiley & Sons, 1991; Vol. 19, pp. 251–353.
[16] Brédas, J. L., Silbey, R. (Eds.), *Conjugated Polymers*. Netherlands: Kluwer Academic Publishers, 1991.
[17] Tsukamoto, J., *Adv. Phys.* **1992**, *41*, 509–546.
[18] Su, W. P., Schrieffer, J. R., Heeger, A. J., *Phys. Rev. Lett.* **1979**, *42*, 1698–1701.
[19] Su, W. P., Schrieffer, J. R., Heeger, A. J., *Phys. Rev. B* **1980**, *22*, 2099–2111.
[20] Su, W. P., Schrieffer, J. R., *Proc. Natl. Acad. Sci. USA* **1980**, *77*, 5626–5629.
[21] Brazovskii, S. A., Kirova, N. N., *Sov. Phys. JETP Lett.* **1981**, *33*, 4–8.
[22] Bishop, A. R., Campbell, D. K., Fesser, K., *Mol. Cryst. Liq. Cryst.* **1981**, *77*, 253–264.
[23] Brédas, J. L., Chance, R. R., Silbey, R., *Mol. Cryst. Liq. Cryst.* **1981**, *77*, 319–332.
[24] Ikehata, S., Kaufer, J., Woerner, T., Pron, A., Druy, M. A., Sivak, A., Heeger, A. J., MacDiarmid, A. G., *Phys. Rev. Lett.* **1980**, *45*, 1123–1126.
[25] Chung, T.-C., Moraes, F., Flood, J. D., Heeger, A. J., *Phys. Rev. B* **1984**, *29*, 2341–2343.
[26] Park, Y. W., Denenstein, A., Chiang, C. K., Heeger, A. J., MacDiarmid, A. G., *Solid State Commun.* **1979**, *29*, 747–751.
[27] Tanaka, M., Watanabe, A., Tanaka, J., *Bull. Chem. Soc. Jpn.* **1980**, *53*, 3430–3435.
[28] Kume, K., Mizuno, K., Mizoguchi, K., Nomura, K., Maniwa, Y., *Mol. Cryst. Liq. Cryst.* **1982**, *83*, 285–290.
[29] Moraes, F., Davidov, D., Kobayashi, M., Chung, T. C., Chen, J., Heeger, A. J., Wudl, F., *Synth. Met.* **1985**, *10*, 169–179.
[30] Kaneto, K., Hayashi, S., Ura, S., Yoshino, K., *J. Phys. Soc. Jpn.* **1985**, *54*, 1146–1153.
[31] Mizoguchi, K., Misoo, K., Kume, K., Kaneto, K., Shiraishi, T., Yoshino, K., *Synth. Met.* **1987**, *18*, 195–198.
[32] Lögdlund, M., Lazzaroni, R., Stafström, S., Salaneck, W. R., Brédas, J. L., *Phys. Rev. Lett.* **1989**, *17*, 1841–1844.
[33] Masubuchi, S., Kazama, S., *Synth. Met.*, **1995**, *74*, 151–158.
[34] Ginder, J. M., Richter, A. F., MacDiarmid, A. G., Epstein, A. J., *Solid State Commun.* **1987**, *63*, 97–101.
[35] Mizoguchi, K., Obana, T., Ueno, S., and Kume, K., *Synth. Met.* **1993**, *55*, 601–606.
[36] Salem, L., *The Molecular Orbital Theory of Conjugated Systems*. New York: Benjamin, 1966.
[37] Kohler, B. E., in: *Conjugated Polymers*: Brédas, J. L. and Silbey, R. (Eds.). Dordrecht: Kluwer Academic Publishers, 1991; pp. 405–434.
[38] Fincher, Jr., C. R., Chen, C.-E., Heeger, A. J., MacDiarmid, A. G., Hasting, J. B., *Phys. Rev. Lett.* **1982**, *48*, 100–104.
[39] Yannoni, C. S., Clarke, T. C., *Phys. Rev. Lett.* **1983**, *51*, 1191–1193.
[40] Villar, H. O., Dupuis, M., Watts, J. D., Hurst, G. J. B., Clementi, E., *J. Chem. Phys.* **1988**, *88*, 1003–1009.
[41] Hirata, S., Torii, H., Tasumi, M., *J. Chem. Phys.* **1995**, *103*, 8964–8979.
[42] Castiglioni, C., Gussoni, M., Zerbi, G., *Synth. Met.* **1989**, *29*, E1–E6.
[43] Cuff, L., Kertesz, M., *Macromolecules* **1994**, *27*, 762–770.
[44] Ohtsuka, H., Master's Thesis, The University of Tokyo, 1993.
[45] Longuet-Higgins, H. C., Salem, L., *Proc. Roy. Soc. A* **1959**, *251*, 172–185.
[46] Pople, J. A., Walmsley, S. H., *Mol. Phys.* **1962**, *5*, 15–20.
[47] Takayama, H., Lin-Liu, Y. R., Maki, K., *Phys. Rev. B* **1980**, *21*, 2388–2393.
[48] Fesser, K., Bishop, A. R., Campbell, D. K., *Phys. Rev. B* **1983**, *27*, 4804–4825.
[49] Onodera, Y., *Phys. Rev. B* **1984**, *30*, 775–785.
[50] Zerbi, G., Gussoni, M., Castiglioni, C., in: *Conjugated Polymers*: Brédas, J. L. and Silbey, R. (Eds.). Dordrecht: Kluwer Academic Publishers, 1991; pp. 435–507.
[51] Horovitz, B., *Solid State Commun.* **1982**, *41*, 729–734.
[52] Vardeny, Z., Ehrenfreund, E., Brafman, O., Horovitz, B., *Phys. Rev. Lett.* **1983**, *51*, 2326–2329.
[53] Ehrenfreund, E., Vardeny, Z., Brafman, O., Horovitz, B., *Phys. Rev. B* **1987**, *36*, 1535–1553.

[54] Castiglioni, C., Navarrete, J. T. L., Zerbi, G., Gussoni, M., *Solid State Commun.* **1988**, *65*, 625–630.
[55] Wilson, E. B., Decius, J. C., Cross, P. C., *Molecular Vibrations*. New York: MacGraw Hill, 1955.
[56] Furukawa, Y., *Springer Ser. Solid-State Sci.* **1992**, *107*, 137–143.
[57] Furukawa, Y., Sakamoto, A., Ohta, H., Tasumi, M., *Synth. Met.* **1992**, *49*, 335–340.
[58] Hirschfeld, T., Chase, B., *Appl. Spectrosc.* **1986**, *40*, 133–137.
[59] Fujiwara, M., Hamaguchi, H., Tasumi, M., *Appl. Spectrosc.* **1986**, *40*, 137–139.
[60] Furukawa, Y., Ohta, H., Sakamoto, A., Tasumi, M., *Spectrochim. Acta* **1991**, *47A*, 1367–1373.
[61] Trotter, J., *Acta Cryst.* **1961**, *14*, 1135–1140.
[62] Hargreaves, A., Rizvi, S. H., *Acta Cryst.* **1962**, *15*, 365–373.
[63] Baudour, J. L., Cailleau, H., Yelon, W. B., *Acta Cryst.* **1977**, *B33*, 1773–1780.
[64] Delugeard, Y., Desuche, J., Baudour, J. L., *Acta Cryst.* **1976**, *B32*, 702–705.
[65] Tabata, M., Satoh, M., Kaneto, K., Yoshino, K., *J. Phys. Soc. Jpn.* **1986**, *55*, 1305–1310.
[66] Satoh, M., Tabata, M., Uesugi, F., Kaneto, K., Yoshino, K., *Synth. Met.* **1987**, *17*, 595–600.
[67] Gillam, A. E., Hey, D. H., *J. Chem. Soc.* **1939**, 1170–1177.
[68] Dale, J., *Acta Chem. Scand.* **1957**, *11*, 650–659.
[69] Balk, P., Hoijtink, G. J., Schreurs, J. W. H., *Rec. Trav. Chim.* **1957**, *76*, 813–823.
[70] Furukawa. Y., Ohtsuka, H., Tasumi, M., *Synth. Met.* **1993**, *55*, 516–523.
[71] Fichou, D., Horowitz, G., Xu, B., Garnier, F., *Synth. Met.* **1990**, *39*, 243–259.
[72] Furukawa, Y., Yokonuma, N., Tasumi, M., Kuroda, M., Nakayama, J., *Mol. Cryst. Liq. Cryst.* **1994**, *256*, 113–120.
[73] Yokonuma, N., Furukawa, Y., Tasumi, M., Kuroda, M., Nakayama, J., *Chem. Phys. Lett. Chem. Phys. Lett.* **1996**, *255*, 431–436.
[74] Deussen, M., Bässler, H., *Chem. Phys.* **1992**, *164*, 247–257.
[75] Sakamoto, A., Furukawa, Y., Tasumi, M., *J. Phys. Chem.* **1994**, *98*, 4635–4640; *J. Phys. Chem.* **1997**, *101*, 1726–1732; Furukawa, Y., Sakamoto, A., Tasumi, M. *Macromol. Symp.* **1996**, *101*, 95–102.
[76] Spangler, C. W., Hall, T. J., *Synth. Met.* **1991**, *44*, 85–93.
[77] Dewar, M. J. S., Trinajstić, N., *Czech. Chem. Commun.* **1970**, *35*, 3136–3189.
[78] Zahradník, R., Čársky, P., *J. Phys. Chem.* **1970**, *74*, 1240–1248.
[79] Furukawa, Y., *Synth. Met.*, **1995**, *69*, 629–632; *J. Phys. Chem.* **1997**, *100*, 15644–15653.
[80] Shacklette, L. W., Chance, R. R., Ivory, D. M., Miller, G. G., Baughman, R. H., *Synth. Met.* **1979**, *1*, 307–320.
[81] Krichene, S., Lefrant, S., Froyer, G., Maurice, F., Pelous, Y., *J. Phys. Colloq.* **1983**, *44*, 733–736.
[82] Krichene, S., Buisson, J. P., Lefrant, S., *Synth. Met.* **1987**, *17*, 589–594.
[83] Furukawa, Y., Ohtsuka, H., Tasumi, M., Wataru, I., Kanbara, T, Yamamoto, T., *J. Raman Spectrosc.* **1993**, *24*, 551–554.
[84] Ohtsuka, H., Furukawa, Y., Tasumi, M., Wataru, I., Yamamoto, T., unpublished work.
[85] Rakovic, D., Bozovic, I., Stepanyan, S. A., Gribov, L. A., *Solid State Commun.* **1982**, *43*, 127–129.
[86] Zannoni, G., Zerbi, G., *J. Chem. Phys.* **1985**, *82*, 31–38.
[87] Buisson, J. P., Krichène, S., Lefrant, S., *Synth. Met.* **1987**, *21*, 229–234.
[88] Ohtsuka, H., Furukawa, Y., Tasumi, M., *Spectrochim. Acta* **1993**, *49A*, 731–737.
[89] Tang, J., Albrecht, A. C., in: *Raman Spectroscopy*: Szymanski, H. (Ed.). New York: Plenum, 1970; Vol. 2, pp. 33–68.
[90] Inagaki, F., Tasumi, M., Miyazawa, T., *J. Mol. Spectrosc.* **1974**, *50*, 286–303.
[91] Warshel, A., Dauber, P., *J. Chem. Phys.* **1977**, *66*, 5477–5488.
[92] Yamaguchi, S., Yoshimizu, N., Maeda, S., *J. Phys. Chem.* **1978**, *82*, 1078–1080.
[93] Peticolas, W. L., Blazej, D. C., *Chem. Phys. Lett.* **1979**, *63*, 604–608.
[94] Kakitani, H., *Chem. Phys. Lett.* **1979**, *64*, 344–347.
[95] Cuff, L., Kertestz, M., *J. Phys. Chem.* **1994**, *98*, 12223–12231.
[96] Takahashi, C., Maeda, S., *Chem. Phys. Lett.* **1974**, *24*, 584–588.
[97] Aleksandrov, I. V., Bobovich, Ya. S., Maslov, V. G., Sidorov, A. N., *Opt. Spectrosc.* **1975**, *38*, 387–389.

[98] Matsunuma, S., Yamaguchi, S., Hirose, C., Maeda, S., *J. Phys. Chem.* **1988**, *92*, 1777–1780.
[99] Furukawa, Y., Ueda, F., Hyodo, Y., Harada, I., Nakajima, T., Kawagoe, T., *Macromolecules* **1988**, *21*, 1297–1305.
[100] Sakamoto, A., Furukawa, Y., Tasumi, M., *J. Phys. Chem.* **1992**, *96*, 3870–3874.
[101] Sakamoto, A., Furukawa, Y., Tasumi, M., *Synth. Met.* **1993**, *55*, 593–598.
[102] Sakamoto, A., Furukawa, Y., Tasumi, M., Noguchi, T., Ohnishi, T., *Synth. Met.* **1995**, *69*, 439–440.
[103] Harada, I., Furukawa, Y., Tasumi, M., Shirakawa, H., Ikeda, S., *J. Chem. Phys.* **1980**, *73*, 4746–4757.
[104] Kuzmany, H., *Phys. Stat. Sol. B* **1980**, *97*, 521–531.
[105] Furukawa, Y., Harada, I., Tasumi, M., Shirakawa, H., Ikeda, S., *Chem. Lett.* **1981**, 1489–1492.
[106] Schügerl, B., Kuzmany, H., *Phys. Stat. Sol. B* **1982**, *111*, 607–617.
[107] Faulques, E., Lefrant, S., *J. Phys. Colloq.* **1983**, *44*, 337–340.
[108] Eckhardt, H., Shacklette, L. W., Szobota, J. S., Baughman, R. H., *Mol. Cryst. Liq. Cryst.* **1985**, *117*, 401–409.
[109] Tanaka, J., Saito, Y., Shimizu, M., Tanaka, C., Tanaka, M., *Bull. Chem. Soc. Jpn.* **1987**, *60*, 1595–1605.
[110] Mulazzi, E., Lefrant, S., *Synth. Met.* **1989**, *28*, D323–D329.
[111] Furukawa, Y., Uchida, Y., Tasumi, M., Spangler, C. W., *Mol. Cryst. Liq. Cryst.* **1994**, *256*, 721–726; Kim, J.-Y., Ando, S., Sakamoto, A., Furukawa, Y., Tasumi, M., *Synth. Met.* **1997**, *89*, 149–152.
[112] Brédas, J. L., Scott, J. C., Yakushi, K., Street, G. B., *Phys. Rev. B* **1984**, *30*, 1023–1025.
[113] Yakushi, K., Lauchlan, L. J., Clarke, T. C., Street, G. B., *J. Chem. Phys.* **1983**, *79*, 4774–4778.
[114] Nechtschein, M., Devreux, F., Genoud, F., Vieil, E., Pernaut, J. M., Genies, E., *Synth. Met.* **1986**, *15*, 59–78.
[115] Chung, T.-C., Kaufman, J. H., Heeger, A. J., and Wudl, F., *Phys. Rev. B* **1984**, *30*, 702–710.
[116] Kaneto, K., Kohno, Y., and Yoshino, K., *Mol. Cryst. Liq. Cryst.* **1985**, *118*, 217–220.
[117] Bradley, D. D. C., Evans, G. P., and Friend, R. H., *Synth. Met.* **1987**, *17*, 651–656.
[118] Voss, K. F., Foster, C. M., Smilowitz, L., Mihailović, D., Askari, S., Srdanov, G., Ni, Z., Shi, S., Heeger, A. J., and Wudl, F., *Phys. Rev. B* **1991**, *43*, 5109–5118.
[119] Hill, M. G., Mann, K. R., Miller, L. L., Penneau, J.-F., *J. Am. Chem. Soc.* **1992**, *114*, 2728–2730.
[120] Stafström, S., in: *Conjugated Polymers*: Brédas, J. L. and Silbey, R. (Eds.). Dordrecht: Kluwer Academic Publishers, 1991; pp. 113–140.
[121] Kivelson, S., Heeger, A. J., *Phys. Rev. Lett.* **1985**, *55*, 308–311.
[122] Mizes, H. A., Conwell, E. M., *Phys. Rev. Lett.* **1993**, *70*, 1505–1508.
[123] Shimoi, Y., Abe, S., *Phys. Rev. B* **1994**, *50*, 14781–14784.
[124] Brédas, J. L., Thémans, B., Fripiat, J. G., André, J. M., Chance, R. R., *Phys. Rev. B* **1984**, *29*, 6761–6773.
[125] Stafström, S., Brédas, J. L., Epstein, A. J., Woo, H. S., Tanner, D. B., Huang, W. S., MacDiarmid, A. G., *Phys. Rev. Lett.* **1987**, *59*, 1464–1467.
[126] Stafström, S., Brédas, J. L., *Phys. Rev. B* **1988**, *38*, 4180–4191.
[127] Brédas, J. L., Beljonne, D., Shuai, Z., Toussaint, J. M., *Synth. Met.* **1991**, *43*, 3743–3746.
[128] Tanaka, C., Tanaka, J., *Synth. Met.* **1993**, *57*, 4377–4384.
[129] Sariciftci, N. S. (Ed.) *Primary Photoexcitations in Conjugated Polymers: Molecular Exciton versus Semiconductor Band Model*. Singapore: World Scientific, 1997, Chapter 17.

5 Vibrational Spectroscopy of Polypeptides

S. Krimm

5.1 Introduction

The polypeptide chain, $(NHCH(R)CO)_n$, is the basic backbone chemical unit of proteins, and knowledge of its possible structures and dynamics is crucial to an understanding of the functions of these important biological macromolecules. In synthetic polypeptides, all the side chains, R, are usually the same; in proteins, they can be any of 20 amino acid residues whose specific sequence is determined by the base sequence in the DNA of the corresponding genome [1]. The unusual property of these chains is their ability to adopt a wide range of three-dimensional structures [2], which accounts for the richness of biological functions that they can perform [1]. This presents a special challenge to mastering the spectroscopy of these polymeric molecules.

While vibrational spectroscopy is not capable of the structural resolution of X-ray diffraction, it nevertheless has some important advantageous features. First, it is not generally limited by physical state: samples can be in the form of powders, crystals, films, solutions, membranous aggregates, etc. Second, a number of different experimental methods probe the structure-dependent vibrational modes of the system: infrared (IR), Raman (both visible and UV-excited resonance), vibrational circular dichroism, and Raman optical activity, many of these with time-resolution capabilities. Finally, in addition to providing structural information, vibrational spectra are sensitive to intra- and intermolecular interaction forces, and thus they also give information about these properties of the system.

To gain such insights, however, requires that we achieve the deepest possible understanding of the vibrational spectrum. Correlations using characteristic band frequencies, particularly with resolution-enhanced spectra [3], can be useful, but it is increasingly being recognized that this is a fundamentally limited approach [4]. For the kind of understanding we seek it is necessary to determine the detailed normal modes of the molecule. Only by achieving an accurate description of these modes can we expect to explain fully all features of the vibrational spectrum and thereby to validate the structure and force field inputs to the normal-mode calculation [5]. This is now beginning to be possible for the polypeptide chain.

In this chapter we review the present state of the calculation of accurate normal modes of polypeptides. Since the burden of such calculations falls primarily on the quality of the force fields, we first review the state of present methods for obtaining such force fields, which include empirical, *ab initio*, and molecular mechanics approaches. We then discuss applications to characterizing the amide modes of the peptide group. Finally, we describe some of the results that have been obtained on characteristic polypeptide conformations, namely extended-chain and helical structures.

5.2 Force Fields

The normal modes of a molecule are determined by its three-dimensional distribution of atomic masses and its intra- and intermolecular force fields. The spectroscopic validation of the structure and these interactions derives from a satisfactory prediction of the frequencies and intensities of the observed IR and Raman bands, with the proviso that the bands are appropriately assigned to the calculated eigenvectors. Because of the complexity of the spectra this part of the process cannot be treated lightly, since more than one band match-up may be possible. A number of methods are available to verify band assignments [5] (isotopic substitution, dichroism, etc.), but since these are usually limited in practice, the success of the normal mode approach will ultimately depend on the reliability of the force fields. We therefore devote part of this section to a discussion of the development of such accurate force fields. This development has proceeded primarily through empirical refinement of force constants but is relying increasingly on inputs from *ab initio* calculations and will probably ultimately be based on molecular mechanics (MM) energy functions.

The development of an empirical force field requires selecting a physical model for the potential, choosing which terms are important, and optimizing the force constant parameters by a least-squares fitting of calculated normal modes to observed bands. (Since there have been summary [5] as well as detailed [6–8] treatments of the methods for obtaining the frequencies (eigenvalues) and forms (eigenvectors) of the normal modes for a given force field, we will not repeat this formalism here but will refer only to those aspects that are relevant to our discussion.) It is useful to recall that while the computational challenge becomes quite demanding for polypeptide structures lacking symmetry, such as globular proteins, for those structures that have helical or translational symmetry, such as the helical and extended-chain forms of synthetic polypeptides, the calculational problem is much simplified. This is because the only modes that can exhibit IR or Raman activity are those in which equivalent vibrations in each helix unit differ in phase by 0, $\pm\xi$, or $\pm 2\xi$, where ξ is the rotation angle about the helix axis between adjacent units, or those in which such vibrations in each translational repeat unit are in phase with one another [9] (otherwise dipole or polarizability derivatives cancel over the chain). The result is that the normal modes of only one unit need to be determined.

Ab initio calculations produce a potential energy surface, from whose second derivatives at the minima one can obtain all of the quadratic (diagonal and off-diagonal) force constants. This comprises a complete force field, but problems exist:

such calculations can only be done for relatively small molecules, and the force constants generally need to be scaled so that calculated frequencies agree with experimental data. Nevertheless, valuable information can be obtained by this approach.

The MM method will be seen to be the most useful one for obtaining force constants that can be used for different conformations of the molecule. In this approach the force constants are obtained from the second derivatives of an assumed potential energy function consisting of quadratic bonded terms and non-quadratic non-bonded terms. Although present MM functions are too crude to be spectroscopically reliable, a new method for deriving such functions holds the promise of providing a reliable vibrational force field for the polypeptide chain.

5.2.1 Empirical Force Fields

The most general force field of a molecule would include anharmonic as well as harmonic terms. However, with the limited experimental information generally available for refining an empirical force field for complex molecules, the harmonic approximation is the only feasible one at present. This means that, for the isolated molecule, we need to know the force constants, $\mathrm{F_{ij}}$, in the quadratic term of the Taylor series expansion of the potential energy, V:

$$V = \frac{1}{2} \sum_{i,j=1}^{n} \mathrm{F_{ij}} r_i r_j \tag{5-1}$$

where r is, for example, an internal displacement coordinate and $n = 3N - 6$, N being the number of atoms in the molecule. If interactions with other molecules are involved, analogous intermolecular interaction energy terms must be included.

Since the maximum number of observable frequencies is n, whereas the general number of $\mathrm{F_{ij}}$s is $n(n+1)/2$, it is necessary to find a physically meaningful model for V that brings the number of empirically determined $\mathrm{F_{ij}}$s into reasonable relation to the number of observable frequencies, even including those of isotopic derivatives. Two main models have been used to accomplish this: the Urey–Bradley force field (UBFF) and the general valence force field (GVFF). Both models incorporate the same diagonal (i.e., $\mathrm{F_{ii}}$) valence-type terms, involving bond stretch, angle bend, and torsion coordinates, but differ in how they treat the off-diagonal (i.e., $\mathrm{F_{ij}}$) terms.

In the unmodified UBFF, elaborated by Shimanouchi [10], off-diagonal terms in V are represented by atom pair 1,3 non-bonded interactions. Because this introduces inherent redundancies in the coordinates, the equilibrium configuration of the molecule represents a state with internal tensions, and therefore linear terms must be included in the force field. By introducing the redundancy condition, the linear terms can be eliminated from V, although the coefficients of the quadratic terms then contain 'intramolecular tension' as well as the valence and non-bonded force constants. The important characteristic of such a UBFF is that only a limited number of valence-type off-diagonal terms are in effect included in the potential. Such simple UBFFs can therefore be physically inaccurate, and as a result they have had to be modified by the explicit inclusion of some additional F_{ij}s [10].

In the GVFF there is no inherent limit to the number of F_{ij}s that are included. There is only the restraint that the number of Fs should not exceed the number of observable frequencies; in fact, it should generally be much smaller so as to over-determine the force field. Since an independent set of coordinates can be chosen, e.g., local symmetry coordinates, and the redundancy conditions explicitly given, there is no need to include linear terms in V, and Eq. (5-1) is the most general representation of the force field. Our discussion will focus on applications of the GVFF.

The refinement of an empirical force field for a macromolecule usually starts with the vibrational analysis of a smaller model system of known structure followed by a transfer of these force constants to the larger system, with possible subsequent force constant adjustments. In the case of the polypeptides, the preferred model system for the peptide group has been *trans*-N-methylacetamide (NMA), $CH_3CONHCH_3$. Synthetic polypeptides of known structure such as β-sheet polyglycine (PG), R = H, and α-helical and β-sheet poly(L-alanine) (PLA), R = CH_3, have served as basic examples of the polypeptide systems.

The force field refinement of the model system introduces important requirements and constraints [5, 7, 8].

1. The types of F_{ij} to be included in V must be selected. In practice, this has been based on prior experiences with the GVFF, but nevertheless the choices are, in general, arbitrary.
2. A starting set of force constant values, from which are calculated a set of normal coordinates, Q_α, and normal frequencies, ν_α, needs to be chosen. These are refined to observed data by a least-squares optimization procedure. We note that a normal mode Q_α can be visualized in terms of the local, usually symmetry (S_i), coordinates since

$$S_i = \sum_\alpha L_{i\alpha} Q_\alpha \qquad (5\text{-}2)$$

and the normal mode calculation provides the elements $L_{i\alpha}$ of the eigenvector matrix. Thus, for a given Q_α the relative S_i are given by the relative $L_{i\alpha}$. Visualization in cartesian coordinates is also possible. An analogous description is given by another characteristic of the normal mode, namely the potential energy distribution (PED). This consists of the relative contributions of each S_i to the total change in potential energy during Q_α, the fractional contribution of a particular S_i being given by

$$F_{ii} L_{i\alpha}^2 / \lambda_\alpha \qquad (5\text{-}3)$$

where $\lambda_\alpha = 4\pi^2 c^2 \nu_\alpha^2$, ν_α in cm^{-1}. (Contributions from F_{ij} elements are typically small and are usually neglected. However, they can be large and negative, which accounts for occasional unusually large contributions in a PED. In such a case, the diagonal elements alone do not give a sufficiently complete description of the normal mode.)

3. Correct band assignments must be made in matching calculated to observed ν_α. This is probably the most important part of the procedure, since acceptable frequency agreement (usually 5–10 cm^{-1}) does not necessarily insure that correct eigenvectors have been obtained. There are a number of properties that help to judge the validity of band assignments: molecular symmetry, which specifies the activity of a mode, and in appropriate situations predicts the IR dichroism (parallel, $\|$, or perpendicular, $\perp$, to the chain axis) and Raman polarization characteristics of the bands; isotopic substitution, which identifies the general nature of a mode, and by the reproduction of specific observed frequency shifts substantiates the correctness of the eigenvectors; overtone and combination bands, which should be consistent with the assignments of fundamentals; and predicted band intensities, which are very sensitive to the eigenvectors. The intensity agreement is particularly important since it provides a test of the eigenvectors that is independent of the frequencies. For example, the IR absorbance, A_α, associated with normal mode Q_α is given by [11]

$$A_\alpha = 42.2547\left(\frac{\partial\vec{\mu}}{\partial Q_\alpha}\right)^2 \tag{5-4}$$

where $\partial\vec{\mu}/\partial Q_\alpha$ is the normal coordinate derivative of the dipole moment, viz., the transition dipole moment (DÅ^{-1} amu$^{-1/2}$), and A_α is in km mol^{-1}. Since from Eq. (5-2)

$$\frac{\partial\vec{\mu}}{\partial Q_\alpha} = \sum_{\mathrm{i}} \mathrm{L}_{\mathrm{i}\alpha}\left(\frac{\partial\vec{\mu}}{\partial S_i}\right) \tag{5-5}$$

we see that, through the eigenvector matrix, the IR intensities sample another physical property of the molecule, namely the dipole derivatives, which arise from the charge re-distributions during the vibrations. Such dipole derivatives can also provide a measure of the coordinate contributions to the IR intensity of a band through a dipole derivative distribution (DDD) [12]. In addition, relative intensities of different symmetry species of a mode can provide information on structure. Thus, in a sample with a molecular chain axis aligned (say mechanically) in a known direction, since the orientation of a specified $\partial\vec{\mu}/\partial Q_\alpha$ with respect to the chain axis can be related mathematically to an experimentally observed dichroic ratio, DR$\equiv A_\alpha(\|)/A_\alpha(\perp)$, a DR measurement can help to establish the $\partial\vec{\mu}/\partial Q_\alpha$ orientation and therefore provide a test of a structural hypothesis.

The refinement of the macromolecular force field from that of the model system follows concepts similar to those described above, with three important additions. The first, and simplest, is that since the structures are different, new kinds of force constants must be introduced. This is not a major problem since initial values, for example for polypeptide side chains, are generally available. Second, the macromolecular system generally experiences inter- as well as intramolecular interactions, and these have to be incorporated into the force field. For polypeptides, these include hydrogen bonds and non-bonded interactions. Third, since for polypeptides in particular we are especially concerned with describing different molecular conformations, attention has to be given to incorporating the conformation dependence of the force constants. This is not easy to do in the framework of an empirical force field.

With respect to the intermolecular interactions, the hydrogen bonds can be treated like the other internal coordinates, i.e., in terms of bond stretching, angle bending, and torsion force constants. However, since hydrogen bonds vary in their geometry, we should ideally know how these force constants depend on the structure of the bond. While preliminary studies have been done for the NH hydrogen bond of the peptide group [13], there is at present no comprehensive description of this relationship.

The important non-bonded interactions are of two kinds, dispersion and transition dipole moment interactions. The former correspond to the so-called van der Waals forces and are generally given by a Lennard–Jones type of potential, viz.

$$V_{nb} = \frac{\mathrm{A_i A}_j}{R_{ij}^{12}} - \frac{\mathrm{B_i B_j}}{R_{ij}^{6}} \tag{5-6}$$

where R_{ij} is the distance between atoms i and j, the exponent of the repulsive term is traditionally 12 (although 9 has been found to be more effective in some cases), the exponent of the attractive term is for basic reasons 6, and $\mathrm{A_i}$ and $\mathrm{B_i}$ are respectively the repulsive and attractive van der Waals parameters of atom i. Such potentials have generally not been added to a GVFF for polypeptides since, in distinction to polyethylene [14], they do not affect midrange and high frequencies significantly. However, they probably are important in describing low-frequency modes.

Transition dipole moment interactions between peptide groups can influence higher frequency modes, as was shown by their effects on the splittings of amide I and amide II modes of α-helix and β-sheet polypeptides [15, 16]. Such transition dipole coupling (TDC) arises from the potential energy of interaction, V_{dd}, between transition moments, $\Delta\vec{\mu}$, in different peptide groups, and is given by [17]

$$V_{dd} = |\Delta\vec{\mu}^{\mathrm{A}}|\,|\Delta\vec{\mu}^{\mathrm{B}}| X_{AB} \tag{5-7}$$

where X_{AB} is a geometrical factor given by

$$X_{AB} = (\hat{\mathrm{e}}_A \cdot \hat{\mathrm{e}}_B - 3\hat{\mathrm{e}}_A \cdot \mathrm{r}_{AB}\hat{\mathrm{e}}_B \cdot \mathrm{r}_{AB})/r_{\mathrm{AB}}^3 \tag{5-8}$$

with ê being the direction of the transition moment and r_{AB} the distance between the centers of the moments. In terms of normal coordinate dipole derivatives,

$$\Delta\vec{\mu} = (4.1058/\nu_\alpha^{1/2})\frac{\partial\vec{\mu}}{\partial Q_\alpha} \tag{5-9}$$

where ν_α is the unperturbed frequency. In view of Eq. (5-5), a calculation of V_{dd} will require a summation over terms in $|\partial\vec{\mu}^A/\partial S_i|\,|\partial\vec{\mu}^B/\partial S_j|$. The force constant associated with V_{dd} is

$$F_{ij}^{dd} = 0.1|\partial\vec{\mu}^A/\partial S_i|\,|\partial\vec{\mu}^B/\partial S_j|X_{ij} \tag{5-10}$$

in mdyn Å^{-1}, and if we take the effect of TDC as a perturbation on ν_α, the frequency shift is given by

$$\Delta\nu_\alpha = (848619/\nu_\alpha)\sum_{ij} \mathrm{L}_{i\alpha}\mathrm{L}_{j\alpha}F_{ij}^{dd} \tag{5-11}$$

Early applications of this theory [18, 19] parametrized $\partial\vec{\mu}/\partial Q$ from observed band splittings. *Ab initio* calculations of these dipole derivatives [20] showed that the empirical TDC parameters were in agreement with *ab initio* as well as IR intensity results [17], thus emphasizing that the TDC mechanism is an important intermolecular interaction in polypeptides, particularly for modes with large $\partial\vec{\mu}/\partial Q$ such as amide I and amide II.

The matter of the conformation dependence of the force field of a polypeptide is one that has received little attention. We know that the differences in vibrational frequencies of different conformers derive mainly from the differences in their three-dimensional structures, with force constant differences playing a secondary role. Nevertheless, certain spectral features cannot be understood without taking into account variations in force constants with conformation (a good example is the *trans* versus *gauche* angle–angle interaction force constant [21]). Independent empirical refinements of α-PLA and β-PLA also demonstrate such force constant differences [5]. The problem is how to express such variations in a general explicit form in V. This is difficult to achieve simply from knowledge of independently refined force fields of different conformers. We return to this question below in our discussion of MM force fields.

Empirical force fields have been developed for NMA, for PG, and for α-PLA and β-PLA. In the case of PLA, it has also been possible to refine an 'approximate' force field in which the CH_3 side chain is replaced by a point mass (of 15 amu) [22]. This is useful in studying modes of the backbone chain that do not couple significantly with side-chain modes.

5.2.2 *Ab Initio* Force Fields

Ab initio molecular orbital calculations, by giving the potential energy surface of a molecule at its minimum energy conformations, provide a complete vibrational force field, i.e., all diagonal and off-diagonal force constants [23]. Since the force constants are given by the second derivatives of the potential in cartesian, and therefore non-redundant, coordinates, the force field is unique. Transformation is then always possible into an internal or local symmetry coordinate basis. Dipole and polarizability derivatives are also obtained from the *ab initio* calculation, and therefore IR and Raman intensities can be predicted.

Aside from the previously mentioned problem of doing such calculations on very large molecules, there are still some limitations at present in directly implementing *ab initio* force field calculations on model systems. At the Hartree–Fock (HF) level, force constants can be 10–30% too large because of basis set limitations and the neglect of electron correlation. This necessitates the empirical scaling of force constants so that calculated frequencies agree with observed. While systematic scaling procedures have been developed [24], there is still variability in the number and kind of scale factors that are used. In any case, a careful comparison of basis sets in terms of their geometry and frequency predictions is still very important. For example, the structure of NMA has non-planar symmetry at the HF/6-31G* level, but it exhibits planar symmetry at the HF/6-31 + G* level, i.e., with the addition of diffuse functions [25]; some well-assigned normal modes of *n*-alkanes cannot be reproduced in the proper order at the scaled HF/6-31G* level but are correctly calculated with a scaled HF/6-31G basis set [26]. With the inclusion of electron correlation, for example by the Møller–Plesset (MP) perturbation method, the force constants are much closer to experimentally compatible values, and such calculations, though very computer-intensive as the molecule gets larger, can provide usable force fields with minimal scaling.

Despite these problems, *ab initio* force fields provide an improved route to avoiding the arbitrariness and incompleteness of empirical force fields. Although such calculations on model systems do not translate into force fields for macromolecules, they can provide important components of such force fields. For example: an *ab initio* force field for glycine hydrogen bonded to two H_2O molecules [27] provided the necessary starting point for refining the carboxyl group part of the empirical force field for glutathione [28]; an *ab initio* force field for diethyl disulfide [29] permitted detailed correlations to be made between SS and CS stretching frequencies and the conformations of disulfide bridges in proteins [30, 31]; an *ab initio* force field for ethanethiol, together with an empirical force field for the peptide group, provided detailed structure–spectra correlations between CS and SH stretch frequencies and the conformation of the cysteine side chains in proteins [32].

Ab initio force fields also demonstrate that force constants vary with conformation, a factor that needs to be taken into account if we are to obtain an accurate description of the normal modes associated with the many possible conformations of the polypeptide chain. This is clearly illustrated by a study of the scaled *ab initio* force fields of four non-hydrogen-bonded conformers of the alanine dipeptide, $CH_3CONHC^{\alpha}H(C^{\beta}H_3)CONHCH_3$ [33, 34]. It was found that diagonal force constants such as the torsion (t) change by very large amounts, and even the stretch (s) and deformation (d) force constants are significantly dependent on the conformation. Off-diagonal force constants can also change significantly with conformation. In principle this is not surprising since we must expect changes in electronic structure with conformation. In terms of conveniently designated components of the energy of the molecule, which of course the *ab initio* calculation does not provide, we can conceive of these changes as arising from changes in the charge distribution with conformation, from changes in the dispersion energies as interatomic distances change, and from any intrinsic valence force constant variations with conformation.

While we can show from *ab initio* calculations that the force field is not independent of molecular conformation, it is much more difficult to specify the explicit nature of this dependence. This is precisely because the method provides only a total energy for a given structure. If we wish to infer the varying contributions of different physical components, we must assume a model for the potential energy and strive to make its behavior mimic the *ab initio* results as closely as possible. It is toward this goal that some present efforts are directed.

5.2.3 Molecular Mechanics Force Fields

An MM energy function, whose second derivatives at a minimum are the spectroscopic force constants, is generally represented as a sum of quadratic and non-quadratic terms. The former encompass the valence-type deformations, such as bond stretching and angle bending, while the latter involve torsions, dispersion interactions, electrostatic interactions, and possible hydrogen-bond terms.

In order to serve as a spectroscopic force field, such an MM function would at least have to reproduce vibrational frequencies to spectroscopic standards, viz., to root-mean-square (rms) errors of the order of 5–10 cm^{-1}. Some early MM potentials came close to this mark [35], but since frequency agreement was not their primary goal (but rather reproduction of structures and energies), subsequent efforts resulted in functions with unacceptable spectroscopic predictability (rms errors of > 50 cm^{-1}). One of the reasons for this was the lack of proper attention to cross-terms in the potential. Initial efforts to improve such potentials have led to only marginally better frequency agreement. For example, MM3 gives an rms error for frequencies below 1700 cm^{-1} of 38 cm^{-1} for butane [36a] (reduced to 22 cm^{-1} in MM4 [36b]) and 47 cm^{-1} for NMA [37]; a Hessian-biased force field [38] gives a similar rms error of 29 cm^{-1} for polyethylene [39]; for CFF93, the rms error for *n*-butane is 25 cm^{-1} [40] and is 34 cm^{-1} for a group of four acyclic and three cyclic hydrocarbons; and a second generation version of AMBER gives 43 cm^{-1} for NMA [41]. Some force fields have been specifically designed to give improved spectroscopic predictability and these [42, 43] indeed show better agreement, although improvement is still desirable: the rms error for the above NMA frequencies is 24 cm^{-1} for SPASIBA [44].

What is clearly needed is an approach that systematically incorporates spectroscopic agreement in the initial stages of the optimization of the MM parameters. This has been achieved by a procedure designed to produce a so-called spectroscopically determined force field (SDFF) [45, 46].

The elements in the SDFF methodology are the following. First, a form is selected for the MM potential, one that is hopefully inclusive enough to incorporate all the physically important contributions. The first implementation of this approach, viz., for linear saturated hydrocarbon chains [47], used a potential of the form

$$V = \frac{1}{2}\sum_i \mathrm{f_{ii}}(q_i - \mathrm{q_{io}})^2 + \sum_{i<j} \mathrm{f_{ij}}(q_i - \mathrm{q_{io}})(q_j - \mathrm{q_{jo}}) + \sum V_{tor} + \sum V_{q,tor} + \sum V_{tor,tor} + \sum V_{nb} \tag{5-12}$$

where the q_is are internal coordinates whose intrinsic (i.e., equilibrium) values are $\mathrm{q_{io}}$ and the $\mathrm{f_{ij}}$ are the intrinsic MM quadratic (i.e., valence-type) force constants. At this stage all possible cross-terms are included. The intrinsic torsion potential is represented by a Fourier cosine series,

$$V_{tor} = \frac{1}{2}\sum_{\mathrm{m}} \mathrm{V_m}(1 \pm \cos \mathrm{m}\chi) \tag{5-13}$$

where $\mathrm{V_m}$ is the barrier to a rotation of periodicity m associated with torsion angle χ. (In general, V_{tor} is a sum over terms in a Fourier series, but in the hydrocarbon case only the $\mathrm{V_3}$ term was found to be necessary.) The non-bonded potential consists of the dispersion term (an exponent of 9 was found to be more suitable for the repulsive term) plus a coulomb term that is usually represented by the interactions between partial charges $\mathrm{Q_i}$ and $\mathrm{Q_j}$ on atoms i and j, i.e., in the case of the hydrocarbons

$$V_{nb} = \frac{\mathrm{A_i A_j}}{R_{ij}{}^9} - \frac{\mathrm{B_i B_j}}{R_{ij}{}^6} + \frac{\mathrm{Q_i Q_j}}{4\pi\kappa\varepsilon_0 R_{ij}} \tag{5-14}$$

where κ is the dielectric constant and ε_0 the permittivity of free space. The summation of V_{nb} is over all atom pairs 1,4 and higher. The $\mathrm{Q_i}$ can be fixed charges or they can include charge fluxes, i.e., $\partial \mathrm{Q_i}/\partial S_\mathrm{j}$ [42]. (Inclusion of charge fluxes also permits the calculation of IR intensities.)

The second element in the procedure is the choice of a spectroscopic force field, i.e., a set of $\mathrm{F_{ij}}$. This could be an empirical force field, but since we wish to provide as complete a description as possible at this stage, a scaled *ab initio* force field is the one of choice. The scaling process also connects the force constants to experimental frequencies and band assignments, and avoids the problems of having incorrect eigenvectors because of using inappropriate basis sets [26].

The most important part is the third element, viz., the ability to make a transformation from the spectroscopic to the MM force field [46]. Since this transformation is analytic, it preserves in the SDFF the frequency as well as structure agreement of the scaled *ab initio* calculation. Although the transformation is initiated by assuming a starting set of V_{nb} parameters, these are subsequently optimized in the refinement procedure [48]. In addition to providing the $\mathrm{f_{ij}}$, the transformation procedure also gives the $\mathrm{q_{io}}$ [46].

The fourth, and also important, element is the application of this transformation to the *ab initio* force fields of a set of conformers of the model molecules for the macromolecular system. In the case of the hydrocarbons, this consisted of the four stable conformers of *n*-pentane and the ten stable conformers of *n*-hexane [26, 47]. At this stage, the non-bonded parameters are optimized under the standard MM assumption that the *intrinsic* force constants (f_{ij}) and *intrinsic* geometry parameters (q_{io}) are the same for all conformers [48]. The relative (*ab initio*) conformer barriers and energies are incorporated in optimizing the V_m and Q_i. At this point, specific forms of the conformation dependence of some f_{ij} can be determined and incorporated in V.

Finally, since it is neither desirable nor necessary to include in V the huge number of f_{ij}, most of which are very small and do not significantly affect the vibrational frequencies, the set of force constants is systematically reduced based on a pre-assigned limit on the frequency error [46]. In this process, the remaining f_{ij} are optimized, usually requiring minimal change, to compensate for the effects of the force constants that were dropped.

There are a number of advantages to the SDFF approach to refining an MM force field. By utilizing a complete (i.e., *ab initio*) spectroscopic force field scaled to experimentally assigned bands, frequency agreement is assured at the start. This not only incorporates highly accurate information about the minima in the potential surface, but it avoids the possibility of biasing the non-bonded parameters to compensate for an incomplete formulation of the f_{ij} terms in V. In addition, since the optimization is done in a nonredundant coordinate basis [49], the uniqueness of the force field is assured; subsequent transformation to a redundant basis of choice is always possible. Finally, since force fields for many different conformers are involved in the optimization, there is a better chance of revealing the nature of specific conformation dependences of the cross-terms.

The application of this procedure to an SDFF for the hydrocarbon chain has given excellent results [47]. With a set of 50 force constant and geometry parameters the observed frequencies of *tt*-pentane and *ttt*-hexane below 1500 cm^{-1} could be reproduced with an rms error of $\sim$11 cm^{-1}. This force field has now been extended to include branched saturated hydrocarbons [50]. An SDFF for the polypeptide chain is now under development [51].

5.3 Amide Modes

Since the peptide group, –CONH–, is the most distinctive component of the backbone of the polypeptide chain, it is not surprising that the normal modes of this group, the so-called amide modes, were first studied in the simplest representative molecule, viz., NMA. It is useful to examine the nature of these modes in NMA and in blocked dipeptides since they have served as important guideposts in the analysis of polypeptide spectra in terms of structure.

5.3.1 *N*-Methylacetamide

Normal mode calculations on NMA have been done with both empirical and scaled *ab initio* force fields. In the empirical force field calculations, the CH_3 group has been treated as a point mass, using a UBFF [52] as well as a GVFF [5], and as an all-atom structure, again with a UBFF [53] and a GVFF [54, 55]. In the *ab initio* calculations, a variety of basis sets and scaling methods have been used: 4-31G, with adjustment of mainly diagonal force constants to liquid-state frequencies of NMA and its isotopomers [56]; 4-31G and 4-31G*, with no scaling [57]; 4-21, with four general scale factors and some approximated force constants [58]; and 4-31G*, with 10 scale factors [59], optimized to matrix-isolated frequencies [60].

All of the above calculations have been on the isolated molecule, even though in many cases comparisons have been made with experimental results on hydrogen-bonded systems (such as the neat liquid). The first normal mode calculation on a hydrogen-bonded system was at 4-31G* on NMA-$(H_2O)_2$, in which one H_2O molecule was hydrogen bonded to the CO group and the other to the NH group, with scale factors optimized to IR and Raman data on the aqueous system [61]. In view of the absence of information on the detailed water structure associated with the NMA molecule in aqueous solution, this was considered to be a satisfactory minimal model for representing the normal modes of hydrogen-bonded NMA. (A 4-31G calculation on NMA-$(H_2O)_3$ has also been done [62].) Subsequently, a 3-21G calculation was done on NMA-$(formamide)_2$, NMA-$(FA)_2$, using six general scale factors [63]. In view of the determination that diffuse functions are necessary to give a planar symmetric peptide group in the isolated molecule [25], the NMA-$(H_2O)_2$ calculations have been repeated at 6-31 + G* [64], and these are discussed below.

Early studies [52, 65, 66] sought to describe the characteristic vibrational modes of the peptide group. In addition to the localized NH s mode, which is subject to Fermi resonance interactions [5], there are other relatively localized modes, labeled amide I–VII (although it was recognized [66] that some of these, because of their delocalization, would be more variable than others). These amide modes have been used as a basis for discussions of peptide spectra, and in Table 5-1 we describe them and compare the PEDs for isolated and aqueous hydrogen-bonded NMA. It should be noted that these results are on fully optimized structures (all frequencies real), in both cases the conformation being N_cC_t. (The symbol indicates that the in-plane *N*-methyl hydrogen is *cis* to the (N)H and the in-plane C-methyl hydrogen is *trans* to the (C)O.)

The amide I mode is primarily CO s, both in NMA and NMA-$(H_2O)_2$. However, the additional contributions to the eigenvector depend very much on the specifics of the calculations (only contributions to the PED of ≥ 10 are given in Table 5-1), as well as, of course, on the absence or presence of hydrogen bonding. For NMA, the next largest contributor in a CH_3 point-mass calculation is CN s [52, 5], whereas in an all-atom calculation it is CCN d [55, 59]. The situation for NMA-$(H_2O)_2$, where the lower frequency is an obvious consequence of hydrogen bonding, is complicated by the experimentally demonstrated coupling of CO s and HOH b [67, 68], which makes the latter the next largest contributor. In NMA-

Table 5-1. Amide modes of isolated and aqueous hydrogen-bonded *N*-Methylacetamide[a].

Mode	ν(obs)[b]	ν(calc)[c]	Potential energy distribution[d]
I	1706	1708	CO s(83) CCN d(11)
	1646	1632	CO s(30) (O)HOH b(30) (H)HOH b(18)
	1626	1626	CO s(38) (O)HOH b(28) (H)HOH b(15)
II	1511	1512	NH ib(51) CN s(28)
	1584	1575	CN s(48) NH ib(36)
III	1265	1264	NH ib(21) CO ib(21) CN s(18) CC s(10) CCH_3 sb(10)
	1313	1305	NH ib(26) CN s(20) CCH_3 sb(20) CO ib(12)
IV	658	648	CC s(36) CO ib(34)
	632	640	CO ib(35) CC s(32)
V	391	391	CN t(118) NH ob(51) CO ob(23)
	(750)[e]	744	CN t(43) H···O b2(28) NH ob(21)
VI	626	627	CO ob(67) CCH_3 or(26) CN t(11)
	(600)[f]	605	CO ob(50) O···H b2(30) CCH_3 or(19) NH ob(10)
VII	–	–	
	195[g]	205	CN t(38) NH ob(21)

[a] First line: NMA; second line(s): NMA-$(H_2O)_2$. 6-31+G* basis set.
[b] NMA: from [59]; NMA-$(H_2O)_2$: from [61] and [68], except where noted.
[c] NMA: from [73]; NMA-$(H_2O)_2$: from [64].
[d] s: stretch, d: deformation, b: bend, ib: in-plane bend, sb: symmetric bend, ob: out-of-plane bend, b2: essentially ob, or: out-of-plane rock, t: torsion. Group coordinates are defined in [59]. Contributions ≥ 10.
[e] Possible value, from [69].
[f] Observed in neat NMA [65].
[g] From [62] (at 206 cm^{-1} from [70]).

$(FA)_2$ this contribution is NH in-plane bend (ib) [63], but such a result is uncertain because of the absence of experimental data on this complex and the resulting lack of optimization of the FA force constants. Nevertheless, we see that even for as localized a mode as CO s the specific nature of the eigenvector can depend on the details of the calculation.

The amide II mode is a mixture of NH ib and CN s, with the former dominating in NMA and the latter, at a higher hydrogen-bonded frequency, dominating in NMA-$(H_2O)_2$, (Table 5-1). It is interesting that the CN s predominance does not appear at 3-21G for NMA-$(FA)_2$ [63], nor at 4-31G [62] or 4-31G* [61] for NMA-$(H_2O)_2$, showing the evident influence of basis set on the predicted eigenvectors. Nor is CN s predominant in polypeptides [5].

The amide III mode is also a mixture of NH ib and CN s, although other coordinates make significant contributions. Since CH_3 symmetric bend (sb) is an important one, CH_3 point-mass calculations [52, 5] distort the relative importance of CC s. As seen from Table 5-1, CO ib is a relatively large contributor in NMA and a small one in NMA-$(H_2O)_2$. (It is not clear why NH ib makes no contribution in NMA-$(FA)_2$ [63].) The position and nature of this mode is variable in polypeptides [5], since NH ib contributes significantly to a number of normal modes in the 1400–1200 cm^{-1} region, with the mixing being dependent on the side-chain structure.

The amide IV mode is a mixture of CO ib and CC s in NMA, although the latter contribution is often replaced by other coordinates in polypeptide structures [5]. This is an example of a peptide group mode that is not in general localized and well defined.

The amide V mode, on the other hand, seems to be a characteristic band of hydrogen-bonded peptide groups, although an unusual feature seems not to have been noted previously. This mode, which is generally a combination of CN t and NH out-of-plane bend (ob), has been assigned to a band near 725 cm^{-1} in neat NMA [65, 66] and possibly ~750 cm^{-1} in aqueous NMA [69]. It is found to have a counterpart, the amide VII mode, calculated at ~200 cm^{-1} [54, 55, 5] and observed near this frequency [62, 70]. However, *ab initio* calculations on an isolated molecule [57, 59] produce only a single CN t, NH ob mode at ~400 cm^{-1}, consistent with an observed Ar-matrix band at 391 cm^{-1} [71]. The increase in the NH ob force constant of an isolated molecule needed to give a hydrogen-bonded frequency in the 700 cm^{-1} region apparently divides the contributions of these coordinates between the two frequency regions. Isolated molecules also exclude the contributions from hydrogen-bond deformation modes (see Table 5-1).

The amide VI mode is predominantly CO ob, with a significant CCH_3 out-of-plane rock (or) contribution. This, of course, cannot be represented in a CH_3 point-mass model [5]. The nature of this mode is also significantly dependent on conformation in polypeptide structures [5].

5.3.2 Blocked Dipeptides

Although the analysis of NMA serves to introduce the representative types of characteristic modes of the peptide group, it does not provide insights into how these modes might be influenced by varying conformational interactions between adjacent groups, such as can occur in polypeptide chains. For this purpose it is useful to examine the modes of a blocked dipeptide, such as *N*-acetyl-L-alanine-*N*-methylamide, the Ala dipeptide. This was first done for just the amide I and amide III modes using a fixed empirical force field and no TDC [72], so the conclusions are necessarily limited. Better insights can be obtained from an *ab initio* study of four non-hydrogen-bonded conformers of this molecule [33].

In Table 5-2 we show the calculated amide I, II, III, and V frequencies and PEDs of the four non-hydrogen-bonded conformers of the Ala dipeptide, which differ mainly in the values of the $\varphi(NC^{\alpha})$, $\psi(C^{\alpha}C)$ torsion angles (see Figure 5-1). (The amide IV and VI modes are no longer distinguishable as relatively pure modes, CO ib and CO ob contributing significantly and mixing with other coordinates in the region of ~830–515 cm^{-1} [33].) It should be noted that in this calculation the 4-21 *ab initio* force constants were scaled with six general scale factors, so the frequencies are not necessarily representative of true experimental values. However, the changes in frequency should be indicative of the effects of changes in conformation.

Table 5-2. Some amide modes of four non-hydrogen-bonded conformers of the alanine dipeptide[a].

Mode	Conformer[b]								
	β_2		α_R		α_L			α'	
I	1719 CO s1(78)[c]	1694 CO s2(81)	1726 CO s1(84)	1679 CO s2(84)	1724 CO s1(79)	1703 CO s2(83)		1720 CO s2(83)	1710 CO s1(82)
II	1550 NH ib2(53) CN s2(27)	1532 NH ib1(61) CN s1(18)	1552 NH ib2(57) CN s2(27)	1526 NH ib1(61) CN s1(18)	1545 NH ib2(61) CN s2(28)	1536 NH ib1(68) CN s1(19)		1540 NH ib2(65) CN s2(25)	1529 NH ib1(68) CN s1(22)
III	1265 CN s1(28) CN s2(12) CO ib(12)	1252 H^α b2(27) CN s2(16) CN s1(14)	1292 CN s2(29) $C^\alpha C$ s(22) NH ib2(22) CO ib2(12)	1253 H^α b1(33) CN s1(24) NH ib1(11)	1265 CN s1(39) CO ib1(17) CC s(14) NH ib1(13)	1247 CN s2(26) H^α b2(22) H^α b1(13) NC s(11)		1295 $C^\alpha C$ s(28) CN s2(27) NH ib2(12) CO ib2(11)	1260 H^α b2(37) CN s1(19) NH ib1(12)
V	672 NH ob2(42) NH ob1(35) CN t2(20) CN t1(17) NC^α t(13)	558 NH ob2(48) NH ob1(38) CN t1(12) CN t2(21) NC^α t(14) CO ib1(13)	657 NH ob2(28) NH ob1(24) CN t2(11)	547 NH ob1(61) CN t1(47) NH ob2(36) NC^α t(28) CN t1(18)	585 CN t2(44) NH ob2(40)	573 NH ob2(40) CN t2(13) $C^\alpha C$ s(12) NH ob1(12) CO ob2(10)	546 NH ob1(113) CN t1(67) NC^α t(65)	605 NH ob2(47) NH ob1(45) CN t2(29) CN t1(21) NC^α t(16) CO ob1(11)	504 NH ob1(71) NH ob2(55) CN t1(53) NC^α t(43) CN t2(38)

[a] From [33].

[b] $\varphi, \psi - \beta_2$: $-134°$, $38°$; α_R: $-92°$, $-6°$; α_L: $61°$, $41°$; α': $-162°$, $-55°$.

[c] s: stretch, ib: in-plane bend, b: bend, ob: out-of-plane bend, t: torsion. Group coordinates are defined in [34]. Contributions ≥ 10.

Keeping in mind that the peptide groups are not identical, because of the different chemical surroundings, we find that amide I frequencies, because of their highly localized nature, generally follow the changes in CO s force constants with conformation [33]. These in turn are determined by the changes in CO bond length (the relationship is essentially linear [33]), which reflect the changing bonded and nonbonded interactions in the molecule with conformation. The larger variation in CO s2 (41 cm^{-1}) than in CO s1 (16 cm^{-1}) is probably a result of its more variable conformational environment adjacent to C^{α}, and therefore the varying (even if small) contributions of other coordinates. It is interesting that in α' the frequency order is reversed compared with the other conformers.

The amide II modes are less sensitive to conformation: the variation in either mode is of the order of 10 cm^{-1}, although the splittings do vary. As with amide I, the different frequencies of the two groups are mainly a reflection of their nonidentical chemical environments.

The latter fact is demonstrated dramatically by the amide III mode: the eigenvectors for the two peptide groups are significantly different, that of the higher frequency mode containing no PED contribution (≥ 10) of H^{α} bend (b) although this coordinate is a major contributor to the lower frequency mode. Nor is NH ib a major contributor, as it was in NMA (in fact, it does not appear in the β_2 modes). The large variations in frequency separation between the two amide III modes reflect the effect of conformation through H^{α} b, since if H^{α} is replaced by D^{α} this variation disappears (the bands are now at 1294 ± 1 and 1266 ± 6 cm^{-1}) [33]. This agrees with previous observations [5] that the amide III mode in polypeptides contains a significant conformation-dependent contribution from H^{α} b. The interaction between H^{α} b and NH ib is illustrated by the fact that in an $(ND)_2$ Ala dipeptide H^{α} b occurs at 1300 and 1296 (β_2), 1320 and 1272 (α_R), 1305 and 1283 (α_L), and 1324 and 1284 (α') cm^{-1} [33]. The H^{α} b coordinate also mixes with CN s to predict significant IR bands at 1277 cm^{-1} in β_2 and 1291 cm^{-1} in α' [33]. Thus, with its added sensitivity to side-chain composition [5], amide III in polypeptides is no longer a simple characteristic NH ib, CN s mode.

The amide V mode, which appears as a single band near 400 cm^{-1} in NMA (see Table 5-1), is split in the nonhydrogen-bonded conformers into at least two bands with predominant NH ob, CN t contributions. (These paired coordinates can make significant contributions to other modes, and each contributes individually to still others.) It might be thought that this splitting (and higher average frequency) is a result of the arbitrary scale factors that were used [33]. However, 4-31G* calculations with scale factors transferred from NMA [59] give similar results [73]. Thus, we must conclude that kinematic as well as potential coupling between the two peptide groups is responsible for the nonlocalized contribution of these coupled coordinates.

The presence of hydrogen bonding obviously introduces a new element into the modes of the Ala dipeptide. This can be examined from the results on the three stable hydrogen-bonded conformers, C5, C7E, and C7A (Figure 5-1) [34]. In Table 5-3 we show their calculated amide I, II, III, and V frequencies and PEDs.

The relative amide I mode frequencies again generally follow the relative values of the CO s force constants [34], although mixing of the two CO s displacements can

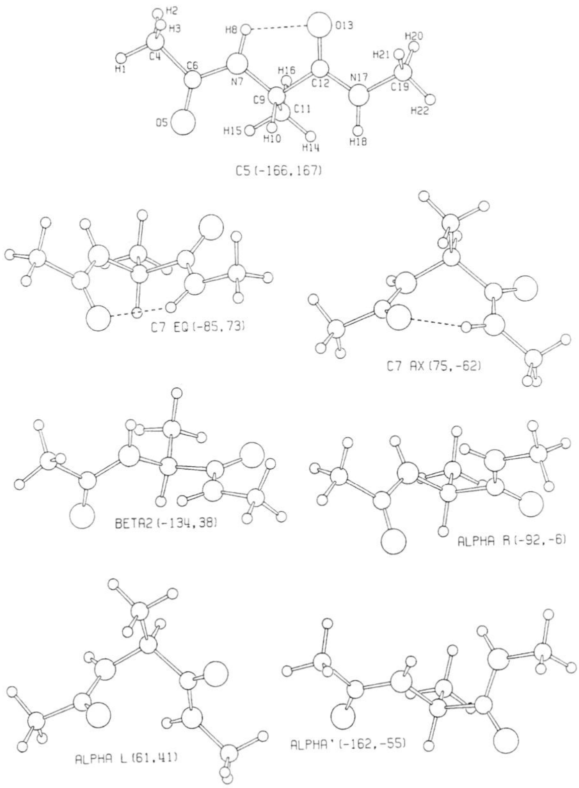

Figure 5-1. Optimized conformations of Ala dipeptide (φ, ψ in parentheses). Peptide groups–1: C6N7, 2: C12N17. Hydrogen bonds are indicated by broken lines. (From [33])

alter this: for C5, even though the CO s1 force constant (11.040) is larger than the CO s2 force constant (10.922), the predominantly CO s1 mode is at a lower frequency (1668 cm^{-1}) than CO s2 (1698 cm^{-1}). This is also counterintuitive in terms of the (admittedly weak) hydrogen bond to CO2, whereas the relative frequencies are in the expected order for C7E and C7A. In distinction to the nonhydrogen-bonded structures, NH ib is now a significant component of the amide I modes of the hydrogen-bonded structures (C7E and C7A). (To a lesser extent, this is also true for the Ala dipeptide in a crystal conformation similar to that of C5 but in which both peptide groups are fully hydrogen bonded [74].)

The amide II mode frequencies generally reflect the influence of hydrogen bonding, which manifests itself primarily through the magnitudes of such force constants as NH ib and CN s. (It should be noted that explicit hydrogen bond coordinates are not included in the coordinate basis [33, 34], so the force constants implicitly reflect the influence of this interaction.) As expected, the higher frequency modes are associated with the peptide group having a hydrogen-bonded NH. In addition, the modes in C5 involve coupled NH ib displacements [34], which makes it difficult in general to infer the explicit dependence of amide II on conformation.

The amide III modes demonstrate an increasing complexity compared with the

Table 5-3. Some amide modes of three hydrogen-bonded conformers of the alanine dipeptide[a].

Mode	Conformer[b]							
	C5			C7E		C7A		
I	1698 CO s2(47)[c] CO s1(34)	1668 CO s1(52) CO s2(32)		1704 CO s2(64) NH ib2(26)	1676 CO s1(70) NH ib1(15)	1699 CO s2(51) NH ib2(23) CO s1(14)	1682 CO s1(55) CO s2(19) NH ib1(15)	
II	1557 NH ib2(49) CN s2(28)	1525 NH ib1(49) CN s1(25) NH ib2(12)		1587 NH ib2(59) CN s2(27) CO s2(21)	1534 NH ib1(60) CN s1(28) CO s1(14)	1594 NH ib2(59) CN s2(28) CO s2(18)	1551 NH ib1(60) CN s1(27) CO s1(14)	
III	1276 CN s1(23) NH ib1(20) CN s2(14) CO ib1(11)	1229 H^{α} b2(40) CN s2(14)		1279 H^{α} b1(28) CN s1(23) NH ib1(16)	1258 CN s2(31) H^{α} b2(20) H^{α} b1(16) NM s(13)	1335 H^{α} b2(30) CN s2(13) H^{α} b1(12)	1325 H^{α} b1(26) CN s2(19) H^{α} b2(18) $C^{\alpha}C$ s(13)	1292 CN s1(37) CO ib(16) MC s(12)
V	693 NH ob1(32) CN t1(23) CO ib2(15) NC^{α} t(14)	683 NH ob1(39) CN t1(25) NC^{α} t(24) CO ob2(14)	585 NH ob2(111) CN t2(57)	754 NH ob2(51) CN t2(40)	577 NH ob1(102) CN t1(76) NC^{α} t(67)	809 NH ob2(51) CN t2(19)	581 NH ob1(116) NC^{α} t(95) CN t1(81)	

[a] From [34].

[b] φ, ψ –C5: −166°, 167°; C7E: −85°, 73°; C7A: 75°, −62°.

[c] s: stretch, ib: in-plane bend, b: bend, ob: out-of-plane bend, t: torsion, d: deformation, M: methyl C. Group coordinates defined in [34]. Contributions ≥10.

nonhydrogen-bonded conformers. While NH ib contributes (at the level of ≥ 10) to the higher frequency C5 and C7E modes, it is absent from the C7A modes. Furthermore, if we ascribe amide III character to H^α b, CN s modes with minimal NH ib, then C7A has such modes at frequencies well over 1300 cm^{-1}, again reflecting the conformation-dependent contribution of H^α b. (The situation is somewhat reversed in the crystalline Ala dipeptide [74], with a calculated mode at 1325 cm^{-1} showing traditional CN s, NH ib character and modes at 1272 and 1258 cm^{-1} having H^α b, CN s character.) Since NH ib can contribute throughout the $\sim$1400–1200 cm^{-1} region, and can mix variably with CN s and H^α bend, it is not surprising that the amide III modes are a complex function of conformation.

The amide V modes are dominated by the effects of hydrogen bonding. Thus, the free amide V mode is at $\sim$580 cm^{-1} in all three conformers, while the bonded modes reflect the increasing hydrogen bond strengths: $\sim$690 cm^{-1} in C5 (the splitting probably due to an accidental resonance with another mode), 754 cm^{-1} in C7E, and 809 cm^{-1} in C7A. Hydrogen bonding thus seems to mask conformational differences in amide V.

These studies on NMA and the Ala dipeptide illustrate that, although it is useful to characterize peptide spectra in terms of supposedly localized amide modes, it is the subtle differences in the detailed nature of these vibrations with varying conformation and hydrogen bonding that provide the spectral clues to differences in structure. This is demonstrated by analyses of tripeptides [28, 75–77], which also involve interactions between two peptide groups. Although done with an empirical force field [5], these analyses show the sensitivity of vibrational spectra to specific structural features. This emphasizes the importance of highly accurate force fields in enabling vibrational analysis to realize its potential for being a powerful tool to study the structures and forces in macromolecules.

5.4 Polypeptides

As will be obvious from our previous discussion, a comprehensive understanding of the vibrational spectra of polypeptides depends on having reliable normal mode analyses to combine with appropriate experimental studies. Such analyses in turn rest on the quality of the force field. At the present time, an SDFF for the polypeptide chain is not available, the most detailed normal mode analyses being based on extensive empirical force fields [5]. These force fields have been tested on tripeptides of known structure with excellent predictive results [28, 75–77], and have proven to be very satisfactory for polypeptide systems. For reasons given in Section 5.2.1, however, we must be careful about a final acceptance of all of their detailed predictions. Acknowledging these reservations, the general features of polypeptide spectra provided by the present detailed empirical force fields can nevertheless be accepted with confidence.

In this section, we describe results of some of the vibrational analyses that have been done on the main secondary conformations found in polypeptides, viz., extended-chain and helical structures. Not all known structures have been the subjects of such analyses, but this approach will certainly be extended to them in the future.

5.4.1 Extended-Chain Polypeptide Structures

5.4.1.1 General Features

Stereochemically acceptable extended-chain (so-called β-sheet) structures were first described by Pauling and Corey from molecular model building [78, 79]. In some cases, specific structures of synthetic polypeptides have been determined from X-ray diffraction studies. These have provided the bases for the vibrational analyses of this class of polypeptide chain conformations.

In these analyses, some structural constraints have had to be imposed in order to achieve a reasonable degree of transferability in the force fields. These constraints have consisted of adopting a standard peptide group geometry, which is probably not a serious problem, and accepting a completely planar peptide group (i.e., the CN torsion angle, ω, equal to 180°), which may be more serious since departures from planarity of 5–10° require relatively little energy (and are not uncommon in proteins) and may have an important effect on some of the modes. The standard geometry of the peptide group is given in Table 5-4, based [80] on the refined X-ray structure of β-PLA [81]. (The dimensions for PGI are slightly different [19].) We therefore assume that chain conformations differ only in their φ, ψ torsion angles.

As Pauling and Corey first pointed out [78, 79], adjacent hydrogen-bonded chains in a β-sheet can occur in two arrangements, parallel and antiparallel (with respect to the direction of the chemical sequence along the chain). In addition, depending on the axial stagger between adjacent chains, the sheet can be 'pleated' or 'rippled', the latter being possible only if all-L and all-D chains alternate in the sheet. (In practice, only PG can satisfy this requirement, since it has not as yet been possible to

Table 5-4. Standard geometry of the peptide group[a].

	Bond length (A°)		Bond angle (degrees)
$C^{\alpha}C$	1.53	$C^{\alpha}CN$	115.4
CO	1.24	$C^{\alpha}CO$	121.0
CN	1.32	CNC^{α}	120.9
NH	1.00	CNH	123.0
NC^{α}	1.47	$NC^{\alpha}C$	109.5
$C^{\alpha}H$	1.07	$NC^{\alpha}H$	109.5
		$CC^{\alpha}H$	109.5
		OCNH	180.0

[a] From [80, 81].

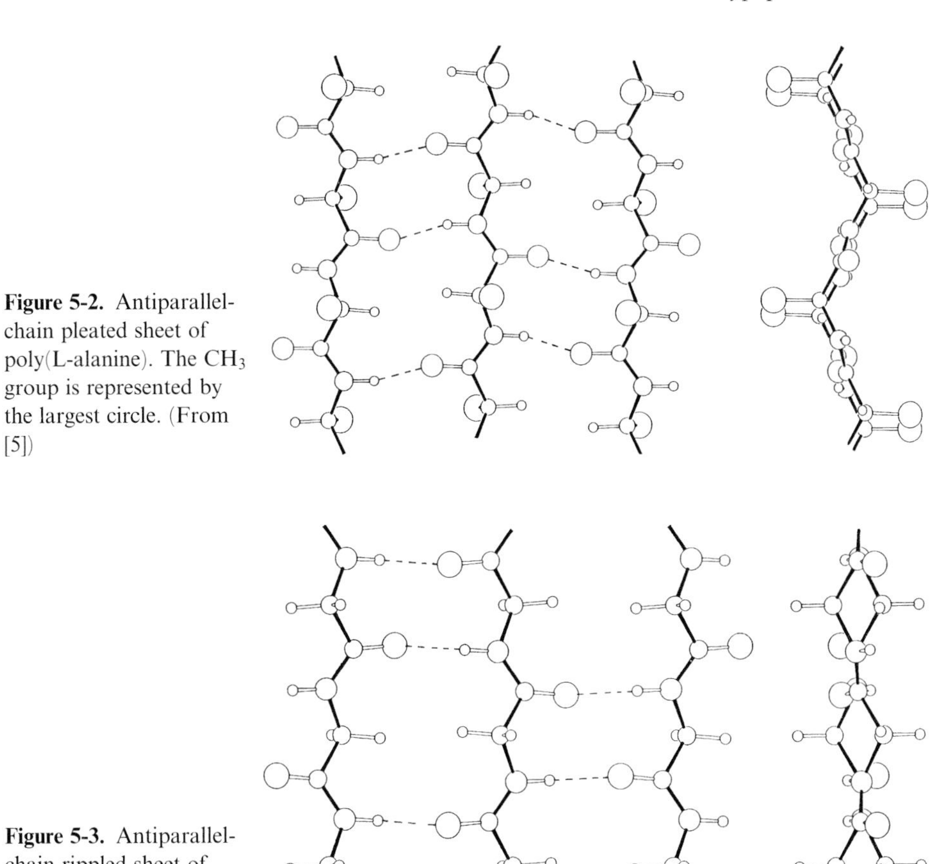

Figure 5-2. Antiparallel-chain pleated sheet of poly(L-alanine). The CH_3 group is represented by the largest circle. (From [5])

Figure 5-3. Antiparallel-chain rippled sheet of polyglycine. (From [5])

produce such a regular alternation with other side-chain-residue polypeptides.) Thus, in principle we can have antiparallel-chain pleated sheet (APPS) (Figure 5-2), antiparallel-chain rippled sheet (APRS) (Figure 5-3), parallel-chain pleated sheet (PPS), and parallel-chain rippled sheet (PRS) extended-chain structures. Their structural parameters are given in Table 5-5 [82].

In describing these structures, and in the vibrational analyses, we assume the sheets to be infinite in two dimensions and therefore calculate only the in-phase unit cell modes. Although modes of only a single sheet are treated, the TDC interactions have been evaluated over a number of sheets (7–11). The effect of finiteness of a single sheet on just the TDC interactions for the amide I mode has been examined [83, 84].

Extended-chain regions are a major component of protein structures. They can occur in mixed antiparallel and parallel arrangements even within a sheet, the sheets can be twisted, and complex 'barrel' arrangements are also found [2]. It is therefore

Table 5-5. Structural parameters of β-sheet structures[a].

Parameter[b]	APPS[c]	APRS[d]	PPS[e]	PRS[f]
a (or $a/2$)	4.73	4.77	4.85	4.80
$b/2$	3.445	3.522	3.25	3.25
φ	−138.4	−149.9	−119.0	−119.0
ψ	135.7	146.5	114.0	114.0
N···O	2.73	2.91	2.74	2.82
H···O	1.75	2.12	1.82	1.75
H–N···O	9.8	31.4	5.5	4.9
N–H···O	164.6	134.4	172.0	172.0

[a] From [82].
[b] Lengths in Å, angles in degrees. a (or $a/2$): lateral separation between adjacent chains. b: axial separation between adjacent units in one chain.
[c] Antiparallel-chain pleated sheet of poly(L-alanine). From [80, 81].
[d] Antiparallel-chain rippled sheet of polyglycine. From [82].
[e] Parallel-chain pleated sheet of poly(L-alanine). From [83].
[f] Parallel-chain rippled sheet of polyglycine. From [83].

of major importance to understand the vibrational dynamics of this type of polypeptide chain conformation.

5.4.1.2 Antiparallel-Chain Structures

Antiparallel-Chain Pleated Sheet

The APPS structure is the predominant one in synthetic polypeptides, and is a prevalent motif in proteins [2]. (Polyglycine is an exception, which we treat below.) An X-ray diffraction study of the simplest representative polypeptide, β-PLA [81], has provided the geometric parameters of the APPS. In addition to those given in Table 5-5, we note that the axial shift between adjacent chains, measured from the position where the hydrogen bond is linear, is 0.27Å (as obtained from a TDC analysis of the amide I mode [80]). The vibrational analysis of β-PLA thus provides the basic spectral characteristics of the APPS.

A PLA sample can be easily oriented, and dichroic IR spectra are therefore readily obtainable. Such spectra [85], together with far-IR spectra [86], are shown in Figure 5-4. Together with spectra of the ND molecule [87, 88], as well as Raman spectra [89] (Figure 5-5), a large amount of spectral information is available, both to optimize an empirical force field and to provide a reasonable description of the normal modes.

The symmetry of the β-PLA structure determines that the normal modes will be distributed among four symmetry species with optical activity and IR dichroism as follows [80]: A–30 modes, Raman; B_1–29 modes, Raman, IR (‖); B_2–29 modes, Raman, IR (⊥); B_3–29 modes, Raman, IR (⊥). The results of the most recent nor-

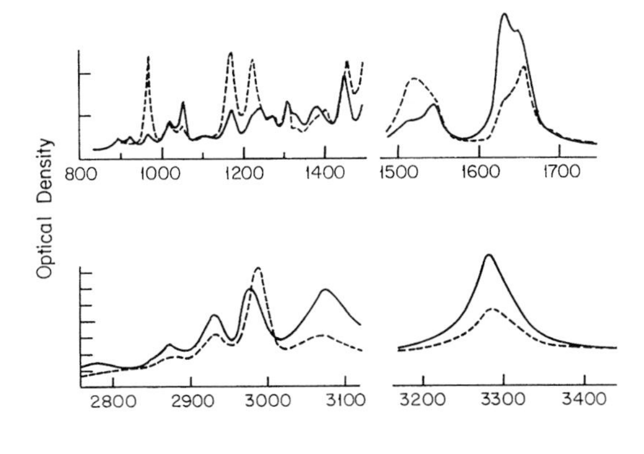

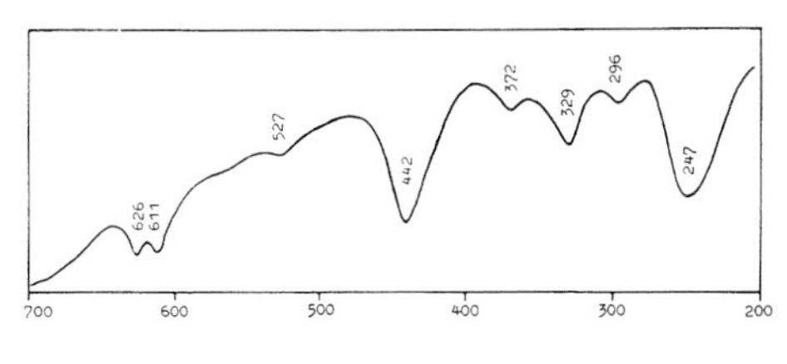

Figure 5-4. Infrared spectra of β-poly(L-alanine). Upper two panels: mid-IR region. (—) Electric vector perpendicular to the stretching direction. (- - -) Electric vector parallel to the stretching direction. (From [85]) Lower panel: far-IR region. (From [86])

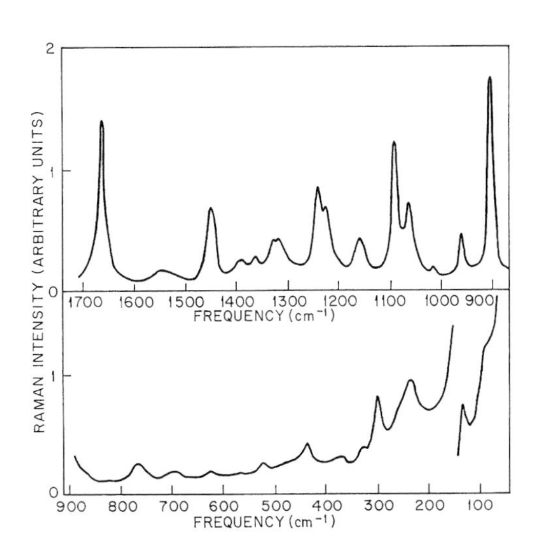

Figure 5-5. Raman spectrum of β-poly(L-alanine). (From [89])

mal mode analysis [22] are given in Table 5-6 (the NH s modes, which are influenced significantly by Fermi resonances [90], and the CH s modes are discussed in an earlier paper [80]).

There are some general points to note about this normal mode analysis. First, the

Table 5-6. Observed and calculated frequencies (in cm^{-1}) of APPS poly(L-alanine).

Observed[a]		Calculated				Potential energy distribution[b]
Raman	IR	A	B_1	B_2	B_3	
					1698	CO s(78) CN s(14)
	1694W ∥		1695			CO s(76) CN s(19)
1669S		1670				CO s(73) CN s(21)
	1632VS ⊥			1630		CO s(70) CN s(21)
					1592	NH ib(57) CN s(21) $C^{\alpha}C$ s(10)
1553VW	1555MW ⊥			1562		NH ib(53) CN s(17) $C^{\alpha}C$ s(14)
1538W		1539				NH ib(48) CN s(22) CO ib(12) $C^{\alpha}C$ s(11)
	1524S ∥		1528			NH ib(41) CN s(26) CO ib(14) $C^{\alpha}C$ s(13)
1451S		1455				CH_3 ab1(44) CH_3 ab2(39)
	1454S ∥		1455			CH_3 ab1(46) CH_3 ab2(37)
					1453	CH_3 ab1(88) CH_3 r1(10)
	1446S ⊥			1452		CH_3 ab1(86) CH_3 r1(10)
1451S		1452				CH_3 ab2(45) CH_3 ab1(43)
	1454S ∥		1452			CH_3 ab2(48) CH_3 ab1(41)
	1446S ⊥			1451		CH_3 ab2(86) CH_3 r2(10)
					1451	CH_3 ab2(89) CH_3 r2(10)
1399W		1402				H^{α} b2(32) CH_3 sb(17) $C^{\alpha}C$ s(14) NH ib(13)
	1402MW ∥		1399			H^{α} b2(33) CH_3 sb(20) NH ib(15) $C^{\alpha}C$ s(11)
	1386W ⊥				1385	CH_3 sb(70) H^{α} b1(16) $C^{\alpha}C^{\beta}$ s(11)
				1383		CH_3 sb(74) H^{α} b1 (16) $C^{\alpha}C^{\beta}$ s(10)
1368W		1372				CH_3 sb(64) H^{α} b2 (16)
	1372MW ∥		1372			CH_3 sb(61) H^{α} b2 (19)
1335W	1330W br ⊥				1332	H^{α} b2(42) NH ib(25) $C^{\alpha}C$ s(12) CH_3 sb(11)
				1317		H^{α} b2(64) NH ib(15)
1311W	1309 sh	{			1305	H^{α} b2(30) CN s(18) CO ib(15)
				1299		NH ib(23) CO ib(18) $C^{\alpha}C$ s(14) CN s(13)
1243S		1236				H^{α} b2(34) NH ib(19) NC^{α} s(19) CN s(13)

1226M	1222S ‖		1231			H^{α} b2(28) NC^{α} s(24) NH ib(18) CN s(11)
		1198				NC^{α} s(36) CH_3 r1(13) $C^{\alpha}C$ s(11) NH ib(10)
					1196	$C^{\alpha}C^{\beta}$ s(33) H^{α} b1(27) NC^{α} s(22) CH_3 sb(11) CH_3 r1(10)
			1195			NC^{α} s(30) NH ib(13) $C^{\alpha}C$ s(12) H^{α} b2(11) CH_3 r1(11)
				1195		$C^{\alpha}C^{\beta}$ s(33) H^{α} b1(26) NC^{α} s(22) CH_3 r1(10) CH_3 sb(10)
1165W		1162				H^{α} b1(56) CH_3 sb(19) $C^{\alpha}C^{\beta}$ s(13)
	1167S ‖		1161			H^{α} b1(54) CH_3 sb(18) $C^{\alpha}C^{\beta}$ s(13) CN s(10)
	1120VW			1125		CH_3 r2(27) H^{α} b1(26) CH_3 sb(11)
					1125	H^{α} b1(26) CH_3 r2(26) CH_3 sb(11)
1092S		1093				CH_3 r2(55) $C^{\alpha}C^{\beta}$ s(21)
			1092			CH_3 r2(54) $C^{\alpha}C^{\beta}$ s(22)
	1084W ⊥			1085		CH_3 r1(27) CH_3 r2(25) $C^{\alpha}C^{\beta}$ s(10) CN s(10)
					1086	CH_3 r1(26) CH_3 r2(26) $C^{\alpha}C^{\beta}$ s(13) CN s(10)
	1052M ⊥				1065	$C^{\alpha}C^{\beta}$ s(23) H^{α} b1(21) CH_3 r1(18) CH_3 r2(17)
				1064		$C^{\alpha}C^{\beta}$ s(22) H^{α} b1(21) CH_3 r2(19) CH_3 r1(17)
1065M		1055				$C^{\alpha}C^{\beta}$ s(50) CH_3 r1(14) H^{α} b1(11)
			1054			$C^{\alpha}C^{\beta}$ s(50) CH_3 r1(13) H^{α} b1(11) CH_3 r2(10)
	966S ‖		970			CH_3 r1(50) NC^{α} s(25)
967M		969				CH_3 r1(50) NC^{α} s(25)
	925M ⊥			918		NC^{α} s(27) CH_3 r1(23) CN s(15)
909VS		913				$C^{\alpha}C$ s(15) CN s(14) CH_3 r2(14) CNC^{α} d(13) CO s(11)
					913	NC^{α} s(30) CH_3 r1(23) CN s(14)
	902 W ‖		904			$C^{\alpha}C$ s(15) CN s(14) CH_3 r2(14) CNC^{α} d(13) CO s(12)
837VW					846	$C^{\alpha}C$ s(34) NC^{α} s(13) CN s(12) $C^{\alpha}C^{\beta}$ s(12)
				844		$C^{\alpha}C$ s(31) NC^{α} s(19) $C^{\alpha}C^{\beta}$ s(14)
					755	CO ib(19) $C^{\alpha}C$ s(15) NC^{α} s(13) C^{β} b2(13) $NC^{\alpha}C$ d(12) CNC^{α} d(11)
775M	778MW			775		CO ib(18) $C^{\alpha}C$ s(16) C^{β} b2(14) NC^{α} s(12) $NC^{\alpha}C$ d(11) CNC^{α} d(11)
				708		CN t(74) NH ob(28) NH···O ib(23)
	706S ⊥ ‖		706			CN t(44) NH ob(41) NH···O ib(19) CO ob(18) H^{α} b1(10)
					705	CN t(74) NH ob(28) NH···O ib(22)
698VW		704				CN t(48) NH ob(41) NH···O ib(19) CO ob(13) H^{α} b1(10)
		668				CO ob(50) CN t(35) C^{β} b1(10)
			662			CO ob(47) CN t(41) C^{β} b1(10)

Table 5-6. *(continued)*

Observed[a]		Calculated				Potential energy distribution[b]
Raman	IR	A	B_1	B_2	B_3	
628VW		624				CO ib(46) $C^{\alpha}C$ s(16) CO ob(13)
	629W ∥		621			CO ib(49) $C^{\alpha}C$ s(16) CO ob(12)
	615W ⊥			626		$C^{\alpha}CN$ d(38) CO ob(18) $NC^{\alpha}C$ d(16) NH ob(16) C^{β} b2(15)
					622	$C^{\alpha}CN$ d(31) CO ob(30) NH ob(22) C^{β} b2(16) $NC^{\alpha}C$ d(14)
					592	CO ob(52) $C^{\alpha}CN$ d(19)
				591		CO ob(62) $C^{\alpha}CN$ d(11) NH ob(11) H^{α} b2(10)
	448M ⊥			447		C^{β} b1(55) NH ob(15)
					447	C^{β} b1(54) NH ob(13)
437W		440				C^{β} b2(75) $NC^{\alpha}C$ d(13)
	432M ∥		440			C^{β} b2(74) $NC^{\alpha}C$ d(12)
	326W ∥		328			$NC^{\alpha}C$ d(25) C^{β} b2(11) CO ob(11) C^{β} b1(10)
332W		327				$NC^{\alpha}C$ d(24) C^{β} b2(11) CO ob(11) C^{β} b1(10)
300M				286		CO ib(33) $NC^{\alpha}C$ d(19) CNC^{α} d(15) C^{β} b2(15) C^{β} b1(13) CO ob(11)
				286		CO ib(33) $NC^{\alpha}C$ d(19) C^{β} b2(15) CNC^{α} d(14) C^{β} b1(13) CO ob(10)
			279			$C^{\alpha}CN$ d(39) $C^{\alpha}C^{\beta}$ t(26) $NC^{\alpha}C$ d(11)
266VWsh		271				$C^{\alpha}CN$ d(38) $C^{\alpha}C^{\beta}$ t(25) $NC^{\alpha}C$ d(17)
				252		C^{β} b2(50)
					252	C^{β} b2(50)
235Msh		240				$C^{\alpha}C^{\beta}$ t(67) $NC^{\alpha}C$ d(11)
	240M ∥		240			$C^{\alpha}C^{\beta}$ t(49) $NC^{\alpha}C$ d(16)
				238		$C^{\alpha}C^{\beta}$ t(91)
					238	$C^{\alpha}C^{\beta}$ t(90)
			211			CNC^{α} d(35) $C^{\alpha}C^{\beta}$ t(23) H···O s(23)
185VW		177				CNC^{α} d(62) $C^{\alpha}CN$ d(15)
					156	H···O s(58) $NC^{\alpha}C$ d(10)
			151			NH ob(39) CO ob(13) NC^{α} s(13) H^{α} b2(10)

135S		147				NH ob(37) CO ob(16) NC^{α} s(12) H^{α} b2(10)
	122Wbr			118		CNC^{α} d(21) NC^{α} C d(18) $C^{\alpha}CN$ d(13) NH ob(12)
			104			NH···O ib(27) CN t(21) C^{β} b1(15)
					103	NH···O ib(33) CN t(21) NH ob(20) $C^{\alpha}C$ t(15) CO···H t(11)
91Msh		91				C^{β} b1(26) NH ob(19) H···O s(14) $C^{\alpha}C$ t(11)
					87	H···O s(33) CNC^{α} d(17) CN t(11) $CN^{\alpha}C$ d(10)
				74		$C^{\alpha}C$ t(24) NC^{α} t(22)H···O s(22) CN t(15)
			41			CO···H ib(21) NH ob(20) C^{β}b1(16) NH···O ib(15) NH···O t(12)
				32		NH···O ib(37) CO···H ib(22) CN t(15)
		31				H···O s(15) CO···H t(15) CO···H ib(14) NH···O ib(14) NH···O t(12) NH ob(11) CN t(10)
		19				CO···H t(35) NH···O t(27) CO···H ib(14) $H^{\alpha}\cdots H^{\alpha}$ s(12)

[a] Data taken from [88]. S: strong, M: medium, W: weak, V: very, sh: shoulder, br: broad, ||: parallel dichroism, ⊥: perpendicular dichroism.

[b] s: stretch, as: antisymmetric stretch, ss: symmetric stretch, b: bend, ib: in-plane bend, ob: out-of-plane bend, ab: antisymmetric bend, sb: symmetric bend, r: rock, d: deformation, t: torsion. Contributions ≥ 10.

empirical force field is capable of giving very acceptable frequency agreement: the rms error between observed and calculated frequencies is 5.9 cm^{-1}. Second, the character of the symmetry species is well predicted: A species modes exhibit only Raman activity, B_1 and B_2 species modes exhibit $\parallel$ and $\perp$ IR dichroism, respectively. Third, while the agreement is somewhat poorer, band shifts on N-deuteration are reasonably well accounted for [88] (for amide I the observed downshift of 4 cm^{-1} is reproduced exactly by the calculation). Finally, TDC is crucial to explaining the large splittings in the amide I and amide II modes: in its absence the amide I modes are calculated at 1670 (A), 1673 (B_1), 1665 (B_2), and 1670 (B_3) cm^{-1}, and the amide II modes are predicted at 1545 (A), 1549 (B_1), 1550 (B_2), and 1554 (B_3) cm^{-1} [5].

With respect to the amide modes, a comparison with the results of the *ab initio* calculation on the hydrogen-bonded Ala dipeptide is of interest, even though the force fields are different. Although localized primarily in CO s, the amide I modes of β-PLA are constrained by symmetry to be coupled modes of the four peptide groups in the unit cell. Interestingly, CN s is now the second major component compared with NH ib for the Ala dipeptide. The coupled amide II modes contain, in addition to NH ib and CN s, $C^{\alpha}C$ s and CO ib contributions not present (at the level of ≥ 10) in the Ala dipeptide. With respect to the strong amide III modes, at 1243 and 1224 cm^{-1}, H^{α} b dominates and NC^{α} s also contributes in addition to NH ib and CN s. As noted above, NH ib is a contributor to many other modes in the 1400–1200 cm^{-1} region. The amide V modes are well accounted for, and in particular the observed absence of any significant dichroism in this band [91] is explained by the near coincidence in frequency of the predicted B_1 and B_3 species modes. Besides the amide modes, local as well as other skeletal modes are very well accounted for.

The excellent agreement obtained for β-PLA would indicate that this force field should be transferable to other APPS polypeptides, of course taking due account of side-chain differences. This has been done for Ca-poly(L-glutamate), Ca-PLG [92], which X-ray and electron diffraction data had indicated to have an APPS structure but which were not extensive enough to provide a definitive conclusion. A proposed model [93] was used as the basis for the vibrational analysis, and the good prediction of the observed bands [92] supports both the model as well as the viability of the force field. The subtle differences between the β-PLA and β-Ca-PLG spectra are accounted for by the normal mode calculation, and emphasize the influence of the side chain on the main-chain modes, particularly on amide III. A similar successful vibrational analysis has been done on an alternating copolymer, APPS poly(L-alanylglycine) [19].

Antiparallel-Chain Rippled Sheet

Although early studies assumed that PG also adopts an APPS structure, electron diffraction studies of 'single crystals' and oriented thin films [94] suggested that extended-chain PG, PGI, has an APRS structure. This was supported by conformational energy calculations [95], and was the basis for a detailed vibrational analysis of the PGI spectra [19, 96], strongly confirming this proposal. (Since the glycine residue is achiral, it can adopt the D or L configuration required of alternate chains in the APRS.) This was aided by extensive IR spectra of isotopic derivatives (NH, CD_2; and ND, CD_2) [97] and by Raman spectra [98], which were important because macroscopic samples of oriented PGI could not be prepared.

The conclusion from vibrational analysis that PGI has an APRS structure was based on a detailed TDC study of the amide I mode splittings for the APPS as compared with the APRS as a function of the axial shift between adjacent chain [19]. No reasonable shift values for the APPS were in agreement with experiment, whereas the APRS gave excellent agreement for a shift of 0 Å, exactly that indicated by the electron diffraction analysis [94]. This structure (Figure 5-3) has a center of inversion and therefore the mutual exclusion rule applies, giving the following symmetry species and activity [19]: A_g–21 modes, Raman; A_u–20 modes, IR(∥); B_g–21 modes, Raman; B_u–19 modes, IR(⊥). The results of the normal mode analysis [96] give an overall rms error of 6.1 cm^{-1}, comparable with that for β-PLA. (The force field for β-PLA was in fact refined from that for PGI [80].) A description of the amide modes of PGI is given in Table 5-7.

The splitting of the amide I modes, somewhat smaller than for the APPS structure, is again due to TDC: in its absence the calculated frequencies are 1684 (A_g), 1677 (A_u), 1676 (B_g), and 1674 (B_u) cm^{-1}. Compared with APPS PLA, $C^{\alpha}CN$ d now makes a noticeable contribution to the eigenvector.

The amide II modes, which are probably at a lower frequency because of the weaker hydrogen bonds, are also split by the TDC interactions: in their absence the calculated modes are at 1534 (A_g), 1535 (A_u), 1559 (B_g), and 1559 (B_u) cm^{-1}. The absence of clearly observed high-frequency modes, particularly an IR band near the expected frequency of 1572 cm^{-1}, raises a very important point: it is necessary to be able to predict intensities as well as frequencies. Intensities require knowledge of dipole derivatives (Eq. (5-4)), but since they also depend on correct eigenvectors (Eq. (5-5)) they provide an independent test of the force field. Such an IR intensity calculation has been done on PGI [20], and the results show that the B_u amide II mode is indeed predicted to be very weak, in agreement with observation.

The amide III mode region is quite complex, with mainly NH ib, CN s contributing near 1300 and 1200 cm^{-1}, where only weak bands are observed. In the 'expected' region (1300–1200 cm^{-1}), the strong bands are due to CH_2 twist (tw)-dominated vibrations. The NH ib contributions now extend to bands about 1400 and below 1200 cm^{-1}.

The amide V mode is at a similar frequency to that in β-PLA, even though the PGI hydrogen bond is weaker. This may be because of the large CN t contribution and/or the somewhat different nature of the modes.

5.4.1.3 Parallel-Chain Structures

Parallel-Chain Pleated Sheet

Although there are no known PPS polypeptides, this general structure is found in proteins [2] and in some tripeptides, in two of which it has been the subject of detailed vibrational analyses [75, 77]. The excellent prediction of observed IR and

Table 5-7. Amide mode frequencies (in cm^{-1}) of APRS polyglycine I.

Observed[a]		Calculated				Potential energy distribution[b]
Raman	IR	A_g	B_u	B_g	B_u	
				1695		CO s(77) CN s(15) C^αCN d(11)
	1685M		1689			CO s(75) CN s(20) C^αCN d(11)
1674S		1677				CO s(74) CN s(21) C^αCN d(11)
	1636S				1643	CO s(69) CN s(22) C^αCN d(11)
				1602		NH ib(56) CN s(19) C^αC s(12)
					1572	NH ib(51) C^αC s(16) CN s(14)
	1517S		1515			NH ib(35) CN s(28) C^αC s(17) CO ib(14)
1515W		1514				NH ib(35) CN s(27) C^αC s(17) CO ib(14)
1410M		1415				CH_2 w(44) CH_2 b(31) NH ib(14)
	1408W		1415			CH_2 w(40) CH_2 b(33) NH ib(13)
				1304		NH ib(30) CO ib(19) CN s(18) C^αC s(16)
	1295W				1286	NH ib(39) C^αC s(17) CO ib(16) CN s(12)
1220W		1213				NC^α s(29) NH ib(23) CH_2 w(18) CH_2 tw(16) CN s(13)
	1214W		1212			NC^α s(29) NH ib(23) CH_2 w(18) CH_2 tw(15) CN s(13)
1162M		1153				NC^α s(50) C^αC s(13) NH ib(12)
				1152		NC^α s(50) C^αC s(14) NH ib(12)
				736		CN t(63) NH···O ib(15) NH ob(11) H···O s(11)
	708S {				718	CN t(75) NH···O ib(19) NH ob(16) H···O s(10)
			718			CN t(79) NH ob(36) NH···O ib(23) H···O s(10)
		702				CN t(79) NH ob(29) NH···O ib(25) H···O s(15)

[a] Data taken from [95]. S: strong, M: medium, W: weak.

[b] s: stretch, b: bend, ib: in-plane bend, ob: out-of-plane bend, d: deformation, w: wag, tw: twist, t: torsion. Contributions ≥10.

Raman bands in these cases, using the same empirical force field as for APPS PLA [5], led to the belief that it would be useful to obtain the frequencies of PPS PLA in order to characterize the vibrational spectrum of this structure. The unperturbed normal modes have been obtained, and the influence of TDC interactions on the amide I and amide II modes, as determined by the intersheet separation, has been assessed [82]. While unperturbed amide I modes occur at 1678 and 1675 cm^{-1}, TDC interactions within one sheet shift these to 1663 and 1642 cm^{-1} and a multisheet array ($11 \times 11 \times 11$ residues) with a separation of 5.3 Å (comparable with that in APPS PLA) leads to modes at 1669 and 1677 cm^{-1}. A similar effect is predicted for amide II: single sheet modes are at 1553 and 1586 cm^{-1}, with the above multisheet separation giving 1549 and 1583 cm^{-1}.

Parallel-Chain Rippled Sheet

A PRS structure is only possible for PG, and is ruled out by the evidence for an APRS structure for this polypeptide. However, the normal modes of PRS PG have also been calculated [82]. It is interesting that, when multisheet TDC coupling is included, the strong Raman amide I mode is predicted to be at a much lower frequency than the strong IR amide I mode, just the opposite of what is observed and predicted for the APRS structure. This demonstrates the power of detailed vibrational spectroscopic analysis in revealing fine details of polypeptide structure.

5.4.2 Helical Polypeptide Structures

5.4.2.1 General Features

The concept of a helical structure as the most general one for a long polymeric chain was developed quite early [99, 100], and possible polypeptide helices having an integral number of residues per turn were described [101]. However, it was only when this constraint was relaxed and well-defined stereochemical criteria invoked [102] that a specific conformation, the α-helix, was discovered [102, 103] that is in fact found in polypeptides and proteins [2]. A number of other helical conformations have been found since then, and their structural parameters, are given in Table 5-8. Vibrational analyses of some of these are described in this section.

Polypeptides are found in both single-stranded intramolecularly hydrogen-bonded helices as well as in conformations in which hydrogen bonds are formed between helices. In the former case, the structure is conveniently described by several parameters: the number of residues per turn, n (positive if right-handed, negative if left-handed); the axial translation per residue, h; and the backbone torsion angles, φ, ψ, and ω. Sometimes the number of atoms in the 'ring' formed by the hydrogen bond, m, is specified [101] (and the residue connectivity via the hydrogen bond can be given; thus, $5 \rightarrow 1$ means the NH group on residue 5 is bonded to the CO group on residue 1). In such a case, the helix will be designated n_m [101].

Among such helices, the α-helix (3.6_{13}) structure has been obtained from a careful analysis of X-ray fiber diffraction patterns of PLA [104]. As is general for helical structures [105], this helix, designated, α_I, has a counterpart, designated α_{II},

Table 5-8. Parameters of some observed helical polypeptide structures.

Structure	n[a]	m[b]	h[c]	φ[d]	ψ[d]	ω[d]	H-bond[e]	Ref.
Intramolecular								
α_I-Helix	3.62	13	1.496	−57.4	−47.5	180	$5 \rightarrow 1$	104
α_{II}-Helix	3.62	13	1.496	−70.5	−35.8	180	$5 \rightarrow 1$	106
3_{10}-Helix	2.99	10	2.01	−45	−30	176	$4 \rightarrow 1$	111
ω-Helix	−4.00	13	1.325	62.8	54.9	175	$5 \rightarrow 1$	112
π-Helix	−4.25	16	1.17	51	74	177	$6 \rightarrow 1$	115
Intermolecular								
Polyglycine II	−3.00		3.10	−76.9	145.3	180		122
Polyproline II	−3.00		3.10	−78.3	148.9	180		128
Polyproline I[f]	3.33		1.90	−83.1	158.0	0		135

[a] Number of residues/turn, + for right-handed and − for left-handed helices.
[b] Number of atoms in hydrogen-bonded 'ring'.
[c] Unit axial translation, in Å.
[d] Dihedral angle, in degrees.
[e] Hydrogen bonding pattern, N–H→O=C.
[f] Cis peptide groups.

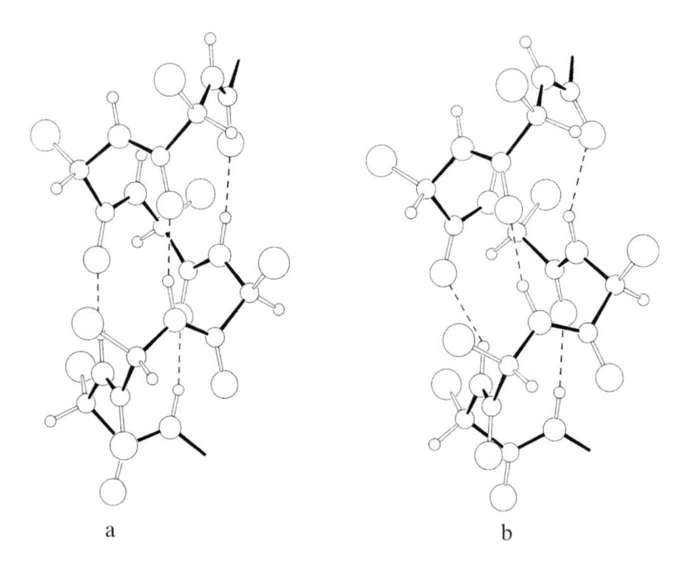

Figure 5-6. Structure of α-helix of poly(L-alanine). (a) α_I-helix; (b) α_{II}-helix. Methyl groups are represented by large circles. (From [5])

which, though having different φ, ψ, has the same n and h, and is comparably allowed energetically [106] (and may be present in bacteriorhodopsin [107]). These structures are shown in Figure 5-6, where it can be seen that the plane of the peptide

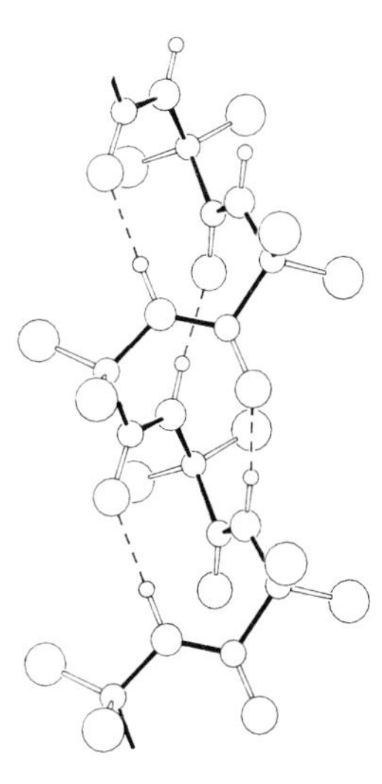

Figure 5-7. Structure of 3_{10}-helix of poly(α-aminoisobutyric acid). Methyl groups are represented by large circles. (From [5])

group in α_{II} is inclined to the helix axis whereas it is almost parallel to the axis in α_I. The α-helix conformation is not only common in polypeptides, and is a basic component of proteins [2], but its topology forms the basis for two-stranded coiled-coil structures such as tropomyosin [108]. The 3_{10}-helix is similar to the α-helix, but has a $4 \rightarrow 1$ hydrogen-bonding pattern. Although segments are found in proteins [2], the only polypeptide in which it has been found, as suggested by electron diffraction [109] and supported by vibrational analysis [110], is poly(α-aminoisobutyric acid), PAIB (Figure 5-7). Its structure, as indicated by conformational energy calculations [111], has a nonplanar peptide group. The ω-helix is a left-handed $5 \rightarrow 1$ helix, -4.0_{13}, proposed from X-ray analysis for the structure of poly(β-benzyl-L-aspartate) [112]. Conformational energy calculations have indicated somewhat different parameters [113], with other calculations suggesting a further variant [114]. The π-helix, said to be present in poly(β-phenethyl-L-aspartate [115], is also a left-handed helix, -4.25_{16}, although a right-handed form with a 4° distorted $NC^{\alpha}C$ angle has been suggested [116]. The ω- and π-helices also have nonplanar peptide groups. Other helical conformations have been proposed, a γ-helix [102, 103] and a 2.2_7-helix [117], but neither of these has as yet been observed in polypeptides. When L and D residues alternate along a chain, which occurs in some transmembrane peptides, a new class of helices results, the so-called β-helices [118, 119]. Such chains can also form double-stranded structures with interchain hydrogen bonds.

The simplest polypeptide structure that has intermolecular hydrogen bonds throughout its lattice is that of PGII, which basically has threefold helical symmetry [120]. (Because of its achiral C^{α}, PGII can be a right- or left-handed helix.) Although the originally proposed structure was modeled with parallel chains, the ease of mechanical conversion of PGII to PGI and the evidence for antiparallel chains in single crystals of PGI led to the suggestion [121] that an antiparallel-chain structure should exist. Such a model was subsequently proposed [122], and a detailed

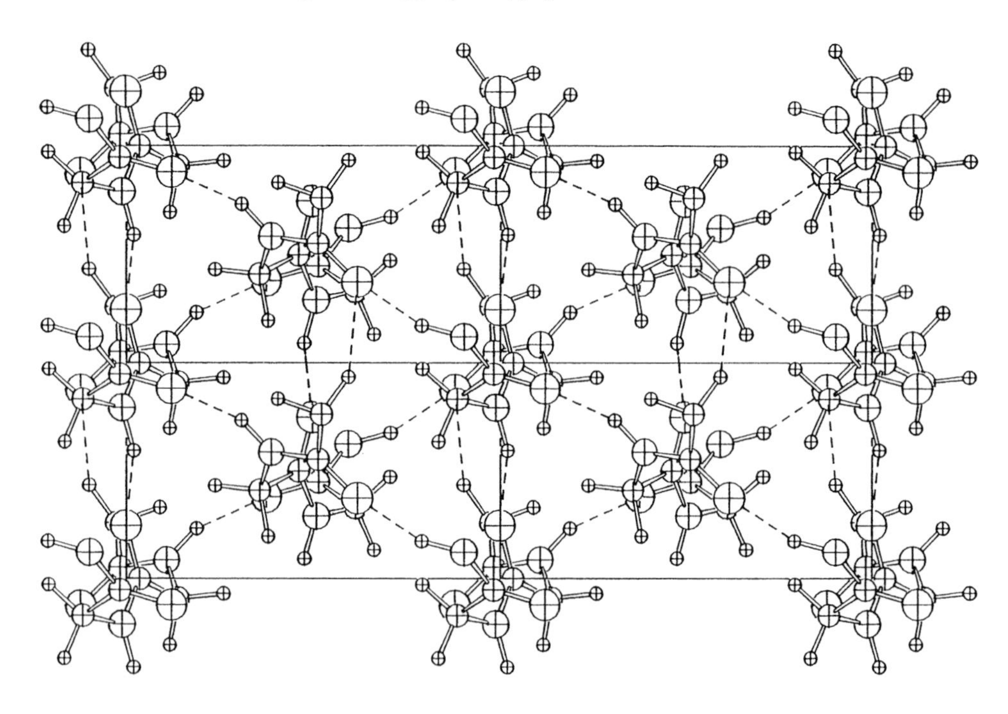

Figure 5-8. Antiparallel-chain structure of polyglycine II. (From [123])

vibrational analysis of the two structures [123] showed that the antiparallel-chain structure (Figure 5-8) was indeed preferred. In these studies, the suggestion that C^{α}–H···O=C hydrogen bonds were present [124], which was supported by spectroscopic studies [125, 126], was convincingly confirmed by normal-mode analysis [123]. The same chain symmetry, in a required left-handed helix, is found in polyproline II, which has trans peptide groups [127, 128], a short contact again suggesting the possibility of a C^{α}–H···O=C hydrogen bond [129]. It is interesting that, from circular dichroism studies [130], a polyproline II type of structure, with 2.5–3.0 residues/turn [131], was indicated to be a preferred local conformation in many 'unordered' polypeptides in solution. This proposal has since received strong support from Raman and normal-mode studies [132] as well as from a more detailed analysis of the circular dichroism results [133]. A triplet of left-handed PGII or polyproline II helices can be thought of as the topological parent of the triple-stranded right-handed coiled-coil structure of collagen [134, 135]. When the peptide groups in polyproline take a *cis* configuration, i.e., polyproline I, the chain adopts a right-handed 10 residue/3 turn helical structure [136].

The variety of helical as well as extended-chain and turn [5] structures emphasizes the flexibility of the polypeptide chain, and thereby the importance of achieving a full understanding of its vibrational dynamics as a function of conformation.

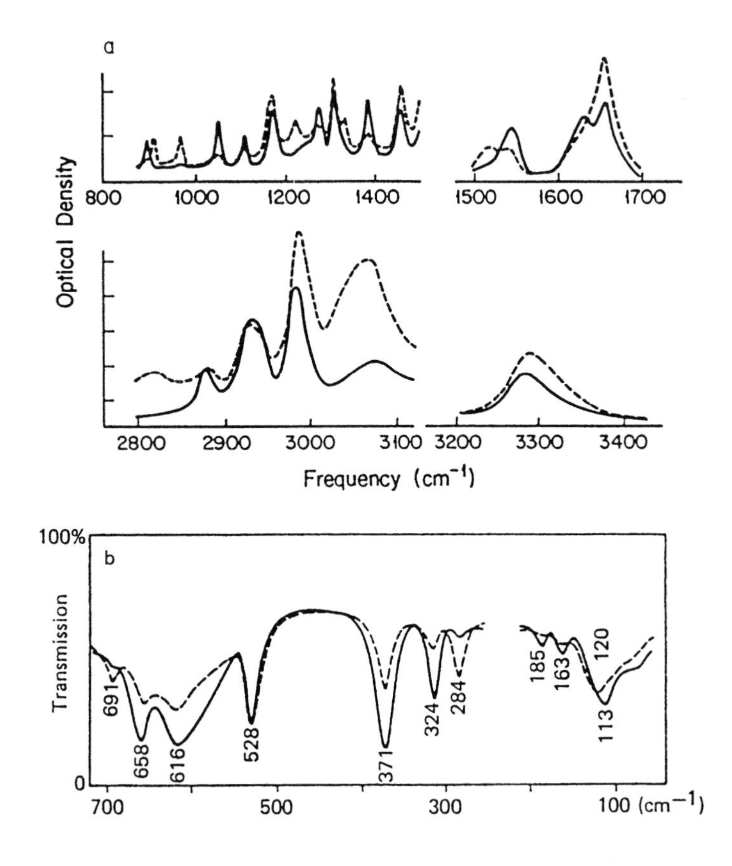

Figure 5-9. Infrared spectra of α-poly(L-alanine). (a) Mid-IR region. (From [85]). (b) Far-IR region. (From [137]). (—) Electric vector perpendicular to stretching direction. (- - -) Electric vector parallel to stretching direction.

5.4.2.2 *α*-Helix

The α_I-PLA structure is well determined from X-ray diffraction [104], and therefore permits a secure refinement of an empirical force field. The helical symmetry results in three optically active modes, defined by the phase difference between similar vibrations in adjacent units of the helix: A(0°) – 28 modes, Raman, IR(‖); E_1 (99.6°) – 29 doubly degenerate modes, Raman, IR (⊥); E_2 (199.1°)–30 doubly degenerate modes, Raman. Extensive IR and Raman studies on α-PLA and α-PLA-ND [5] have provided good experimental data. Polarized IR spectra [85, 137] are shown in Figure 5-9 and Raman spectra [138] in Figure 5-10. The force field for the normal-mode calculation [106] was refined using that of *β*-PLA [22] as a starting point. The frequencies and PEDs [106] are given in Table 5-9; the rms frequency error is 7.6 cm^{-1}.

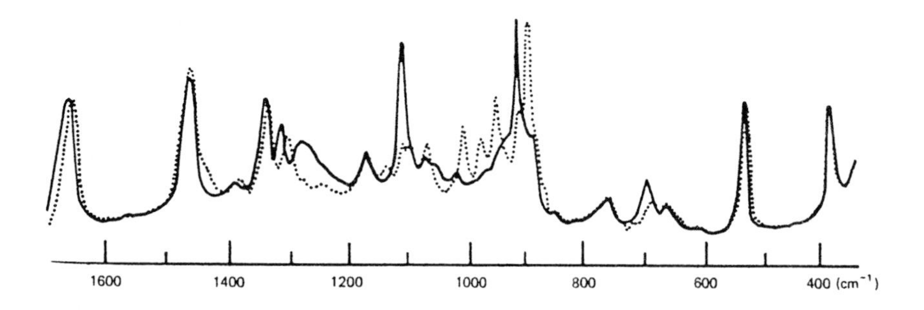

Figure 5-10. Raman spectra of α-poly (L-alanine). (—) Undeuterated. (···) N-deuterated. (From [138])

The observed and calculated amide I mode splittings agree quite well, and are much smaller than for β-PLA. (The specific A-E_2 splitting, due to TDC, remains to be verified; this will depend on having good polarized Raman spectra.) The mode now contains less CN s and more C^{α}CN d than β-PLA, and a change in the small (<10) NH ib contribution is largely responsible for the observed 8 cm^{-1} downshift (calculated 10 cm^{-1}) on N-deuteration [106]. The calculated unperturbed amide II modes at 1529 (A), 1532 (E_1), and 1536 (E_2) cm^{-1} again show how important TDC is in explaining the observed 29 cm^{-1} A-E_1 splitting. The NH ib contribution to amide III now dominates over H^{α} b as compared with β-PLA, and CN s does not appear. The increase in these frequencies, from the 1240–1220 cm^{-1} region in β-PLA to the 1280–1260 cm^{-1} region in α-PLA, must be due to the difference in conformation (and the resulting change in interaction between H^{α} b and NH ib), since the relevant force constants are in fact smaller in the α-helix than in the β-sheet [106]. The lower amide V frequencies in α-PLA, 658 and 618 cm^{-1}, than in β-PLA, 706 cm^{-1}, are undoubtedly a result of the weaker hydrogen bond in the helical structure (N···O = 2.86 Å in α-PLA versus 2.73 Å in β-PLA).

Other synthetic polypeptides adopt the α_I-helix structure, and it is of interest to see how their normal modes are influenced by a change in the side chain. A polypeptide that has been analyzed with the same main chain force field as α-PLA is α-poly(L-glutamic acid), α-PLGH [139]. Some of its modes are compared with those of α_I-PLA in Table 5-10. The amide I mode is hardly affected. The observed amide II mode A-E_1 splitting increases somewhat, from 29 cm^{-1} in α-PLA to 40 cm^{-1} in α-PLGH; since the eigenvectors are identical [139], this may reflect a very subtle change in structure that alters the TDC interactions. The amide III modes are somewhat more affected, and this is a result of the side chain: $C\gamma H_2$ tw now contributes to the E_2 and E_1 modes. Analogous changes influence skeletal modes in the 1200–900 cm^{-1} region, in the case of the CN s, CNC^{α} d modes even altering the frequency order of the species. The amide V modes are also shifted by such changes in eigenvectors, and these have an even greater impact on the low-frequency skeletal modes. The constellation of such changes may well be characteristic of the side chain.

Table 5-9. Observed and calculated frequencies (in cm^{-1}) of α_I-poly(L-alanine).

Observed[a]		Calculated			Potential energy distribution[b]
Raman	IR	A	E_1	E_2	
	1658VS ∥	1657			CO s(82) CN s(10) C^{α}CN d(10)
1655S			1655		CO s(82) CN s(11) C^{α}CN d(10)
				1645	CO s(83) CN s(12) C^{α}CN d(10)
				1540	NH ib(46) CN s(31) CO ib(12) C^{α}C s(11)
1543VW	1545VS ⊥		1538		NH ib(46) CN s(33) CO ib(11) C^{α}C s(10)
	1516Msh	1519			NH ib(45) CN s(34) CO ib(11)
1458S	1458VS ∥ ⊥	1452			CH_3 ab1(47) CH_3 ab2(41) CH_3 r1(10)
			1452		CH_3 ab2(54) CH_3 ab1(31)
				1452	CH_3 ab2(59) CH_3 ab1(26)
		1451			CH_3 ab2(45) CH_3 ab1(41) CH_3 r2(10)
			1451		CH_3 ab1(57) CH_3 ab2(31)
				1451	CH_3 ab1(62) CH_3 ab2(26)
1377W	1381S ⊥	1379	1379	1379	CH_3 sb(100)
				1349	H^{α} b1(34) NH ib(17) C^{α}C s(17) $C^{\alpha}C^{\beta}$ s(10)
1338Msh			1345		H^{α} b1(25) H^{α} b2(20) C^{α}C s(17) NH ib(12)
1326S	1328sh ∥	1334			H^{α} b2(58) C^{α}C s(16)
				1314	H^{α} b2(81)
1308M	1307 ∥ ⊥		1308		H^{α} b2(56)H^{α} b1(23)
		1301			H^{α} b1(62)
1278W				1287	NH ib(23) NC^{α} s(18) H^{α} b1(18)
1271W	1270M ⊥		1278		NH ib(28) H^{α} b2(15) NC^{α} s(14) H^{α} b1(11)
1261W	1265Msh	1262			NH ib(40) H^{α} b2(28)
1167M	1170S ∥ ⊥	1178			NC^{α} s(32) CH_3 r1(20) $C^{\alpha}C^{\beta}$ s(6)
			1167		$C^{\alpha}C^{\beta}$ s(32) NC^{α} s(21) CH_3 r1(15) H^{α} b1(10)
				1162	$C^{\alpha}C^{\beta}$ s(41) NC^{α} s(16) H^{α} b1(12) CH_3 r1(12)
		1115			$C^{\alpha}C^{\beta}$ s(64) CH_3 r2(15)
1105S	1108S ⊥		1103		$C^{\alpha}C^{\beta}$ s(39) CH_3 r2(18)
				1094	$C^{\alpha}C^{\beta}$ s(26) CH_3 r2(21)
				1043	CH_3 r1(47) H^{α} b1(20)
1050W	1051VS ⊥		1037		CH_3 r1(39) H^{α} b1(22) CH_3 r2(15)
1017W	1016W ∥	1026			CH_3 r2(28) H^{α} b1(25) CH_3 r1(24)
970W	968M ∥	982			CH_3 r1(34) NC^{α} s(29) CH_3 r2(17) C^{α}C s(15)
			955		CH_3 r2(32) NC^{α} s(20) CH_3 r1(15)
940VW				951	CH_3 r2(41) NC^{α} s(16)
908VS	909M ∥	910			CNs (23) CNC^{α} d(16) CO ib(12) CH_3 r2(11) CO s(10)
	893S ⊥		901		CN s(18) C^{α}C s(12) CO ib(12) $C^{\alpha}C^{\beta}$ s(10)
882W				896	$C^{\alpha}C^{\beta}$ s(19)C^{α}C s(14) CN s(14) CO ib(12) NC^{α} s(11)
773VW	774M		780		CO ob(42) C^{β} b1(10) CN t(10)
				767	CO ob(52) C^{β} b1(10)
756W		754			CO ob(38) CN t(30)

Table 5-9. (*continued*)

Observed[a]		Calculated			Potential energy distribution[b]
Raman	IR	A	E_1	E_2	
693M	691Wsh $\parallel$	700			NC$^{\alpha}$C d(36) C$^{\alpha}$CN d(26)
				675	CN t(32) CO ib(18) C$^{\alpha}$C s(14)
662W	658S $\perp$		660		CN t(37) NH ob(21) NC$^{\alpha}$C d(12)
				637	CN t(59) NH ob(43) NH···O ib(13)
	618S $\perp$		608		CN t(47) NH ob(23) CO ob(15) CO ib(12)
		589			CN t(68) NH ob(36) CO ob(26) NH···O ib(11)
530VS	526S $\parallel$ $\perp$	537			CO ib(29) C$^{\alpha}$CN d(21) C$^{\alpha}$C s(17) C$^{\beta}$ b2(17) NH ob(11)
			522		NC$^{\alpha}$C d(31) C$^{\alpha}$C s(14) C$^{\alpha}$CN d(11) CO ib(11)
				492	NC$^{\alpha}$C d(37) CO ib(17) C$^{\alpha}$C s(16)
375S	375S $\perp$		374		CO ob(16) NH ob(16) C$^{\beta}$ b2(15) C$^{\alpha}$CN d(15) CO ib(15) CNC$^{\alpha}$ d(10)
	~366Msh	367			CO ob(21) C$^{\beta}$ b1(17) NC$^{\alpha}$C d(14) NH ob(13) C$^{\beta}$ b2(11)
				366	C$^{\beta}$ b2(49) C$^{\alpha}$CN d(22) C$^{\beta}$ b1(16)
328W	324S $\perp$		326		C$^{\beta}$ b2(42) C$^{\beta}$ b1(19) CO ib(16)
310S				310	CNC$^{\alpha}$ d(30) CO ib(20) C$^{\beta}$ b1(20) CO ob(17)
294M	290M $\parallel$	307			C$^{\beta}$ b2(34) CO ib(29) CNC$^{\alpha}$ d(15)
260M	259Wsh	264			C$^{\alpha}$C$^{\beta}$ t(35) C$^{\beta}$ b2(14)
	240W		245		C$^{\alpha}$C$^{\beta}$ t(91)
				244	C$^{\alpha}$C$^{\beta}$ t(95)
	223VW	230			C$^{\alpha}$C$^{\beta}$ t(63)
209VW				205	C$^{\alpha}$CN d(27) C$^{\beta}$ b2(25) C$^{\beta}$ b1(12) CO ob(10)
189M	188M $\perp$		197		C$^{\alpha}$CN d(15) C$^{\beta}$ b2(12) CO ob(12)
165M	163M $\perp$		155		CNC$^{\alpha}$ d(33) C$^{\alpha}$CN d(19) C$^{\beta}$ b1(14) NH ob(14) NC$^{\alpha}$C d(12)
159S				151	NH ob(43) CNC$^{\alpha}$ d(20) NC$^{\alpha}$C d(11)
	120S $\parallel$	136			CNC$^{\alpha}$ d(34) C$^{\alpha}$CN d(21) NC$^{\alpha}$C d(16) C$^{\beta}$ b1(12)
	113Msh $\perp$		96		NH ob(24) C$^{\alpha}$C t(20) NC$^{\alpha}$ t(18) CN t(18) H···O s(10)
87W	84W	94			CN t(27) NC$^{\alpha}$ t(23) C$^{\alpha}$C t(22) H···O s(15) NH ob(10)
				87	NH ob(33) NH···O ib(20) C$^{\beta}$ b1(12) NC$^{\alpha}$C d(10) CNC$^{\alpha}$ d(10)
				49	NH ob(38) NC$^{\alpha}$ t(24) H···O s(23) CN t(15) C$^{\beta}$ b1(13)
			40		NH ob(58) C$^{\beta}$ b1(20) H···O s(11)
				38	C$^{\alpha}$C t(53) H···O s(18)

[a] S: strong, M: medium, W: weak, V: very, sh: shoulder, $\parallel$: parallel dichroism, $\perp$: perpendicular dichroism.

[b] s: stretch, as: antisymmetric stretch, ss: symmetric stretch, b: bend, ib: in-plane bend, r: rock, d: deformation, t: torsion. Constributions ≥ 10.

Table 5-10. Comparison of some modes of α_I-poly(L-glutamic acid) and α_{II}-poly(L-alanine) with those of α_I-poly(L-alanine).

Mode[a]	α_I-PLA[b]		α_I-PLGH[c]		α_{II}-PLA	
	Observed	Calculated	Observed	Calculated	Observed[d]	Calculated[b]
I	1658 1655	1657(A) 1655(E_1)	1653 1652	1657(A) 1655(E_1)	1667 1661	1667(A) 1660(E_1)
II	1545 1516	1538(E_1) 1519(A)	1550 1510	1537(E_1) 1517(A)	1547	1540(E_1) 1515(A)
III	1338 1278 1270 1265	1345(E_1) 1287(E_2) 1278(E_1) 1262(A)	1340 1296 1283	1326(E_1) 1299(E_2) 1287(E_1) 1263(A)		1346(E_1) 1281(E_2) 1272(E_1) 1260(A)
NC^{α} s, $C^{\alpha}C^{\beta}$ s	1170	1178(A) 1167(E_1)	1118	1129(A) 1129(E_1)		1175(A) 1169(E_1)
CN s, CNC^{α} d	909 893	910(A) 901(E_1)	928 924	929(E_1) 922(A)		909(A) 900(E_1)
V	658 618	660(E_1) 608(E_1)	670 618	678(E_1) 626(E_1)		666(E_1) 615(E_1)
CO ib, $C^{\alpha}CN$ d	528	537(A)	565	549(A)		536(A)
$NC^{\alpha}C$ d	528	522(E_1)		515(E_1)		518(E_1)
CO ob, C^{β} b	375 366	374(E_1) 367(A)	409	402(E_1) 368(A)		376(A) 369(E_1)
C^{β} b	292	307(A)	318	334(A)		302(A)
$C^{\alpha}C^{\beta}$ t	260	264(A)	280	265(A)		274(A)

[a] I, II, III, V: amide modes, in cm^{-1}; s: stretch, b: bend, ib: in-plane bend, ob: out-of-plane bend, d: deformation, t: torsion.
[b] From [106].
[c] From [139].
[d] From [140].

Small changes in backbone structure, such as are represented by the α_{II}-helix, also lead to predictable spectral changes, as seen from Table 5-10. The amide I mode, found about 10 cm^{-1} higher in (putative [107]) α_{II}-helices [140], is expected to be particularly affected because of the weaker hydrogen bond resulting from the tilted peptide group. The other higher-frequency modes do not change very much, but there are some possible characteristic changes in the low frequency region: the frequency order of the symmetry species of the CO ob, C^{β} b modes inverts, and the frequency separation between C^{β} b and $C^{\alpha}C^{\beta}$ t is expected to decrease significantly in α_{II}. Thus, vibrational spectra can be sensitive to even very small changes in structure.

In addition to frequencies, dichroic properties can be altered by small structural changes. Since the peptide group in the α_{II}-helix is more inclined to the helix axis than it is in the α_I-helix, the DRs of amide modes are expected to differ significantly between the two structures. To calculate the DR of, for example, amide I of an α-helix we need to know $(\partial\vec{\mu}/\partial Q_I)_{\|}$ and $(\partial\vec{\mu}/\partial Q_I)_{\perp}$, which involves knowing the L_{iI} and $\partial\vec{\mu}/\partial S_i$ (see Eq. (5-5)). Since the L_{iI} are known from the normal mode calculation and the $\partial\vec{\mu}/\partial S_i$ have been obtained from *ab initio* calculations [63], it becomes possible to synthesize a band intensity profile for a given (even finite) α-helix structure. Such a calculation has been done for the α_I-helix and the α_{II}-helix for amide I and amide II by computing the IR intensity profiles for the two polarization components of these two bands [141]. The results show that the DRs are significantly different enough to permit identification of these structures from this property.

The ability to predict a band intensity profile opens up an important additional dimension in vibrational analysis. It means that we will be able to relate subtle spectral differences to small structural changes with a greater degree of confidence. Thus, it has been possible to confirm that an observed three-component contour of the amide I band of tropomyosin is indeed expected for a coiled-coil α-helix [142]. Extensions to understanding the normal modes of proteins become possible [143]. The systematic incorporation in an SDFF of dipoles and dipole fluxes to calculate IR intensities [144] will finally bring to the vibrational analysis of polymeric molecules the completeness and flexibility needed to make it a much more powerful structural tool.

5.4.2.3 3_{10}-Helix

The 3_{10}-helix represents a different topology than the α-helix, being $4 \rightarrow 1$ rather than $5 \rightarrow 1$, and is therefore of importance in studying the vibrational dynamics of polypeptide helices. At present, its only clear identification has been in PAIB, so its characterization from this polypeptide may lack some generality. On the other hand, the excellence of the experimental data [110] makes its vibrational analysis secure.

Since the structure of PAIB has not been determined in detail by diffraction methods, the normal-mode studies [110] were based on a 3_{10} structure obtained from conformational analysis [111] (Figure 5-7). The vibrational studies compared experimental data with predictions for two helices, and the results clearly favored the 3_{10}- over the α-helix. The threefold screw symmetry of this structure results in E_1 and E_2 species modes reducing to doubly degenerate E species modes. The main chain force field was the same as that for α_I-PLA, with additional force constants refined for the $(CH_3)_2$ group. Some modes from the full analysis [110] are compared with those of α_I-PLA in Table 5-11.

The observed amide I frequencies are slightly lower for the 3_{10}-helix than for the α_I-helix because of the slightly stronger hydrogen bond ($N \cdots O = 2.83$ Å versus 2.86 Å, respectively). The increased splitting is undoubtedly due to the different TDC interactions as a function of conformation. This probably also accounts for the large change in the A–E splitting of the amide II modes. The nominal amide III modes (they contain no CN s and NH ib dominates only in the 1312 cm^{-1} mode)

Table 5-11. Comparison of some modes of 3_{10}-poly(α-aminoisobutyric acid) and α_I-poly(L-alanine).

Mode[a]	3_{10}-PA1B[b]			α_I-PLA
	Observed		Calculated	Calculated
	Raman	IR		
I		1656 ∥	1665(A)	1657(A)
	1647		1661(E)	1655(E_1)
II		1545 ⊥	1547(E)	1538(E_1)
	1531		1533(A)	1519(A)
III	1339		1346(E)	1345(E_1)
	1313		1312(A)	
				1287(E_1)
				1278(E_2)
	1280	1280 ∥	1287(A)	1262(A)
CN s, CNC$^\alpha$ d	908	905 ∥	905(A)	910(A)
V		694 ⊥	701(E)	660(E_1)
		680 ∥	676(A)	
				608(E)
				589(A)
NC$^\alpha$C d	594	595 ∥	594(A)	522(E_1)
C$^\alpha$CN d, CO ib	568		557(A)	537(A)

[a] I, II, III, V: amide modes in cm^{-1}; s: stretch, d: deformation, ib: in-plane bend.
[b] From [110]. ∥: parallel dichroism, ⊥: perpendicular dichroism.
[c] From [106].

are well accounted for, and exhibit a quite different pattern from that of the α_I-helix. This is undoubtedly due in part to the different involvement of side-chain structures in the two polypeptides. Since no such major difference is seen in the amide V eigenvectors, the large frequency differences between the two helices must be due to differences in hydrogen-bonding geometry and/or main-chain conformation. (The disappearance of the 694 and 680 cm^{-1} bands on N-deuteration [110] unambiguously confirms their assignments to amide V.) The skeletal NC$^\alpha$C d and C$^\alpha$CN d, CO ib modes have reversed frequency order (and in one case changed species) in 3_{10}-PAIB, and again, this is probably partly due to the involvement of side-chain coordinates in the eigenvectors. Future studies of, for example, a 3_{10}-helix of PLA should help to establish the possible generality of these results.

5.4.2.4 Polyglycine II

The PGII structure is the simplest one that is representative of polypeptide helices without intramolecular hydrogen bonding. In this case, the antiparallel chains of essentially threefold helices [122, 123] are completely intermolecularly N–H···O=C hydrogen bonded (Figure 5-8), with strong spectroscopic evidence [123] for additional C^{α}–H···O=C hydrogen bonds. The basis for this conclusion is of interest since it demonstrates the sensitivity of the spectroscopic technique to subtle features of the structure.

The vibrational analysis [123] was based on comparing the normal modes of a parallel-chain and an antiparallel-chain structure of PGII. In both structures all N–H···O=C hydrogen bonds can be made. However, whereas in the parallel-chain arrangement all CH_2 groups participate equivalently in C^{α}–H···O=C hydrogen bonds (implying that all C=O groups have bifurcated hydrogen bonds), in the antiparallel-chain arrangement only every third CH_2 group can be involved in such interactions, and only between like-directed chains. The result is that the chain no longer has strict threefold symmetry, and as a result the degeneracy of the E species modes is broken. This leads to the expected, and observed, presence of a larger number of bands than predicted by a strictly threefold-symmetric chain structure, and also of bands that occur in regions where no modes are predicted for the parallel-chain structure. Such symmetry-based conclusions are powerful constraints on structural possibilities.

The antiparallel-chain structure (Figure 5-8) has only a screw axis of symmetry, and its modes are distributed as follows: A-62 modes, Raman, IR; B-61 modes, Raman, IR. As for PGI, extensive IR spectra on isotopic derivatives (NH, CD_2; ND, CH_2; and ND, CD_2) were available [97] as well as Raman spectra on the N-deuterated molecule [98]. The PGI force field [96] was used as a starting point, and refinement required small adjustments in 10 of ~70 intramolecular force constants [123]. Amide mode frequencies are given in Table 5-12; the overall rms frequency error is 5.4 cm^{-1}.

The hydrogen bond in PGII is stronger than that in PGI (N···O = 2.69 Å vs 2.91 Å), so it is not surprising that the average amide I mode frequency is lower in PGII (the unperturbed modes are at 1651.0 ± 1.7 cm^{-1}). The very different splittings are due to the different TDC interactions for the two structures. The somewhat higher amide II modes are also consistent with stronger hydrogen bonds. The amide III modes with significant NH ib, CN s contributions are found near 1280 cm^{-1}, strongly mixed with CH_2 tw, quite different from the situation in PGI. The isolated NH ib contribution is now found in the 1400–1300 cm^{-1} region. The amide V modes in PGII are divided between a higher frequency group near 750 cm^{-1} and a lower frequency group near 670 cm^{-1}. This is a much larger separation than found for other helical structures, and may be a consequence of the interchain hydrogen bonding.

5.5 Summary

The main message conveyed by the body of present vibrational spectroscopic studies of peptides and polypeptides is that detailed information on their structures

Table 5-12. Amide mode frequencies (in cm^{-1}) of antiparallel-chain polyglycine II.

Observed[a]		Calculated		Potential energy distribution
Raman	IR	A	B	
		1656		CO s(71) CN s(20) C^{α}CN d(10)
1654VS	1655W		1654	CO s(74) CN s(20) C^{α}CN d(10)
			1653	CO s(72) CN s(20) C^{α}CN d(10)
		1651		CO s(73) CN s(19) C^{α}CN d(10)
		1649		CO s(73) CN s(20) C^{α}CN d(10)
	1640VS		1645	CO s(74) CN s(19) C^{α}CN d(10)
1560W	1560W	1565	1565	NH ib(59) CN s(18)
	1550S	1555	1552	NH ib(52) CN s(20) C^{α}C s(10)
		1548	1548	NH ib(54) CN s(21) C^{α}C s(12)
1383MS	1377M	1383	1380	CH_2 w(58) C^{α}C s(15) NH ib(13)
		1350	1350	CH_2 w(46) CH_2 b(16) NH ib(16) C^{α}C s(14)
1334VW	1332VW	1344	1345	CH_2 w(50) NH ib(16) CH_2 b(14) C^{α} C s(13)
		1303	1304	CH_2 tw(31) NH ib(14) CH_2 w(13) CN s(12)
1283M	1283M	1290	1290	CH_2 tw(29) CH_2 w(23) NH ib(13) CN s(12)
752VW	751W	760	759	CN t(20) NH ob(19) NH$\cdots$O ib(14) NC^{α}C d(13) CO ib(12)
742VW		740		CN t(53) NH$\cdots$O ib(22) NH ob(18)
	740M		738	CN t(49) NH$\cdots$O ib(20) NH ob(16) CO ib(11)
			678	CN t(77) NH ob(27) NH t(11) CO ob(10)
673M		674		CN t(83) NH ob(28) NH t(11) NH$\cdots$O ib(10)
		664	661	CN t(88) NH ob(30) NH t(14)

[a] Data taken from [123]. S: strong, M: medium, W: weak, V: very.
[b] s: stretch, b: bend, ib: in-plane bend, ob: out-of-plane bend, d: deformation, w: wag, tw: twist, t: torsion. Contributions ≥ 10.

and interactions can be obtained if experimental data are interpreted by careful normal-mode analysis. Although the systematic application of extensive empirical force fields has revealed the validity of this approach [5], the full power of this technique remains to be exploited. Our discussion has concentrated, because of space limitations, on the amide modes, but of course the sensitivity to conformation and forces resides in the entire spectrum. The key to extracting this information will be the physical reliability of the force fields used in the normal-mode calculations. We have come a substantial distance toward this goal with existing force fields [5], but improvements can be made and some are already in sight. These presently involve empirical force fields, such as the development of an ab initio-based α_I-PLA force field [145] utilizing better IR [145] and polarized Raman [146] data as well as an understanding of helix vibrations [147] that avoids the weak-coupling and perturbation approximations [9]. However, the ultimate goal will be achieved when we have a conformation-dependent force field as represented by an SDFF for the polypeptide chain [51].

Acknowledgments

The author's contributions to the development of this field have been supported by the National Science Foundation through grants from its Biophysics and Polymers Programs. Helpful discussions with Noemi G. Mirkin and Kim Palmö are gratefully acknowledged.

5.6 References

[1] Stryer, L., *Biochemistry*. Fourth Edition. New York: W. H. Freeman & Co., 1995.
[2] Richardson, J., *Adv. Protein Chem.* **1981**, *34*, 167–339.
[3] Surewicz, W. K., Mantsch, H. H., *Biochim. Biophys. Acta* **1988**, *952*, 115–130.
[4] Surewicz, W. K., Mantsch, H. H., Chapman, D., *Biochemistry* **1993**, *32*, 389–394.
[5] Krimm, S. and Bandekar, J., *Adv. Protein Chem.* **1986**, *38*, 181–364.
[6] Herzberg, G., *Molecular Spectra and Structure. II. Infrared and Raman Spectra of Polyatomic Molecules*. New Jersey: Van Nostrand-Reinhold, 1945.
[7] Wilson, E. B., Decius, J. C., Cross, P. C., *Molecular Vibrations*. New York: McGraw-Hill, 1955.
[8] Califano, S., *Vibrational States*. New York: Wiley, 1976.
[9] Higgs, P. W., *Proc. Roy. Soc. London Ser. A* **1953**, *220*, 472–485.
[10] Shimanouchi, T., in: *Physical Chemistry: An Advanced Treatise*: Eyring, H., Henderson, D., Jast, W. (Eds.). New York: Academic Press, 1970; Vol. 4, pp. 233–306.
[11] Person, W., Zerbi, G., Eds., *Vibrational Intensities in Infrared and Raman Spectroscopy*. Amsterdam: Elsevier, 1982.
[12] Qian, W., Krimm, S., *J. Phys. Chem.* **1993**, *97*, 11578–11579.
[13] Cheam, T. C., Krimm, S., *J. Mol. Struct.* **1986**, *146*, 175–189.
[14] Tasumi, M., Krimm, S., *J. Chem. Phys.* **1967**, *46*, 755–766.
[15] Krimm, S., Abe, Y., *Proc. Natl. Acad. Sci. U.S.A.* **1972**, *69*, 2788–2792.
[16] Moore, W. H., Krimm, S., *Proc. Natl. Acad. Sci. U.S.A.* **1975**, *72*, 4933–4935.
[17] Cheam, T. C., Krimm, S., *Chem. Phys. Lett.* **1984**, *107*, 613–616.
[18] Abe, Y., Krimm, S., *Biopolymers* **1972**, *11*, 1817–1839.
[19] Moore, W. H., Krimm, S., *Biopolymers* **1976**, *15*, 2439–2464.
[20] Cheam, T. C., Krimm, S., *J. Chem. Phys.* **1985**, *82*, 1631–1641.
[21] Schachtschneider, J. H., Snyder, R. G., *Spectrochim. Acta* **1963**, *19*, 117–168.
[22] Dwivedi, A. M., Krimm, S., *J. Phys. Chem.* **1984**, *88*, 620–627.
[23] Hehre, W. J., Radom, L., v. R. Schleyer, P., Pople, J. A., *Ab Initio Molecular Orbital Theory*. New York: John Wiley & Sons, 1986.
[24] Fogarasi, G., Pulay, P., in *Vibrational Spectra and Structure*: Durig, J. R. (Ed.). Amsterdam: Elsevier, 1985; Vol. 14, pp. 125–219.
[25] Mirkin, N. G., Krimm, S., *J. Mol. Struct. (Theochem)* **1995**, *334*, 1–6.
[26] Mirkin, N. G., Krimm, S., *J. Phys. Chem.* **1993**, *97*, 13887–13895.
[27] Qian, W., Krimm, S., *J. Phys. Chem.* **1994**, *98*, 9992–10000.
[28] Qian, W., Krimm, S., *Biopolymers* **1994**, *34*, 1377–1394.
[29] Zhao, W., Bandekar, J., Krimm, S., *J. Mol. Struct.* **1990**, *238*, 43–54.
[30] Qian, W., Zhao, W., Krimm, S., *J. Mol. Struct.* **1991**, *250*, 89–102.
[31] Qian, W., Krimm, S., *Biopolymers* **1992**, *32*, 1025–1033.
[32] Qian, W., Krimm, S., *Biopolymers* **1992**, *32*, 1503–1518.
[33] Cheam, T. C., Krimm, S., *J. Mol. Struct. (Theochem)* **1990**, *206*, 173–203.
[34] Cheam, T. C., Krimm, S., *J. Mol. Struct. (Theochem)* **1989**, *188*, 15–43.

[35] Lifson, S., Stern, P. S., *J. Chem. Phys.* **1982**, *77*, 4542–4550.
[36] (a) Lii, J.-H., Allinger, N. L., *J. Am. Chem. Soc.* **1989**, *111*, 8566–8575. (b) Allinger, N. L., Chen, K., Lii, J.-H., *Comp. Chem.* **1996**, *17*, 642–668.
[37] Lii, J.-H., Allinger, N. L., *J. Comp. Chem.* **1991**, *12*, 186–199.
[38] Dasgupta, S., Goddard III, W. A., *J. Chem. Phys.* **1989**, *90*, 7207–7215.
[39] Karasawa, N., Dasgupta, S., Goddard III, W. A., *J. Phys. Chem.* **1991**, *95*, 2260–2272.
[40] Hwang, M. J., Stockfisch, T. P., Hagler, A. T., *J. Am. Chem. Soc.* **1994**, *116*, 2515–2525.
[41] Cornell, W. D., Cieplak, P., Bayly, C. I., Gould, I. R., Merz, Jr., K. M., Ferguson, D. M., Spellmeyer, D. C., Fox, T., Caldwell, J. W., Kollman, P. A., *J. Am. Chem. Soc.* **1995**, *117*, 5179–5197.
[42] Miwa, Y., Machida, K., *J. Am. Chem. Soc.* **1988**, *110*, 5183–5189.
[43] Derreumaux, P., Dauchez, M., Vergoten, G., *J. Mol. Struct.* **1993**, *295*, 203–221.
[44] Derreumaux, P., Vergoten, G., *J. Chem. Phys.* **1995**, *102*, 8586–8605.
[45] Palmö, K., Pietilä, L.-O., Krimm, S., *Computers Chem.* **1991**, *15*, 249–250.
[46] Palmö, K., Pietilä, L.-O., Krimm, S., *J. Comp. Chem.* **1991**, *12*, 385–390.
[47] Palmö, K., Mirkin, N. G., Pietilä, L.-O., Krimm, S. *Macromolecules* **1993**, *26*, 6831–6840.
[48] Palmö, K., Pietilä, L.-O., Krimm, S., *Computers Chem.* **1993**, *17*, 67–72.
[49] Palmö, K., Pietilä, L.-O., Krimm, S. *J. Comp. Chem.* **1992**, *13*, 1142–1150.
[50] Palmö, K., Mirkin, N. G., Krimm, S., *J. Phys. Chem. A* **1998**, *102*, 6448–6456.
[51] Palmö, K., Mirkin, N. G., Krimm, S., to be published.
[52] Miyazawa, T., Shimanouchi, T., Mizushima, S., *J. Chem. Phys.* **1958**, *29*, 611–616.
[53] Jakeš, J., Schneider, B., *Coll. Czech. Chem. Commun.* **1968**, *33*, 643–655.
[54] Jakeš, J., Krimm, S., *Spectrochim. Acta* **1971**, *27A*, 19–34.
[55] Rey-Lafon, M., Forel, M. T., Garrigou-Lagrange, C., *Spectrochim. Acta* **1973**, *29A*, 471–486.
[56] Sugawara, Y., Hirakawa, A. Y., Tsuboi, M., *J. Mol. Spectrosc.* **1984**, *108*, 206–214.
[57] Sugawara, Y., Hirakawa, A. Y., Tsuboi, M., Kato, K., Morokuma, K., *J. Mol. Spectrosc.* **1986**, *115*, 21–33.
[58] Balázs, A., *J. Mol. Struct. (Theochem)* **1987**, *153*, 103–120.
[59] Mirkin, N. G., Krimm, S., *J. Mol. Struct.* **1991**, *242*, 143–160.
[60] Ataka, S., Takeuchi, H., Tasumi, M., *J. Mol. Struct.* **1984**, *113*, 147–160.
[61] Mirkin, N. G., Krimm, S., *J. Am. Chem. Soc.* **1991**, *113*, 9742–9747.
[62] Williams, R. W., *Biopolymers* **1992**, *32*, 829–847.
[63] Cheam, T. C., *J. Mol. Struct. (Theochem)* **1992**, *257*, 57–73.
[64] Mirkin, N. G., Krimm, S., *J. Mol. Struct.*, **1996**, *377*, 219–234.
[65] Miyazawa, T., Shimanouchi, T., Mizushima, S., *J. Chem. Phys.* **1956**, *24*, 408–418.
[66] Miyazawa, T., in: *Polyamino Acids, Polypeptides, and Proteins*: M. A. Stahmann (Ed.). Madison: University of Wisconsin Press, 1962; pp. 201–217.
[67] Chen, X. G., Schweitzer-Stenner, R., Krimm, S., Mirkin, N. G., Asher, S. A., *J. Am. Chem. Soc.* **1994**, *116*, 11141–11142.
[68] Chen, X. G., Schweitzer-Stenner, R., Asher, S. A., Mirkin, N. G., Krimm, S., *J. Phys. Chem.* **1995**, *99*, 3074–3083.
[69] Song, S., Asher, S. A., Krimm, S., Bandekar, J., *J. Am. Chem. Soc.* **1988**, *110*, 8547–8548.
[70] Miyazawa, T., *Bull. Chem. Soc. Japan* **1961**, *34*, 691–696.
[71] Fillaux, F., de Lozé, C., *Chem. Phys. Lett.* **1976**, *39*, 547–551.
[72] Hsu, S. L., Moore, W. H., Krimm, S., *Biopolymers* **1976**, *15*, 1513–1528.
[73] Mirkin, N. G., Krimm, S., private communication.
[74] Cheam, T. C., *J. Mol. Struct.* **1993**, *295*, 259–271.
[75] Bandekar, J., Krimm, S., *Biopolymers* **1988**, *27*, 885–908.
[76] Sundius, T., Bandekar, J., Krimm, S., *J. Mol. Struct.* **1989**, *214*, 119–142.
[77] Qian, W., Bandekar, J., Krimm, S., *Biopolymers* **1991**, *31*, 193–210.
[78] Pauling, L., Corey, R.B., *Proc. Natl. Acad. Sci. U.S.A.* **1951**, *37*, 729–740.
[79] Pauling, L., Corey, R. B., *Proc. Natl. Acad. Sci. U.S.A.* **1953**, *39*, 253–256.
[80] Moore, W. H., Krimm, S., *Biopolymers* **1976**, *15*, 2465–2483.
[81] Arnott, S., Dover, S. D., Elliott, A., *J. Mol. Biol.* **1967**, *30*, 201–208.
[82] Bandekar, J., Krimm, S., *Biopolymers* **1988**, *27*, 909–921.

[83] Chirgadze, Yu. N., Nevskaya, N. A., *Biopolymers* **1976**, *15*, 607–625.
[84] Chirgadze, Yu. N., Nevskaya, N. A., *Biopolymers* **1976**, *15*, 627–636.
[85] Elliott, A., *Proc. Roy. Soc. London Ser. A* **1954**, *226*, 408–421.
[86] Itoh, K., Katabuchi, H., *Biopolymers* **1972**, *11*, 1593–1605.
[87] Masuda, Y., Fukushima, K., Fujii, T., Miyazawa, T., *Biopolymers* **1969**, *8*, 91–99.
[88] Dwivedi, A. M., Krimm, S., *Macromolecules* **1982**, *15*, 186–193.
[89] Fanconi, B., *Biopolymers* **1973**, *12*, 2759–2776.
[90] Krimm, S., Dwivedi, A. M., *J. Raman Spectrosc.* **1982**, *12*, 133–137.
[91] Itoh, K., Nakahara, T., Shimanouchi, T., Oya, M., Uno, J., Iwakura, Y., *Biopolymers* **1968**, *6*, 1759–1766.
[92] Sengupta, P. K., Krimm, S., Hsu, S. L., *Biopolymers* **1984**, *23*, 1565–1594.
[93] Keith, H. D., Padden, F. J., Giannoni, G., *J. Mol. Biol.* **1969**, *43*, 423–438.
[94] Lotz, B., *J. Mol. Biol.* **1974**, *87*, 169–180.
[95] Colonna-Cesari, F., Premilat, S., Lotz, B., *J. Mol. Biol.* **1974**, *87*, 181–191.
[96] Dwivedi, A. M., Krimm, S., *Macromolecules* **1982**, *15*, 177–185.
[97] Suzuki, S., Iwashita, Y., Shimanouchi, T., Tsuboi, M., *Biopolymers* **1966**, *4*, 337–350.
[98] Small, E., Fanconi, B., Peticolas, W. L., *J. Chem. Phys.* **1970**, *52*, 4369–4379.
[99] Huggins, M. L., *Chem. Revs.* **1943**, *32*, 195–218.
[100] Crane, H. R., *Sci. Mon.* **1950**, *70*, 376–389.
[101] Bragg, L., Kendrew, J. C., Perutz, M. F., *Proc. Roy. Soc. London Ser. A.* **1950**, *203*, 321–357.
[102] Pauling, L., Corey, R. B., Banson, H. R., *Proc. Natl. Acad. Sci. U.S.A.* **1951**, *37*, 205–211.
[103] Pauling, L., Corey, R. B., *Proc. Natl. Acad. Sci. U.S.A.* **1951**, *37*, 235–240.
[104] Arnott, S., Dover, S. D., *J. Mol. Biol.* **1967**, *30*, 209–212.
[105] Miyazawa, T., *J. Polym. Sci.* **1961**, *55*, 215–231.
[106] Dwivedi, A. M., Krimm, S., *Biopolymers* **1984**, *23*, 923–943.
[107] Krimm, S., Dwivedi, A. M., *Science* **1982**, *216*, 407–408.
[108] Whitby, F. G., Kent, H., Stewart, F., Stewart, M., Xie, X., Hatch, V., Cohen, C., Phillips, Jr., G. N., *J. Mol. Biol.* **1992**, *227*, 441–452.
[109] Malcolm, B. R., *Biopolymers* **1983**, *22*, 319–321.
[110] Dwivedi, A. M., Krimm, S., Malcolm, B. R., *Biopolymers* **1984**, *23*, 2025–2065.
[111] Prasad, B. V. V., Sasisekharan, V., *Macromolecules* **1979**, *12*, 1107–1110.
[112] Bradbury, E. M., Brown, L., Downie, A. R., Elliott, A., Fraser, R. D. B., Hanby, W. E., *J. Mol. Biol.* **1962**, *5*, 230–247.
[113] McGuire, R. E., Vanderkooi, G., Momany, F. A., Ingwall, R. T., Crippen, G. M., Lotan, N., Tuttle, R. W., Kashuba, K. L., Scheraga, H. A., *Macromolecules* **1971**, *4*, 112–124.
[114] Nambudripad, R., Bansal, M., Sasisekharan, V., *Int. J. Peptide Protein Res.* **1981**, *18*, 374–382.
[115] Saski, S., Yasumoto, Y., Uematsu, I., *Macromolecules* **1981**, *14*, 1797–1801.
[116] Low, B., Grenville-Wells, H. J., *Proc. Natl. Acad. Sci. U.S.A.* **1953**, *39*, 785–801.
[117] Donohue, J., *Proc. Natl. Acad. Sci. U.S.A.* **1953**, *39*, 470–478.
[118] Urry, D. W., *Proc. Natl. Acad. Sci. U.S.A.* **1971**, *68*, 672–676.
[119] Lotz, B., Colonna-Cesari, F., Heitz, F., Spach, G., *J. Mol. Biol.* **1976**, *106*, 915–942.
[120] Crick, F. H. C., Rich, A., *Nature* **1955**, *176*, 780–781.
[121] Krimm, S., *Nature* **1966**, *212*, 1482–1483.
[122] Ramachandran, G. N., Ramakrishnan, C., Venkatachalam, C. M., in: *Conformation of Biopolymers*: Ramachandran, G. N. (Ed.) New York: Academic Press, 1967; Vol. 2, pp. 429–438.
[123] Dwivedi, A. M., Krimm, S., *Biopolymers* **1982**, *21*, 2377–2397.
[124] Ramachandran, G. N., Sasisekharan, V., Ramakrishnan, C., *Biochim. Biophys. Acta* **1966**, *112*, 168–170.
[125] Krimm, S., Kuroiwa, K., Rebane, T., in: *Conformation of Biopolymers*: Ramachandran, G. N. (Ed.) New York: Academic Press, 1967; Vol. 2, pp. 439–447.
[126] Krimm, S., Kuroiwa, K., *Biopolymers* **1968**, *6*, 401–407.
[127] Cowan, P. M., McGavin, S., *Nature* **1955**, *176*, 501–503.
[128] Arnott, S., Dover, S. D., *Acta Crystallogr.* **1968**, *B24*, 599–601.
[129] Sasisekharan, V., *Acta Cryst.* **1959**, *12*, 897–903.

[130] Tiffany, M. L., Krimm, S., *Biopolymers* **1968**, *6*, 1379–1382.
[131] Krimm, S., Mark, J. E., *Proc. Natl. Acad. Sci. U.S.A.* **1968**, *60*, 1122–1129.
[132] Sengupta, P. K., Krimm, S., *Biopolymers* **1987**, *26*, S99–S107.
[133] Woody, R. W., *Adv. Biophys. Chem.* **1992**, *2*, 37–79.
[134] Ramachandran, G. N., Kartha, G., *Nature* **1955**, *176*, 593–595.
[135] Rich, A., Crick, F. H. C., *J. Mol. Biol.* **1961**, *3*, 483–506.
[136] Traub, W., Shmueli, U., in: *Aspects of Protein Structure:* Ramachandran, G. N. (Ed.) New York: Academic Press, 1963; pp. 81–92.
[137] Itoh, K., Shimanouchi, T., *Biopolymers* **1970**, *9*, 383–399.
[138] Frushour, B. G., Painter, P. C., Koenig, J. L., *J. Macromol. Sci. Rev. Macromol. Chem.* **1976**, *C15*, 29–115.
[139] Sengupta, P. K., Krimm, S., *Biopolymers* **1985**, *24*, 1479–1491.
[140] Rothschild, K. J., Clark, N. A., *Biophys. J.* **1979**, *25*, 473–488.
[141] Reisdorf, Jr., W. C., Krimm, S., *Biophys. J.* **1995**, *69*, 271–273.
[142] Reisdorf, Jr., W. C., Krimm, S. *Biochemistry*, **1996**, *35*, 1383–1386.
[143] Krimm, S., Reisdorf, Jr., W. C., *Faraday Discuss.* **1994**, *99*, 181–197.
[144] Palmö, K., Krimm, S., *J. Comp. Chem.* **1998**, *19*, 754–768.
[145] Lee, S.-H., Krimm, S., *Biopolymers* **1998**, *46*, 283–317.
[146] Lee, S.-H., Krimm, S., *J. Raman Spectrosc.* **1998**, *29*, 73–80.
[147] Lee, S.-H., Krimm, S., *Chem. Phys.* **1998**, *230*, 277–295.

Index